T0227677

BIOMARKERS
Biochemical, Physiological, and Histological Markers of Anthropogenic Stress

Edited by

Robert J. Huggett, Virginia Institute of Marine Science
Richard A. Kimerle, Monsanto Company
Paul M. Mehrle, Jr., ABC Laboratories
Harold L. Bergman, University of Wyoming

Proceedings of the Eighth Pellston Workshop
Keystone, Colorado, July 23-28, 1989
Sponsored by The American Petroleum Institute, Battelle Columbus Division,
Eastman Kodak Company, Monsanto Company, The Procter & Gamble
Company, US Air Force, US Army, US EPA, US Fish and Wildlife Services,
US Navy, and Roy F. Weston, Inc.

SETAC Special Publications Series

Series Editors

Dr. C.H. Ward
Department of Environmental Science and Engineering, Rice University
Dr. B.T. Walton
Environmental Sciences Division Oak Ridge National Laboratory
Dr. T.W. LaPoint
The Institute of Wildlife and Environmental Toxicology
Clemson University

Publication sponsored by the Society of Environmental Toxicology
and Chemistry (SETAC) and the SETAC Foundation for Environmental Education, Inc.

CRC Press
Taylor & Francis Group
Boca Raton London New York

CRC Press is an imprint of the
Taylor & Francis Group, an **informa** business

Library of Congress Cataloging-in-Publication Data

Catalog record is available from the Library of Congress.

A Library of Congress record exists under LC control number: 91043878

ISBN 13: 978-1-315-89117-0 (hbk)
ISBN 13: 978-1-351-07027-0 (ebk)

The SETAC Special Publication Series

The SETAC Special Publications series was established by the Society of Environmental Toxicology and Chemistry to provide in-depth reviews and critical appraisals on scientific subjects relevant to understanding the impacts of chemicals and technology on the environment. The series consists of single- and multiple-authored/edited books on topics selected by the SETAC Board of Directors for their importance, timeliness, and their contribution to multidisciplinary approaches to solving environmental problems. The diversity and breadth of subjects covered in the series will reflect the wide range of disciplines encompassed by environmental toxicology, environmental chemistry, and hazard/risk assessment. Despite this diversity, the goals of these volumes are similar; they are to present the reader with authoritative coverage of the literature, as well as paradigms, methodologies, controversies, research needs, and new development specific to the featured topics. All books in the series are peer reviewed for SETAC by acknowledged experts.

The SETAC Special Publications will be useful to environmental scientists in research, research management, chemical manufacturing, regulation, and education, as well as to students considering careers in these areas. The series will provide information for keeping abreast of recent developments in familiar areas and for rapid introduction to principles and approaches in new subject areas.

Biomarkers: Biochemical, Physiological, and Histological Markers of Anthropogenic Stress, is the sixth volume to be published in this series. It presents the proceedings of the Biomarkers Workshop, the eighth Pellston Workshop, held July 23–28, 1989, in Keystone, Colorado. The workshop was organized to bring selected professionals together to discuss the use of biomarkers for use in assessing exposure and effect of toxicants. This volume presents critical analyses of using biochemical, physiological, and histological changes to estimate exposure of organisms to toxicants or resultant effects of exposure to toxicants. It represents a large body of pioneering work and is a significant contribution.

Thomas W. LaPoint
Editor, SETAC Special Publications
The Institute of Wildlife and Environmental Toxicology
Clemson University

Preface

This publication is the result of the eighth Pellston Workshop, the last four of which were sponsored by the Society of Environmental Toxicology and Chemistry (SETAC). The focus is on biomarkers and their present and future utility to determine either exposure to or effects of anthropogenic stress. Each chapter is based on discussion papers generated before and during a workshop held in Keystone, Colorado in July 1989. Each chapter was peer reviewed by at least two experts in that particular field.

The reader may find several topics discussed in various chapters. Because the subject matter itself has no sharp boundaries, it was necessary to include many overlapping topics for the integrity of the chapters.

Acknowledgments

This publication was made possible by financial support from The American Petroleum Institute, Battelle Columbus Division, Eastman Kodak Company, Monsanto Company, The Proctor & Gamble Company, U.S. Air Force, U.S. Army, U.S. Environmental Protection Agency, U.S. Fish and Wildlife Service, U.S. Navy, and Roy F. Weston, Inc. Their commitment to advancing environmental sciences is appreciated.

Sincere thanks is expressed to the authors who gave many hours of their time to generate this document. Special recognition is also given to Janice Dickson, Scott Dyer, Aida Farag, Marveen Fishman, Kathryn Gallagher, Jennifer Gundersen, K. D. Howard, and Phyllis Howard, who labored to make the workshop a success.

The editors acknowledge the excellent editorial support provided by Dean Premo of Whitewater Associates and by Phyllis Howard and Shirley Sterling of the Virginia Institute of Marine Science, The College of William and Mary.

Robert J. Huggett is professor of Marine Science at the Virginia Institute of Marine Science, the School of Marine Science of The College of William and Mary.

Dr. Huggett received his MS degree from the Scripps Institution of Oceanography, The University of California, San Diego and his PhD from The College of William and Mary specializing in Marine Chemistry.

He has authored or co-authored over 70 research articles on subjects ranging from the fate and transport of trace metals and anthropogenic organic substances to environmental risk assessments. He is a member of the Society of Environmental Toxicology and Chemistry (SETAC) and was a Board member from 1988–1990. He is presently a member of the Environmental Protection Agency's Science Advisory Board serving on the Environmental Processes and Effects Committee and the Executive Committee.

Dr. Richard Kimerle is a Senior Environmental Science Fellow with the Monsanto Company in St. Louis, Missouri. His academic training at the University of Missouri and Oregon State University is in aquatic ecology and water pollution biology. His industrial experience of over 24 years includes chemical safety testing and ecological risk assessment of products, effluents, and hazardous waste sites. He is an active member in the Society of Environmental Toxicology and Chemistry (SETAC) and the EPA's Science Advisory Board.

Dr. Paul Mehrle, Jr. is Vice President, Business and Program Development, for ABC Laboratories, Inc., Columbia, Missouri. Prior to joining ABC Laboratories in 1990, he was with the research and development program of the U.S. Department of Interior, Fish and Wildlife Service, for 20 years. He directed the laboratory and field toxicology research programs at the Service's National Fisheries Contaminant Research Center in Columbia, Missouri.

Dr. Mehrle received his undergraduate degree in biology from Southwestern at Memphis (Rhodes College), Memphis, Tennessee, in 1967. He earned his MS in zoology in 1969 and PhD in biochemistry in 1971 from the University of Missouri.

Dr. Mehrle has directed extensive research programs in environmental toxicology during his career. His research has focused on the biochemical mechanisms of chemical contaminants in aquatic organisms and development of biomarkers for use in hazard assessments. Dr. Mehrle has published over 60 scientific papers.

Dr. Mehrle is a member of the Society of Environmental Toxicology and Chemistry (SETAC), and is currently serving on the society's Board of Directors. He was elected Secretary/Treasurer of SETAC, and appointed Editor of *Environmental Toxicology and Chemistry*. He is a member of American Society for Testing and Materials (ASTM), American Chemical Society, American Fisheries Society, and Sigma Xi. He has served on numerous editorial boards of scientific journals and provided advisory services and scientific consultations to government agencies and private industries.

Dr. Paul Mehrle, Jr. is Vice President, Business and Program Development, for ABC Laboratories Igor, Columbia, Missouri. Prior to joining ABC Laboratories in 1989, he was with the research and developmental program of the U.S. Department of Interior Fish and Wildlife Services for 20 years. He directed the laboratory and field technology research programs at the Service's National Fisheries Contaminant Research Center in Columbia, Missouri.

Dr. Mehrle received his undergraduate degree in zoology from Southwestern at Memphis (Rhodes College), Memphis, Tennessee in 1967. He earned the MS in zoology in 1968 and PhD in ecophysiology in 1972 from the University of Missouri. Dr. Mehrle has directed extensive research in environmental toxicology and fisheries research. His research has resulted in the basic developments related to ... He has authored over 80 scientific papers.

Dr. Mehrle maintains active membership in several professional societies and has served in the society's board of directors and the Society (SETAC). He was a Toxicology Terminology SETAC, and appointed liaison of Environmental Toxicology and Chemistry. He is a member of American Society for Testing and Materials (ASTM), American Chemical Society, American Fisheries Society, and Sigma Xi. He has served on numerous editorial boards of scientific journals and provided advisory services and scientific consultations to government agencies and private industries.

Harold L. Bergman is professor of Zoology and Physiology and Director of the Red Buttes Environmental Biology Laboratory at the University of Wyoming.

Dr. Bergman received BA and MS degrees in biology at Eastern Michigan University and a PhD in fisheries biology at Michigan State University. Prior to joining the University of Wyoming faculty in 1975, he was a fishery biologist at the Great Lakes Fishery Laboratory, U.S. Fish and Wildlife Service, Ann Arbor, Michigan, and Research Associate in the Environmental Sciences Division at Oak Ridge National Laboratory, Oak Ridge, Tennessee.

He has authored or co-authored over 60 research articles on diverse topics related to his principal research interests in fish physiology and aquatic toxicology. Professional society memberships include the American Association for the Advancement of Science, American Fisheries Society, North American Benthological Society, Sigma Xi, and the Society of Environmental Tosicology and Chemistry (SETAC), which he served as President in 1984-1985.

Contents

Introduction

Robert J. Huggett, Richard A. Kimerle, Paul M. Mehrle, Harold L. Bergman, Kenneth L. Dickson, James A. Fava, John F. McCarthy, Rodney Parrish, Philip B. Dorn, Vic McFarland, Garet Lahvis

Over the past decade, an increasing emphasis has been placed on the use of biochemical, physiological, and histological changes as well as aberrations in organisms to estimate either exposure to chemicals or resultant effects. Many environmental scientists have adopted the term *biomarker* to refer to these changes. Biomarker, plus the less commonly used terms *bio-indicator* and *biocriteria*, specifies the exposure or effect measurements in organisms, including plant or animal microorganisms from the laboratory or the field. Some environmental scientists and managers use biomarkers very broadly to include measurements of exposure or effect at any level of biological organization. This includes suborganismal, organismal, population, community, or ecosystem. The most common usage of the term, however, has been for biochemical, physiological, or histological indicators of either exposure to, or effects of, xenobiotic chemicals at the suborganismal or organismal level. And it is this meaning that has been adopted for the purposes of this publication.

Environmental scientists have found many potentially useful tools for detecting either exposure to, or effects of, chemicals, thanks to increased knowledge from basic research on the mechanisms of action and the identification of various chemical residues in organisms. These tools have been adopted increasingly in laboratory studies to understand mechanisms of action or early indicators of pathology. Biomarkers provide general indications of exposure or effect, such as certain physiological or histological pathologies. Other biomarkers provide very specific indications, such as detection of a particular DNA adduct. These tools have been used in field monitoring studies to understand the temporal and spatial extent of environmental contamination and effects. This accumulated knowledge has led to the kinds of applications that are discussed here.

The rate at which biomarkers are being used is increasing, and there is a strong impetus for carefully evaluating them. Their proper use has many potential benefits,

1

but their improper application or interpretation have severe liabilities. In order to objectively evaluate candidate biomarkers, criteria should be developed to evaluate their strengths and weaknesses. Some of these criteria are listed below.

General Indicators — If the biomarker responds to a variety of different chemicals, it may be useful as a general indicator of exposure to contaminant mixtures and might be particularly useful in monitoring or screening.

Relative Sensitivity — Two concepts are embedded in this criterion: (1) how sensitive is the biomarker compared to conventional endpoints such as lethality, reproductivity, or growth impairment, and (2) how sensitive is the biomarker relative to other candidate biomarkers?

Biological Specificity — Certain biomarkers may have greater applicability to certain groups of organisms because of their different metabolic capabilities. For example, aromatic hydrocarbon metabolism is different in bivalve molluscs than in fish or mammals.

Chemical Specificity — Some biomarkers respond to effect or exposure while others react to, and diagnose, specific chemicals or classes of chemicals.

Clarity of Interpretation — How clear-cut is the endpoint as an indicator of exposure or effects of anthropogenic stress? Can the endpoints be distinguished from responses to natural physiological or natural environmental stresses?

Time to Manifestation of Endpoint — The response time course of different biomarkers can vary widely from nearly instantaneous to years, as required for manifestation of some cancerous lesions. Depending on the objective of a particular study, rapid or slow manifestations may be desirable.

Permanence of Response — The persistence of endpoints may vary widely among biomarkers. Depending on the scale of temporal integration desired in a monitoring study, reversibility may be an important consideration. Is the biomarker a transient or a permanent manifestation of exposure and/or effect?

Inherent Variability (reliability) — Variability in biomarker response may arise from two sources: (1) from physiological or environmental influences that modulate the organism's response to a chemical, and (2) from the inherent biological response to a defined exposure. The first source of variability is addressed by developing a fundamental understanding of how nonchemical stresses affect the biomarker responses. This criterion affects the second source of variability by evaluating candidate biomarker with respect to its inherent variability and its reliability as an indicator of exposure or effects of chemicals.

Linkage to Higher Level Effects — Biomarkers of effects at the lower levels of biological organization (molecular or biochemical responses) are most meaningful when the biological significance of these responses can be clearly linked to effects at higher levels of organization (e.g., growth or reproduction). The lack of linkage may be the result of the biomarker's state of development or of the current understanding of its implications to higher levels of organization. The lack of clear-cut linkages does not invalidate the application of the biomarker, but it may limit its predictive potential and thus constrict the use of the biomarker in decisions concerning environmental health. Is the biomarker predictive of higher level (i.e., population, community, and/ or ecosystem) effects?

Applicability to Field Conditions — Some biomarkers may be measured only in a laboratory setting, while others have utility for field evaluations. Does the biomarker have proven or potential for making field measurements?

Validation in the Field — Has the candidate biomarker been successfully used in a field evaluation?

Methodological Considerations — When biomarkers are applied to evaluations of chemical contamination, it is important to understand the limitations of the methods. Important considerations from a practical perspective are precision and cost. *Precision* refers to the analytical reproducibility of the method. The *cost and ease of the assay* may be an important consideration if the biomarker is to be widely applied in biomonitoring studies.

Status of Method's Utility — Is the dependability, validity, and significance of the biomarker sufficiently established to permit current use in environmental monitoring and resource management?

The authors of this volume have considered some or all of these criteria in their individual chapters. The six chapters that follow address the following types of biomarkers: DNA alterations, proteins, metabolites, immunological, histopathological, and physiological-nonspecific.

CHAPTER 1

Physiological And Nonspecific Biomarkers

Foster L. Mayer, Donald J. Versteeg, Michael J. McKee, Leroy C. Folmar,
Robert L. Graney, Delbert C. McCume, and Barnett A. Rattner

Physiological and nonspecific biomarkers have been used extensively in the laboratory to document and quantify both exposure to, and effects of, environmental pollutants. As monitors for exposure, biomarkers have the advantage of quantifying only biologically available pollutants. As measures of effects in the laboratory, biomarkers can integrate the effects of multiple stressors and can assist in elucidating mechanisms of effects (i.e., mode of action). These laboratory studies will be critical to validate biomarkers as methods to instantaneously assess the condition of a population, community, or ecosystem.

Despite their utility in the laboratory, biomarkers have not been extensively applied in actual environmental assessments of effluents, nonpoint source pollution, and effects of land use practices. The lack of extensive use of biomarkers has many causes, including:

- the relative ease and simplicity of analytical chemistry methods for many environmental contaminants
- the disproportionate emphasis on development of biomarkers rather than their application
- the present difficulty in causally linking biomarker effects with specific environmental alterations

- the present lack of a direct linkage between biomarker effects and relevant population, community, and ecosystem-level effects

This chapter discusses the utility of physiological biomarkers and lists the criteria to be used in selecting biomarkers to address specific ecological questions. After a general discussion of this topic, several specific biomarkers of exposure and effects are described individually. For each of these biomarkers, research needs are identified for further development and for evaluation of their ecological relevance.

INTRODUCTION

During the past 20 years, toxicological effects in the aquatic environment have been assessed from the suborganismal to the ecosystem level of organization (Larsson et al. 1985; Miller 1981). Research, primarily at the population, community, and ecosystem levels, is being used to monitor environmental effects, conduct hazard assessments, and make regulatory decisions. The predictive utility of research at these levels of biological organization is limited because ecologically important effects (e.g., death or impaired organismal function) have already occurred by the time they can be detected.

Recently, as the discipline of environmental toxicology has matured and environmental regulation has become more complex, biomarkers at the suborganismal levels of organization (biochemical, physiological, and histological) have been considered to be viable measures of responses to stressors (Wedemeyer and McLeay 1981; Goldstein 1981). Biomarkers can allow for rapid assessment of organism health. As tools to monitor biological function and health, they are quantifiable biochemical, physiological, or histological measures that relate in a dose- or time-dependent manner the degree of dysfunction that the contaminant has produced.

Biomarkers can be indicators of either exposure or effects. The best current use of biomarkers is in understanding exposure of organisms to biologically available environmental pollutants. However, the greatest future utility of biomarkers may be in the in situ quantification of effects and diagnosis of cause.

Research from the cellular to the organismal level has not been utilized extensively in situ to address environmental questions. This is unfortunate since information unavailable by other methods can be obtained by these means in a timely and cost-effective manner. Current methods need additional research attention, especially in the area of field validation, before they can be used extensively in hazard assessment and regulatory toxicology. For new biomarkers, a sound research strategy for the development and implementation of biomarkers must be prepared and shared in order to direct and focus future research efforts.

The remainder of this chapter concentrates on methods for assessing the health of an organism or population. We review the response of organisms to stressors, discuss the rationale for using biomarkers, review problems associated with the field use of biomarkers for effects, and discuss experimental considerations in the design of field exposures.

Background

Frequently, the focus of biomarker research is to quantify the condition or health of organisms in situ. Health of an organism can be defined as the residual capacity to withstand stress; the more stressed, the less capable is the organism of withstanding further stress (Bayne et al. 1985). Stress has been variously defined, and several excellent discussions are available on it relative to ecotoxicology (Lugo 1981; Pickering 1981). Various uses of stress have become so entrenched that a consensus definition is unlikely (Pickering 1981). Therefore, as discussed by Pickering (1981), it is essential that terminology be specifically defined to avoid confusion among the various terms. In this discussion, we conform to the following concept of stress. Brett (1958) defined stress at the individual level of ecological organization as "a state produced by an environmental or other factor which extends the adaptive responses of an animal beyond the normal range, or which disturbs the normal functioning to such an extent that the chances of survival are significantly reduced." This definition is essentially consistent with the stressor/stress concept described by Selye (1956) and Fitch and Johnson (1977). A stressor is any condition or situation that causes a system to mobilize its resources and increase its energy expenditure (Lugo 1981). Stress is the response of the system to the stressor via increase in energy expenditure (Lugo 1981).

Seyle (1950a,b; 1976) envisaged stressors to produce specific reactions in an organism, which when sufficient in intensity, would stimulate a systemic stress response resulting in the general adaptation syndrome or GAS (Figure 1). Mazeaud et al. (1977) further extended the GAS concept to population level responses (Figure 2). Pickering (1981) provided an excellent critique of Selye's concept as it relates to quantification of stress in organisms and ecosystems. A modern deviation from Selye's original treatise is that the general adaptation syndrome (GAS) is itself a specific response based on neuroendocrine stimulation associated with awareness (Mason 1975a,b). Therefore, caution should be exercised in extending Selye's original concept to assume that biomarkers of GAS are always reliable indicators of chronic stress. For example, Schreck and Lorz (1978) reported reductions in survival of coho salmon (*Oncorhynchus kisutch*) without stimulation of GAS as evidenced by plasma cortisol levels. The reader is referred to Pickering (1981) for a more detailed critique.

Chronic toxicant effects may occur without stimulation of the GAS. Lugo (1981) discussed the importance of energy costs of stress and emphasized the significance of energy drains during exposure to chronic stressors. It is conceivable that reductions in survival, growth, and reproduction might occur via an energetic route without stimulation of GAS. This emphasizes the importance of selecting biomarkers that are intimately related with survival, growth, and reproduction as potential monitors for in situ population effects.

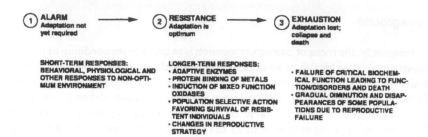

Figure 1.   The general adaptation syndrome of Selye (1976).

For additional publications related to stress, the reader is referred to Cairns et al. 1984, Casillas and Smith 1977, Donaldson 1981, Harvey et al. 1984, Johansson-Sjobeck et al. 1978, Lowe-Jinde and Niimi 1984, Mazeaud et al. 1977, Munck et al. 1984, Nolan and Duke 1983, Sibly and Calow 1986, Siegel 1980, Turner and Bagnara 1976, Versteeg and Giesy 1986, and Wedemeyer et al. 1984.

Biomarker Selection and Development

Biomarkers of chemical and physical stressor effects are attractive alternatives to more traditional measures for several reasons. Under field conditions, organisms are exposed to a multiplicity of chemical and physical stressors, against a background of naturally occurring seasonal fluctuations that, in and of themselves, are potentially stressful to the organism. Biomarkers have the potential to act as integrative measures at the suborganismal level to indicate adverse conditions, whether natural or not, preceding population-level effects. In addition, since the toxicological response to a chemical is caused by the interaction between the toxicant and biochemical receptor, biochemical responses would be expected to be the most immediate. That is, these responses would occur before responses are observed at higher levels of organization. Therefore, biomarkers should respond more rapidly than the whole organism.

Biomarkers can be categorized into nonspecific and specific as discussed previously. Biomarkers indicative of a nonspecific response to stress include any measure that is altered by exposure to a variety of stressors. Some of these nonspecific biomarkers (e.g., RNA/DNA, radiolabeled amino acid or nucleotide incorporation, and adenylate energy charge) give direct information about the growth rate or potential of an organism. Nonspecific biomarkers can integrate the simultaneous impacts of multiple toxicants or environmental factors on the organism, since all types of stressors can affect these endpoints. While these measures cannot be used to identify the specific toxicant causing an effect, many environmental situations consist of multiple stressors causing effects in an interactive manner.

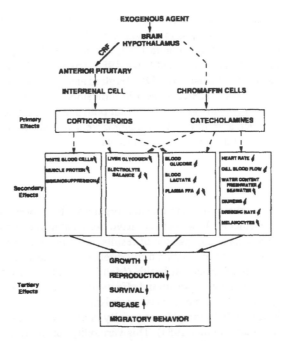

Figure 2.   The physiological, biochemical, and population level responses of an organism to stress, corticotropic releasing hormone (CRF), adrenocorticotropic hormone (ACTH), and free fatty acid (FFA). Modified from Mazeaud et al. 1977.

Specific indicators of stress are of two types: organ- and toxicant-specific. Organ-specific biomarkers include organ function tests, histopathology, organ-specific enzymes, and isoenzymes. Organ function tests include p-aminohippurate uptake by the kidney (Miller 1981) and bromosulfothalein and indocyanine green clearance by the liver (Patton 1978; Gingerich and Weber 1979). To date, these tests have not been used extensively in environmental toxicology; however, they are potentially useful for understanding toxicant-induced alterations in organ function.

Organ-specific biomarkers often depend on detection of certain enzymes at increased concentrations in a few organs. These enzymes appear in blood when those organs are damaged and are indicative of the presence and extent of damage. Lactate dehydrogenase, transaminases, creatinine phosphokinase, lysosomal enzymes, and alkaline phosphatase are examples of enzymes used as organ-specific indicators of toxic effects. Organ-specific isozymes such as lactate dehydrogenase and creatinine phosphokinase, may provide additional information on the specific organ affected in fish. Relatively few studies have used these methods to identify specific organ damage (Lockhart et al. 1975; Versteeg 1985).

Toxicant-specific biomarkers can involve quantification of certain enzyme activities or biomolecules in a tissue. They indicate exposure and possibly effects due to a single chemical or a related group of compounds. Examples include inhibition of acetylcholinesterase activity in brain tissue by organophosphorus pesticides (Coppage et al. 1975; Grue et al. 1983; Lowe-Jinde and Niimi 1984), cytochrome P_{450}-monooxygenase (Gooch and Matsumura 1983; Rattner et al. 1989), metallothionein and metal binding proteins (Nolan and Duke 1983; Dixon and Sprague 1981), and amino-levulinic acid dehydratase (ALAD) (Berglind and Sjobeck 1985; Finley et al. 1976; Hodson et al. 1984). Toxicant-specific indicators have been correlated with exposure of aquatic organisms to environmental pollutants in the laboratory and the field. To date, however, alterations in these biomarkers have shown poor correlation with population, community, or ecosystem effects.

Methods

Methods for biomarkers in environmental organisms have been largely derived from mammalian medicine. Few methods have been optimized for use with feral organisms, which has made comparison of results generated with different methods difficult. Further, use of the mammalian interpretation of a test result may lead to inaccurate assessments of organism health (Mehrle and Mayer 1985). The use of biomarkers for environmental exposure and effects presents several unique problems, namely variability and selection of the test organism.

Results obtained with field samples are usually more variable than those from the laboratory (Lockhart and Metner 1984). This variability has many causes including increased genetic heterogeneity, diversity of age and size classes, variability in exposure to chemical and environmental stressors (e.g., dissolved oxygen, suspended solids, predation, etc.), variable food quality, effects of sampling and handling, and a host of other factors. To increase the effective use of biomarkers in the field, we should design and implement studies to increase our understanding of the impact these factors have on biomarker response and develop a reliable baseline of normal values (Barnhardt 1969; Mitz and Giesy 1985a,b; Versteeg 1985).

Selection of the organism is critical in developing a monitoring program based on biomarkers. Factors to be considered include:

1. exposure of caged or transplanted organisms versus sampling indigenous ones
2. knowledge of the organism's biochemistry and physiology
3. size of the organism and reproductive state
4. the ease of sampling, which includes availability of sufficient numbers and ages
5. trophic level
6. sensitivity
7. ecosystem/societal importance

Considering the first factor, cage or transplant studies give a high degree of control and allow direct interpretation of the effect of exposure. However, they cannot take into account the effects of long-term adaptation or acclimation to current environmental conditions. In sampling indigenous biota, care must be taken to select appropriate control or reference sites. Even then, alterations in a given biomarker associated with adaptation or acclimation may result in statistically but not biologically significant changes. Biologically significant changes must occur if the biomarkers are useful as indicators of change at any level. A monitoring program using multiple biomarkers or a suite of measurements would better enable the identification of stressed populations.

Criteria for Field Use of Biomarkers

Choice of the appropriate biomarker for monitoring toxicant exposure and effects requires consideration of a variety of factors (Livingstone 1985; Versteeg et al. 1988). We established five criteria for the selection and development of useful biomarkers:

1. The biomarker should be relatively easy to measure, allowing quantification of multiple individuals. Parameters that are expensive, time consuming, and require sophisticated expertise to measure will not be generally useful in quantifying effects on field populations.
2. The biomarker should respond in a dose or time-dependent manner to the toxicant so the magnitude of the exposure or effect can be determined.
3. The biomarker should be sensitive but applied appropriately. There is no need to develop measures of acutely lethal concentrations of a compound. At present, the gain in sensitivity of biomarkers over traditional endpoints of survival, growth, and reproduction in chronic studies is negligible (Mayer et al. 1986).
4. The variability due to other factors (i.e., season, temperature, sex, weight, and handling) should be understood and within acceptable limits.
5. The measure must have biological significance. Only biomarkers that can be linked to important biological processes and for which changes can be interpreted should be used.

Not all of these criteria need to be fulfilled for a biomarker to be useful. However, these criteria will assist in biomarker selection, study planning, implementation, and data interpretation.

Research Needs for Biomarker Use

Many biomarkers have been developed and utilized successfully to quantify toxicant effects in the laboratory; however, those techniques have not been used to address important environmental problems. To be of greatest use, biochemical and physiological measures of contaminant exposure and effects should have application in the field. To achieve this goal, researchers have several general tasks before them:

*   Accumulate basic knowledge of biochemical and physiological "norms."
*   Define biomarker variability.
*   Develop biomarkers that discriminate the acclimation response from toxic effects.
*   Conduct field evaluation.

•   Identify relationships between biomarker response and important, ecologically relevant effects.

Once information is obtained in each of these areas, biomarkers may have regulatory application. However, regulatory uses must not be pushed beyond the science.

Accumulate basic knowledge of biochemical and physiological "norms." Mayer (1983) and Mehrle and Mayer (1980) have discussed the paradox of clinical tests in aquatic toxicology: the need for increased understanding of the biochemical toxicology of aquatic organisms is impeded by the lack of knowledge of the basic biochemistry of the organisms, including "normal" physiological ranges (Mehrle and Mayer 1980). This paradox greatly limits the applicability of many biomarkers. Developing a strong background data base for specific organisms is essential. Besides establishing the "normal" range of a particular indicator, one must also establish the statistical confidence in determining that the parameter is out of the normal range. This is dependent on the variance of the parameter as well as the sample size. We advocate determining the parameter standard deviation and using power analysis to establish the required sample size to demonstrate a statistically significant effect at a given power (Type I and Type II error) (Giesy and Allred 1985; Wedemeyer and Yasutake 1977). Few studies provide information on the range of values for a parameter. However, Wedemeyer and Nelson (1975) provide the basis for calculating both Gaussian (parametric) and nonparametric tolerance intervals to estimate "normal" ranges for biochemical and physiological parameters in aquatic organisms.

Define biomarker variability. Variability in the measurement of responses to environmental contaminants is a problem throughout environmental toxicology. This variability has three sources: organismal, environmental (including toxicant effects; see Dively et al. 1977 for discussion), and methodological. Biomarkers can succeed only when these variability issues are resolved. Within a single species, several parameters such as diet, genetic strain, sex, age, environmental temperature, and sampling techniques can affect the "normal" range of biomarkers, such as enzyme activities and substrate concentration (Barnhart 1969). Also, activities of enzymes can vary greatly among different tissues within the same organism (Bouck 1980) and among different species. For this reason, we suggest that representative species be selected for intensive study in several different habitats. For instance, the large volume of information currently available on the rainbow trout (*Salmo gairdneri*) makes it such a species. Mammalian toxicologists have reduced variability through standardized and optimized procedures and selective breeding of test organisms. Since little can presently be done to reduce the organismal variability, environmental toxicologists must reduce biomarker variability through proper selection of biomarkers, research on experimental protocols and factors affecting variability, and application of appropriate statistical methods (Adams et al. 1985). Some biomarkers may be needlessly rejected due to high variability; in fact, decreased variability can indicate significant effects on the genetic pool of a population, and possibly, less plasticity in responding to biotic and abiotic stresses. Variability needs to be better defined and not always considered a criterion for rejection.

Develop biomarkers that discriminate the acclimation response from toxic effects. Xenobiotic-induced alterations at the suborganismal level of organization are not biologically relevant if changes are fully compensated for by the organism and do not result in adverse effects (Livingstone 1982). Therefore, it is critical to differentiate between the acclimation (homeostatic) response and toxic effects (Figure 3). Certain specific alterations in an organism's physiology occur after stimulus by natural environmental stressors, exposure to nontoxic doses (i.e., exposure duration ¥ concentration) of toxicants, and exposure to toxic doses. The first two responses may not result in serious effects on the individual or population and should be differentiated from the third. Biochemical and physiological changes caused by environmental cues and low doses of toxicants have survival value to the organism and are considered acclimation. The difficulty in discriminating acclimation response from the toxic response may have several causes. Exposure to a low-level stress can cause an alteration in a biochemical parameter without a change in the population due to the acclimation response of the organism (Figure 2). As the stress is increased and toxicity occurs, the biochemical response may either increase as the organism attempts further acclimation or decrease, representing exhaustion of the acclimation response. If the biochemical response increases as toxicity occurs, it may be difficult to statistically discriminate between a response in the acclimation range and a slightly increased response in the toxic range. It may, however, be possible to distinguish between the "normal" state and the acclimated state. If the biochemical response then decreases as toxicity occurs, levels of the biochemical parameter may occur within the acclimation and toxic ranges that are equal. Resolution of this dilemma will involve appropriate selection of the biomarker.

Conduct field evaluation. The utility of many biomarkers in the field is unknown since testing and extensive evaluation of methods is lacking. While field research is more difficult than laboratory research, environmental relevance should be of greater concern to the scientist. We must apply our procedures to field situations and either confirm or reject each technique in this arena.

Identify relationships between biomarker response and important, ecologically relevant effects. For a biomarker to be applied in the field, it should be correlated with a significant effect such as survival, growth, or reproduction. A significant effect is defined as any effect causing a biologically or sociologically relevant effect on population size or condition. Clinical measures of pathological effects are successful in humans because a large body of correlative information exists. Because this type of information is lacking for many feral organisms, development of biomarkers for assessing toxicant exposure and effects is more difficult than they are for mammalian species.

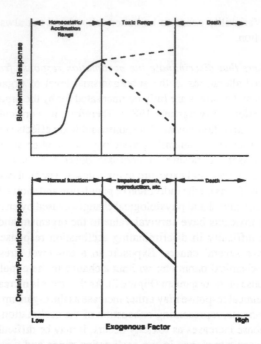

Figure 3.   A comparison of the biochemical and organismal response of an organism to a stressor demonstrating the difficulty in discriminating between a homeostatic response of a biochemical parameter and a "toxic" response. From Versteeg et al. Aquatic Toxicology and Hazard assessment ASTM STP 971, (Philadelphia: American Society for Testing Materials, 1988) pp 289–306. With permission.

SUMMARY

Biomarkers can be important tools for the environmental scientist/manager. Like any tool, biomarkers have advantages and disadvantages that must be considered in selection, application, and interpretation. Much additional research is needed to provide a greater number of calibrated and environmentally useful biomarkers.

Perhaps the most challenging task will be to develop biomarkers of effects that differentiate toxicity from exposure. Fortunately, toxicologists appear to be reaching agreement on the critical endpoints to be quantified in the laboratory in single species population-level toxicity tests (Woltering 1984; Suter et al. 1987; Mayer et al. 1986). Comparing biomarkers against these tests will simplify evaluation and verification of biomarker application.

The remainder of this chapter discusses selected physiological and nonspecific biomarkers expected to be useful in quantifying ecologically relevant exposure and effects. Biomarkers discussed in these sections are organized into the following categories:

- direct enzyme inhibition
- energetics
- endocrine
- blood chemistry
- growth rate
- gross indices
- lysosomal membrane stability
- free amino acids

DIRECT ENZYME INHIBITION

Quantification of enzyme activity in plants and animals can serve as a valuable biomarker of pollutant exposure and effect. Toxicants can inhibit enzymes at very specific sites (e.g., esteratic site of acetylcholinesterase) or effects can be evoked by less specific interactions with various moieties (e.g., sulfhydryl groups). Some of these responses are quite toxicant specific (e.g., organophosphorus insecticide inhibition of acetylcholinesterase), while others are far more generalized (e.g., stimulation of peroxidase activity in plants by ozone, sulfur dioxide, and hydrogen fluoride). Changes in enzyme activity often precede changes in condition and function, and onset of altered activity in plants and animals can be rapid. Changes in isozyme patterns can provide valuable clues to the affected organ, tissue, cell, or subcellular compartment. These techniques are precise, and may be used to screen large numbers of samples at a fraction of the cost of determining residue burdens.

Four types of biomarkers representative of direct enzyme inhibition are acetylcholinesterase, delta-aminolevulinic acid dehydratase, adenosine triphosphatase, and certain plant enzymes.

Acetylcholinesterase

Knowledge of the inhibition of acetylcholinesterase (AChE) accumulated concurrently with the development of organophosphorus and carbamate compounds (including nerve gas and pesticides). Its inhibition is linked directly with the mechanism of toxic action (irreversible or reversible binding to the esteratic site, and potentiation of cholinergic effects). AChE is responsible for hydrolyzing acetylcholine into choline and acetic acid (O'Brien 1967). As an indicator of exposure to these compounds, AChE and nonspecific cholinesterase (ChE) activities in blood and tissues emerged as a diagnostic tool in the biomedical area. Subsequently, quantification of this enzyme was applied to laboratory and field studies with both lower and higher vertebrates to assess exposure to organophosphorus and carbamate insecticides (Weiss 1958, 1961, 1965; Weiss and Gakstatter 1964; Williams and Sova 1966; Holland et al. 1967; Morgan et al. 1973; Coppage and Braidech 1976; Ludke et al. 1975; Grue et al. 1983).

Carbamate and organophosphorus can be compounds less hazardous to the environment than organochlorine chemicals because of their short half-life, and because they do not persist in animal tissues. Nevertheless, some of these chemicals are extremely toxic for short periods after application, when they or their active metabolites inhibit AChE and disrupt neurotransmitter processes in the central nervous system. Death from paralysis can often result.

For vertebrates, acetylcholinesterase is vital for normal neural functioning of the sensory, integrative, and neuromuscular systems. In salmonids, for instance, inhibition of this enzyme alters respiration (Klaverkamp et al. 1977), swimming (Matton and Lattam 1969; Post and Leisure 1974), feeding (Wildish and Lister 1973; Bull and McInerney 1974), and social interaction (Symons 1973). In birds, alterations in behavior, endocrine function, thermoregulation, reproduction, and tolerance to non-contaminant environmental stressors have been reported (Grue et al. 1983). For invertebrates, ChE-inhibiting compounds are lethal and have profound population effects. In other vertebrates, ChE-inhibition can impair certain physiological functions and may be lethal. However, to date, only individual and local population impacts have been documented. Such local population effects may serve to delineate impacted areas.

Distribution of AChE and ChE in tissues is ubiquitous, although measurements are generally confined to whole blood, plasma, and brain tissue. Broad generalizations regarding the type of cholinesterase in a given tissue are usually not valid, and one cannot always predict the type, importance, or substrate specificity of the ChE by its location. There are two classical types: acetylcholinesterase and butyrylcholinesterase, which are found in mammalian erythrocytes and serum, respectively (O'Brien 1967). In mammals, AChE preferentially hydrolyzes acetylcholine and butyrylcholinesterase (BChE) preferentially hydrolyzes butyrylcholine. Some of these differences do not apply to insect ChE (e.g., fly-head ChE hydrolyzes butyrylcholine readily, but in other respects behaves like mammalian AChE). In some species, such as ducks, ChE is propionylcholine specific. Nervous tissue has ChEs with a variety of substrate specificities. The main ChE in lake trout (*Salvelinus namaycush*) and the crayfish *Procambarus clarkii* is AChE, while in the crab *Uca pugnax* it is BChE (Guilbault et al. 1972). Physiological norms of brain cholinesterase activity have been established for several species of birds and mammals (Westlake et al. 1983; Hill 1988).

In addition to anticholinesterase insecticides, a few other contaminants, including mercury, have been shown to cause inhibition. Depression of AChE activity occurs in cases of severe infection, anemia, malnutrition, and liver disease. Responses in blood and plasma are quite sensitive, well below signs of overt intoxication. There are some distinct structure-activity relationships associated with the degree of inhibition evoked by anticholinesterases. Inherent variability among species and tissues necessitates inclusion of concurrent controls in laboratory and field evaluations.

History of Use

The relationship between brain AChE activity and organophosphorus and carbamate exposure appears to be close in birds regardless of age, sex, or species (Grue et al.

1983), but such relationships are essentially unknown in aquatic invertebrates and fish. Inhibition of brain AChE in fish by organophosphorus insecticides has been used to detect pollution of natural waters by these compounds (Williams and Sova 1966; Holland et al. 1967), but the interpretation of the relationship between inhibition to exposure and death is controversial (Nicholson 1967; Cox 1968; Gibson et al. 1969). Death of fish after exposure to organophosphorus pesticides in the laboratory was reported at brain activities ranging from 5.4 to 92% of normal (Weiss 1958, 1961), but fish have survived at activities as low as 10 to 20% of normal (Weiss 1961; Gibson et al. 1969).

Potential Value

AChE is a useful biomarker of exposure and could be developed for use in exposure quantification of effects.

Research Needs

Further research is indicated to define the functions of various cholinesterases with different substrate specificities and to establish the significance of inhibition among representative species.

Delta-Aminolevulinic Acid Dehydratase

Delta-aminolevulinic acid dehydratase (ALAD) is a cytosolic enzyme found in many tissues and active in the synthesis of hemoglobin by catalyzing the formation of porphobilinogen, a precursor of heme (Hammond and Belile 1980). Lead exposure causes a dose-dependent decrease in erythrocyte ALAD activity in fish, birds, and mammals through direct inhibition of the enzyme (Johansson-Sjobeck and Larsson 1979; Scheuhammer 1987). Blood lead concentrations have been directly correlated with ALAD inhibition (Hodson et al. 1977; Hodson et al. 1980; Finley et al. 1976). Because toxic lead concentrations based on ALAD inhibition are similar to laboratory-derived maximum acceptable toxicant concentration (MATC) values, Hodson et al. (1977) advocated this method to relate exposure and toxicity. ALAD is not the rate-limiting enzyme in heme synthesis. Toxic effects of lead in fish have not been causally related to ALAD inhibition, and compensatory mechanisms in fish are speculated to counteract whatever effect lead has on hemoglobin production (Johansson-Sjobeck and Larsson 1979).

History of Use

Field studies have demonstrated ALAD inhibition in fish from lead polluted lakes, in passerines nesting near highways, and in waterfowl ingesting spent lead shot. Often,

ALAD is maximally inhibited before other signs of toxicity become apparent, indicating that ALAD is a good monitor of lead exposure but not necessarily toxicity (Haux et al. 1986).

Potential Value

Assays of ALAD are relatively simple, inexpensive, accurate, and precise. It is recommended that ALAD be used as a measure of lead exposure.

Research Needs

No research is indicated at this time.

Adenosine Triphosphatase (ATP)

Mg^+-activated, Na^+/K^+-adenosine triphosphatase (Na^+/K^+-ATPase) is a membrane-bound protein that transports Na^+ and K^+ ions across the cell using the energy gained from the conversion of ATP to ADP (adenosine diphosphate)(Haya and Waiwood 1983). This enzyme is primarily responsible for maintaining the transmembrane Na^+ and K^+ gradients. Effects on enzyme activity may occur due to a reduction in ATP availability or a direct toxic effect on the enzyme. In vitro exposure of Na^+/K^+-ATPase to a variety of organochlorine compounds and metals inhibits this enzyme (Cutkomp et al. 1972; Desaiah and Koch 1975; Tucker 1979; Watson and Beamish 1980).

History of Use

In vivo, acute exposures of aquatic organisms to metals have resulted in variable effects on gill Na^+/K^+-ATPase. Acute exposure of the rainbow trout to acidic water containing aluminum decreased Na^+/K^+-ATPase activity (Saunders et al. 1983; Staurnes et al. 1984) while exposure to mercury or methylmercury did not alter gill Na^+/K^+-ATPase (Lock et al. 1981). Cadmium exposure of the American lobster (Tucker 1979) and zinc exposure of trout (Watson and Beamish 1980) and salmon (Lorz and McPherson 1976) have all resulted in increased ATPase activity in gill tissue.

Potential Value

The activity of Na^+/K^+-ATPase in fish gill, and possibly other animal tissues, may represent a useful nonspecific biomarker as it is easily quantified and is affected by a variety of toxicants. However, the ecological significance of changes in ATPase activity is difficult to assess due to the large reserve capacity of ATPase activity. In addition, ATPase activity may either increase or decrease in response to environmental stress, and separating normal variation from toxicant-induced alterations may be extremely difficult.

Research Needs

The majority of research has focused on changes in enzyme activity under controlled laboratory exposures. The sensitivity of this endpoint as a nonspecific indicator requires further elucidation. Field evaluation research, incorporating natural stress and varying pollutant exposures, would provide an assessment of the utility of this enzyme.

Plant Enzymes

Activities of enzymes in foliar tissues of terrestrial plants have been studied with respect to the mechanisms of action of air pollutants and the influence of environmental or biological factors upon them (Wellburn et al. 1976).

History of Use

Some enzymes, such as peroxidase and its isozymes, have been used as nonspecific markers of a general metabolic shift in response to toxicant exposure. Some enzymes have been chosen for their importance in one or more metabolic pathways, such as ribulose-1,5-bisphosphate carboxylase/oxygenase and its role in photosynthetic carbon dioxide assimilation. Other enzymes have been selected because of high sensitivity to a particular pollutant, such as enolase due to its sensitivity to fluoride and position in glycolysis. The toxic interaction of two or more air pollutants has also been studied by monitoring the activities of several enzymes (Wellburn 1984). Finally, batteries of enzymes have been used to study genetic changes in populations of herbaceous and woody species exposed to heavy metals, sulfur dioxide, or other pollutants (Scholz et al. 1989).

Potential Value

Among the plant enzymes having possible value as biomarkers for air pollutants are those that mediate processes that detoxify the pollutant or its products and whose activities are increased by exposure to the pollutant. One group includes superoxide dismutase and enzymes of the ascorbate glutathione system (e.g., ascorbate peroxidase, glutathione [ascorbate] dehydrogenase, glutathione reductase, and glutathione synthetase) for detoxification of oxygen radicals arising from, or associated with, the action of sulfur dioxide or ozone (see reviews by Alscher et al. 1987; Heath 1988.) A second group includes nitrate reductase (reduces nitrate to nitrite), nitrite reductase (reduces nitrite to ammonium), and glutamine/glutamate synthetase (a portal for the assimilation of ammonia into amino acids), which are involved in the metabolism of oxides of nitrogen. Nevertheless, changes in the activities of nitrate or nitrite reductase in response to oxides of nitrogen are determined by several factors, such as species, cultivar, and stage of development of the plant; light intensity; and form and level of nitrogen in the soil (Wellburn 1990).

Research Needs

Most information has come from controlled exposures to pollutants under experimental conditions that exclude the other environmental stresses capable of producing changes in the activities of these enzymes. Until more is known as to the specificity and sensitivity (in the sense of Bishop et al. 1975) of changes in enzyme activity, their utility as biomarkers in natural populations under ambient conditions is problematic.

ENERGETICS

There are a multitude of energetic responses that organisms exhibit in response to toxicant-induced stress. Some of these responses can be directly related to the general adaptation syndrome (GAS) (Selye 1976), which is based on the concept that energy is required to maintain homeostasis. Other responses are the direct result of toxicants interfering with key metabolic pathways. Acute responses, which are generally controlled by hormones, are considered to be the initial response to the stressor and often involve the increase of energy-related substrates in the plasma/hemolymph. Chronic stress can initiate compensatory physiological adjustments such that changes in energy metabolism may be required to maintain normal physiological/biochemical functions. Growth, reproduction, and maintenance metabolism demand the majority of an organism's energy expenditure, so any increased maintenance requirement may result in reduced growth or reproduction. Consequently, a comparative assessment of the energy status of an organism can be indicative of its overall condition.

The use of energetics has potential utility in both field monitoring and laboratory toxicity assessment programs. Field utility is often based on one's ability to separate toxic effects from normal variations or "noise" caused by accessory factors such as temperature, diet, salinity, and reproductive status. Under laboratory conditions, the most useful applications of energetic measurements may be as short-term indicators of chronic toxicity.

Since energetics covers a broad range of potential biomarkers, specific background and rationale are necessary for each endpoint. Biomarkers to be discussed as energetic-related endpoints are adenylate energy charge (AEC), energy reserves (proximate analysis and whole body calorimetry), enzymes of intermediary metabolism, scope for growth, and oxygen:nitrogen ratios. Many of the principles associated with energetics not only apply to animals but also terrestrial plants subjected to the chronic stress of air pollution.

Adenylate Energy Charge

Adenylate energy charge (AEC) can be considered a measure of the metabolic energy available to an organism from the adenylate pool (Atkinson 1977; Knowles 1977; Haya and Waiwood 1983). AEC is a direct calculation based on the measured concentrations of adenosine triphosphate (ATP), adenosine diphosphate (ADP), and adenosine monophosphate (AMP):

$$AEC = \frac{ATP + \frac{1}{2}ADP}{ATP + ADP + AMP}$$

Based on increased use of energy by organisms under stress, a decrease in the concentration of ATP (the primary source of "biochemical" energy) may be expected. High values for AEC (0.8 to 0.9) are expected for organisms operating under optimum conditions, and lower values are found for organisms existing in "suboptimum" or stressful conditions (Giesy et al. 1983; Ivanovici 1979). Since ATP is a direct modulation of the activity of key enzymes, changes in the concentration of ATP can directly influence key metabolic pathways. For example, shifts toward gluconeogenesis will occur when ATP levels are high, and glycogenolysis will occur when AMP levels are high.

There are a number of properties of AEC that make it a potential biomarker for sublethal effects in organisms (Giesy et al. 1983; Ivanovici 1979) as follows:

1. The adenylate energy system is ubiquitous among organisms.
2. Adenylates are important in the regulation of catabolic and anabolic processes such that they are integrative measures of metabolism.
3. AEC values can be related to the physiological state of the organism.
4. Relative to other biomarkers, variation among organisms is not great.
5. AEC represents a general, nonspecific indicator of stress.

History of Use

AEC has been evaluated extensively in a variety of organisms exposed to various stressors and/or conditions. In many cases, the changes observed have correlated fairly well with the general "condition" of the organism (Chapman et al. 1971; Ivanovici 1979; Montague and Dawes 1975; Haya and Waiwood 1983; Reinert and Hobreiter 1984) Organisms such as algae and bacteria were the focus of much of the earlier work (Atkinson 1977; Knowles 1977). Recent research has focused on higher organisms, specifically molluscs and crustaceans. Anoxia has been shown to have a significant effect on AEC in aquatic organisms, and this has been reviewed extensively by Haya and Waiwood (1983). Diet, reproductive condition, age, season, and geographical location have all been shown to influence AEC of crustaceans and molluscs (Giesy et al. 1981; Dickson and Giesy 1981; 1982). AEC in aquatic invertebrates (crustaceans and molluscs) exposed to metals and organic compounds has been shown to decrease (Giesy et al. 1981; Haya et al. 1980; Skjoldal and Bakke 1978; Ivanovici and Wiebe 1982; Giesy et al. 1983). Relatively few studies have been conducted on the effects of chemicals on AEC in fish (Reinert and Hobreiter 1984; Heath 1984). In general, the work on molluscs indicates AEC to be more stable and less variable than other biomarkers, and thus potentially more useful as an in situ indicator.

As for plants, no statistically significant changes in AEC were found in leaves of corn, tomato, or bean exposed to gaseous hydrogen fluoride (McCune et al. 1970), nor in bean leaves exposed to ozone, although the levels of adenylates were increased (Pell and Brennan 1973). In perennial rye grass, sulfur dioxide had no effect and nitrogen dioxide increased AEC, although the combination of both pollutants decreased AEC (Wellburn et al. 1981).

Potential Value

Although AEC has been shown to be altered in a large number of organisms exposed to various stressors, its potential as a laboratory or field biomarker is still unknown. As previously stated, since AEC is a highly regulated pool of biomolecules, the sensitivity of the pool relative to chronic stress is questionable. Often organisms are close to death before the pool significantly decreases (Skjoldal and Bakke 1978). In situations where laboratory tests indicate sensitivity of AEC to sublethal concentrations, it is unknown whether this sensitivity will be realized under field conditions in which the natural variability can be high. Organism selection is also very important, since stress tolerance can influence AEC (Giesy et al. 1983). Molluscs are the recommended organism for AEC because factors such as handling stress and inherent variability do not appear to be as important (Ivanovici and Wiebe 1982; Giesy et al. 1983; 1981).

Research Needs

The relationship between changes in AEC and corresponding reductions in growth and/or reproduction of the organism must be addressed. Ivanovici and Wiebe (1982) have examined the data for a variety of organisms and concluded that organisms with values between 0.5 and 0.7 had slower growth rates and did not reproduce; however, these organisms were able to recover. If values dropped below 0.5, the organisms never recovered from the stressor. These ranges of values have not been validated under field conditions. Research is needed to determine if, under field conditions, decreases in AEC can be separated from background variability and related to growth and reproductive effects.

Energy Reserves

Both physical and chemical stressors can cause changes in the concentration of stored energy reserves in fish and invertebrates. During times when energy intake exceeds the maintenance, growth, and reproductive requirements of an organism, energy is "stored" as either glycogen or lipids. During times of increased energy demand, such as stress or reproduction, these energy stores are mobilized. During severe stress, proteins can also be used as an energy source, although protein is not synthesized and stored for this purpose.

Whole-body or tissue-specific analysis of these three types of energy reserves has been used extensively as a general indicator of the energetic status of an organism. Such changes are generally indicative of long-term, sublethal exposure to a stressor and are normally not manifested under acute exposure conditions. Extensive background and control data is required to interpret the significance of energy levels in field-collected organisms. As with all energetics related biomarkers, a large array of factors (e.g., diet, reproductive status, sex, age, location, and season) will influence the energy reserve of the organisms and must be carefully considered in any interpretation of energetics data.

History of Use

Glycogen. Glycogen is a branched polymer of glucose residues and, for most animals, represents the readily mobilizable storage form of glucose. Both increases and decreases in glycogenolysis can occur due to toxicant-induced stress, which results in either depletion or accumulation of glycogen. In most cases, the increased energy demand associated with stress results in a depletion of glycogen reserves (Bhagyalakshmi et al. 1984a; Thomas et al. 1981). Since glycogen storage and mobilization is restricted to certain tissues, observed alterations in glycogen concentrations are often tissue specific. In mussels, the hepatopancreas and mantle are the primary storage organs for glycogen and generally the first to be utilized (Bayne 1973a,b). In vertebrates, the liver is the primary glycogen storage location (Grizzle and Rogers 1976). Glycogen content can be influenced by diet (Bayne 1973a,b), reproductive condition, and season (Newell and Bayne 1980).

The glycogen content of fish tissue is affected by both acute and chronic exposure to metals (Arillo et al. 1982; Gill and Pant 1983; Sastry and Sunita 1983; Chaudhry 1984) and organic compounds (Verma et al. 1983; Srivastava and Gupta 1981; Murty and Devi 1982; Pant and Singh 1983). In the large majority of cases, the glycogen content was found to decrease. Very similar results have been found for invertebrates (Thomas et al. 1981; Coglianese and Neff 1982; Bhagyalakshmi et al. 1983b; Graney and Giesy 1986b). The depletion of glycogen reserves has been usually attributed to the increased energy demand associated with chemical-induced stress.

Lipids. Lipids provide an essential, readily available energy source for a large number of aquatic invertebrates (Holland 1978; Voogt 1983), fish (Miller et al. 1976; Glass et al. 1975), and birds. Their importance as a primary energy source vs a secondary one varies with species and season (Morris 1971; Gardner and Riley 1972). As the organism develops, it may alter its primary storage form. For example, benthic invertebrate larvae often store neutral lipids; as they develop, their primary form of energy storage may shift to glycogen (Holland 1978). Similarly, during molting and/ or reproductive cycles, the mobilization and utilization of different energy substrates may shift (Barclay et al. 1983; O'Connor and Gilbert 1969; Frank et al. 1975).

The tissue distribution of lipids may also vary greatly (Holland 1978; Love 1970). In fish, the lipid content of dark muscle is much greater than that of white muscle. Given this, the effects of stressors on lipid is most meaningful to the analysis of specific tissues rather than whole body content. As with invertebrates, the total and tissue-specific lipid content of fish can be influenced by a variety of factors, including diet (Phillips et al. 1956; Fletcher 1984), reproductive condition (Love 1970), salinity (Daikoku et al. 1982), and temperature (Hansen and Abraham 1983).

In most cases, lipid content of fish decreased during toxicant-induced stress (Murty and Devi 1982; Rao and Rao 1984b; Dixon 1975; Dey et al. 1983). However, the decreases observed in total lipids are often accompanied by "no effect" or increases in specific components of the lipid constituents (i.e., triglycerides, fatty acids, and phospholipids). Decreases in lipid content have also been noted in freshwater (Graney and Giesy 1986b) and marine invertebrates (Lee et al. 1981; Capuzzo et al. 1984) exposed to toxicants.

Protein. Whole body protein concentrations are influenced by a variety of environmental factors (Claybrook 1983). Obviously, a large percentage of an organism's body is composed of structural proteins. As noted, structural proteins are not intended as stored energy reserves. Under conditions of stress (i.e., starvation), many organisms and particularly invertebrates will mobilize proteins as an energy source (Florkin and Scheer 1970; Gilles 1970; Bayne 1973b). During this mobilization, energy is obtained via the oxidation of amino acids. Glycogen and/or lipids are generally utilized first; however, depending on season, reproductive status of the organism, and the length of stressful conditions, proteins can be an important energy source (Bayne 1973a).

During toxicant-induced stress, protein catabolism was increased in oysters exposed to naphthalene (Riley and Mix 1981), in freshwater crabs exposed to sumithion (Bhagyalakshmi et al. 1983a) and in amphipods exposed to pentachlorophenol (Graney and Giesy 1986b). Increased proteolysis to meet the energy demands of stress were assumed to cause the decline in structural proteins.

Potential Value

Glycogen. The measurement of glycogen reserves in either whole-body homogenates or individual tissues represents a crude, although often useful measurement of the relative energy status of an organism at a particular point in time. Since an organism's energy reserve can directly influence growth and reproductive potential, measurement of such reserves can theoretically be predictive of higher level effects. It is extremely simple and inexpensive to measure important attributes of any biomarker. However, the influence of accessory factors present under field conditions often hinders the interpretation of glycogen levels relative to toxicant-induced stress.

Lipids. As with glycogen, total lipid content represents a relatively crude measurement of energy reserves. Influencing accessory factors must be quantified and understood before lipids can be used as an in situ biomarker. A considerable amount of information is lost by conducting whole organism analysis since the tissue distribution can often provide critical information. Since lipids are closely coupled with growth and reproduction, a direct link may often be found between decreased lipid levels and decreased growth and reproduction. Lipids are most useful as sublethal indicators of long-term stress, although changes in specific lipid components may occur during acute stress.

Protein. As an individual biomarker, measurement of whole organism protein content has little utility. Protein is the last source of energy to be utilized in stressed organisms and thus will not be a very sensitive biomarker. Mobilization and utilization of glycogen and lipids will normally occur before protein catabolism. Since protein content is easily measured and often used to standardize other endpoints, however, information is often available and, when available, should be evaluated among the battery of biomarkers.

Research Needs

Establishing baseline conditions for glycogen is essential to be able to differentiate effects. Field validation research is required for the use of glycogen, lipids, and protein.

Whole Body Calorimetry

The whole-body caloric content of organisms can be calculated either by conversion of energy reserves (i.e., glycogen, lipid, and protein) to caloric equivalents or by direct whole-body calorimetry. The latter is simpler and more direct; however, information on the contribution of the various storage components is lost. Given the crude nature of whole-body proximate analysis, the loss of information may not influence one's ability to evaluate the energetic status of the organism.

History of Use

This technique has been used to evaluate toxicant-related stress in fish (Krueger et al. 1968).

Potential Value

Although crude, whole-body calorimetry may represent a very useful biomarker to be used with other endpoints. It is a simple, inexpensive method of estimating the energetic content of an organism at a particular time.

The influence of accessory factors under field conditions may complicate interpretation of changes. Thus, whole-body calorimetry may be indicative of severe stress only and may not represent a very sensitive indicator. Given the importance of energetics in an organism's growth and reproduction, changes in caloric content can indicate higher level effects.

Research Needs

Work needs to be conducted under field conditions to determine the actual sensitivity of the method and ability to separate toxicant-related effects from background variability. Additional research is needed to determine the usefulness of this biomarker for invertebrates.

Enzymes of Intermediary Metabolism

Basically, there are four different processes that suggest enzymes could respond to specific or nonspecific chemical stress: (1) direct enzyme inhibition, (2) enzyme induction by specific classes of chemicals, (3) elevation of serum enzymes via tissue damage, and (4) alterations in enzyme activity as a result of changes in metabolic pathways or fluxes. This subsection will deal with the fourth category.

The catalytic function of enzymes in biochemical reactions renders them essential for the normal function of all organisms. Enzyme activity is generally regulated such that specific substrates or entire pathways may be homeostatically adjusted to compensate for endogenous or exogenous changes. The complexity of these pathways is enormous; in aquatic invertebrates, the regulatory control of the various enzymes is only beginning to be understood (Zammitt and Newsholme 1976). There are dozens of enzymes that could be used as sublethal indicators of stress. The selection of the appropriate enzymes for any given purpose is enhanced by knowing the type of stressor; the functional rationale for the use of the enzymes; and the physical, chemical, and biological properties of the ecosystem.

Specific enzymes regulating a variety of metabolic pathways can be altered as a result of stress-related homeostatic adjustments induced by toxicant exposure. These changes may involve (1) shifts in catabolic pathways, such as glycogenolysis, glycolysis, and lipolysis, for maximization of energy production to meet demands created by the stressor; (2) changes in anabolic processes, such as protein synthesis for tissue repair; (3) maintenance of redox balance via reoxidation of glycolytically produced NADH; and (4) anaerobically induced shifts in key branchpoint enzymes. Measurement of the rate-limiting or regulatory enzyme controlling these processes can be a useful indicator of the metabolic status of the organism. It is essential that the enzyme being measured is rate limiting. Little quantitative information can be derived by measuring enzymes that catalyze reversible, near-equilibrium reactions; these enzymes are not saturated, and changes in substrate level may alter flux through the pathway without affecting enzyme activity (Crabtree and Newsholme 1975).

Choice of the appropriate enzyme, tissue, and organism to be used in a comprehensive monitoring program can be extremely difficult and requires consideration of a multitude of factors. Since a single enzyme will not normally provide adequate information regarding an organism's condition (except for specific inhibition by a toxicant), the selection of a series of enzymes is recommended. Profiling a suite of enzyme activities will maximize one's ability to determine abnormal processes. Inherent variability of the enzyme needs to be considered since seasonal and molt cycle fluctuations, along with the organism's developmental state and diet, can influence the resolution at which stressed and unstressed populations can be separated (Barnhart 1969; Livingstone 1981). In addition, the tissue-specific distribution of many enzymes also necessitates the selection of appropriate tissues for monitoring each enzyme and requires the development of consistent techniques for sampling tissues for enzyme analysis, especially in smaller organisms.

A number of specific, rate-limiting enzymes have been used as general indicators of stress in aquatic organisms:

- Glycogen phosphorylase is responsible for the breakdown of glycogen and is activated during times of increased energy demand.
- Hexokinase catalyzes the first step of the glycolytic sequence.
- Phosphofructokinase represents the major control point in glycolysis and increases during increased energy demand.
- Pyruvate kinase catalyzes the terminal reaction of glycolysis.
- Isocitrate dehydrogenase is a key regulatory enzyme of the citric acid cycle.
- Glutamate dehydrogenase catalyzes the oxidative deamination of glutamate.
- Aspartate aminotransferase catalyzes the reversible transfer of an amino group from glutamate to oxaloacetate.
- Glucose-6-phosphate dehydrogenase is the key regulatory enzyme of the pentose-phosphate shunt and is essential for the regeneration of reducing power (NADPH).
- Lactate dehydrogenase reduces pyruvate to lactate and is important in redox maintenance.

The previous list does not represent all the enzymes that have been investigated, but is representative of those that may be used. A more thorough review has been provided by Bayne et al. (1985).

History of Use

A considerable amount of research has been conducted on the effects of toxicants on enzyme activities in fish and invertebrates. Glycogen phosphorylase activity has been altered during toxicant exposure in fish (Simon et al. 1983; Morata et al. 1982;

Dragomirescu et al. 1975), although changes in activity have not always been observed (Rush and Umminger 1978). Elevations in phosphorylase activity have been noted in invertebrates exposed to pesticides (Bhagyalakshmi et al. 1983b). Hexokinase, a potential indicator of glycolytic flux, has received little attention as a potential biomarker, although changes have been observed in marine molluscs (Widdows et al. 1982) and fish (Bostrom and Johansson 1972; Sastry and Subhadra 1985). Phosphofructokinase has been shown to both increase (Widdows et al. 1982) and decrease (Blackstock 1980) in invertebrates under toxicant stress. Decreases in pyruvate kinase (PK) activity have been observed in invertebrates and fish exposed to either metals or organic compounds (Blackstock 1980; Phelps et al. 1981; Gould 1981; Bostrom and Johansson 1972); however, PK activity has also been shown to be unaffected by toxicant stress (Widdows et al. 1982; Gould 1980). Isocitrate dehydrogenase activity has been shown to increase (Widdows et al. 1982), decrease (Bhagyalakshmi et al. 1984a), or remain unchanged (Gould and Grieg 1983) in toxicant-stressed organisms. Similarly, changes in the activity of glutamate dehydrogenase and aspartate aminotransferase have not always been consistent (Blackstock 1980; Bhagyalakshmi et al. 1984b; Sastry and Subhadra 1985; Mitz and Giesy 1985b; Phelps et al. 1981; Heitz et al. 1974; Thurberg et al. 1977; Blackstock 1978). The activity of glucose-6-phosphate dehydrogenase has been shown to increase in mussels (Widdows et al. 1982) and decrease in mussels and lobsters (Gould and Grieg 1983; Phelps et al. 1981) during stress. Increases in lactate dehydrogenase activity have been reported for invertebrates (Gould 1980; Phelps et al. 1981). For all of the enzymes just discussed, accessory factors (i.e., diet, season, temperature, sex, and reproductive condition) have been shown to have considerable influence on enzyme activity (McWhinnie et al. 1972; Blackstock 1978; Head and Gabbott 1979; Gould 1981; Frank et al. 1975; Livingstone 1981).

Potential Value

· The measurement of a single enzyme involved with energy metabolism provides only limited information. Measuring a battery of enzymes can provide insight into the metabolic status of the organism. For example, measuring glycogen phosphorylase, hexokinase, and phosphofructokinase can provide information on both carbon flux through the glycolytic pathway and the relative sources of the carbon (i.e., glucose vs. glycogen) (Zammitt and Newsholme 1976). Since both increases and decreases in enzyme activity are possible and not always predictable, a battery of enzymes necessarily enhances interpretation.

A primary limitation in this biomarker is the current lack of data on the key metabolic pathways in aquatic organisms. In addition, the variability observed in field-collected organisms prevents reliable measurement of the sensitivity of enzyme alteration to toxicant-related stress. Together, these drawbacks make interpretation and extrapolation to higher levels of organization extremely difficult.

Research Needs

Before metabolic enzymes can be used as effective in situ biomarkers, a considerable amount of research is needed on their basic biochemistry. Enzyme alterations need to be related to the function of the organism.

Scope for Growth

Scope for Growth (SFG) is an integrative measure of the energy status of an organism at a particular time. It is based on the concept that energy in excess of that required for normal maintenance will be available for the growth and reproduction of the organism. By measuring critical components of energy metabolism, the relative energy balance of the organism can be estimated. Examination of the energy balance equation (Winberg 1971) reveals three critical components that must be measured to calculate Scope for Growth:

$$P = A - (R + U)$$

where

  P = Scope for Growth or energy available for growth and
        reproduction (gamete production)
  A = energy absorbed from food
  R = energy lost via respiration
  U = energy lost via excretion

All energetic measurements are converted to energy equivalents (joules per hour). SFG can increase or decrease depending on a variety of physiological and environmental factors including changes in feeding rate and food availability, alterations in efficiency of food digestion and absorption, and either increases or decreases in respiration and excretion rates (Bayne et al. 1985).

The large majority of work on SFG has been conducted with marine bivalves (Bayne et al. 1985); some work has also been conducted with fish (Warren and Davis 1967). Use as a field monitor requires sampling of organisms and transporting them to a laboratory for measurement of the critical parameters. Whole organism respiration, or oxygen consumption, and nitrogen excretion represent lost pathways easily measured via static or continuous flow methods. Calculation of the energy available from food requires measurement of the amount of food ingested and the gut absorption efficiency of that food. Bayne et al. (1985) have reviewed the methods available for measuring the endpoints necessary to calculate SFG.

History of Use

A considerable amount of work has been conducted on the influence of extrinsic and intrinsic factors on SFG. Temperature (Widdows and Bayne 1971; Widdows 1978), salinity (Stickle and Sabourin 1979), dissolved oxygen (Bayne et al. 1985), food concentration (Griffiths and King 1979), body size (Widdows 1978), and reproductive conditions (Bayne et al. 1985) have all been shown to influence SFG in marine molluscs. Laboratory tests have been conducted that indicate that petroleum hydrocarbons reduce SFG in marine crustaceans and molluscs (Gilfillan and Vandermeulen 1978; Widdows et al. 1982). Similar effects have been observed in molluscan field studies, whose results have often been correlated with general environmental degradation (Widdows et al. 1981; Gilfillan et al. 1976; Gilfillan and Vandermeulen 1978; Bayne et al. 1979). The key to the interpretation of the field studies is to have a large data base on the "normal" condition of the organisms and/or a good reference site for comparison with the test organisms.

Potential Value

SFG has considerable potential as an in situ indicator of toxicant-induced stress. It is an easily measured, general indicator of organism health. As with all biomarkers, it is crucial to have a good understanding of the factors influencing SFG. Bivalve mollusks represent well-studied organisms in SFG work. Their sedentary nature and relative ease of handling make them perhaps the most useful organism for this purpose. Indigenous populations can be used, but the best approach may be to use transplanted organisms, especially when working with bivalve mollusks. Although transplanting has several advantages and can help to reduce the variability associated with specific accessory factors, care must be taken to ensure that exotic species are not introduced into new habitats. Most of the research shows that SFG can be a fairly sensitive indicator of effects and general ecosystem condition. Since the endpoints within the SFG calculation measure factors directly linked to the growth and reproduction of the organism, SFG has an extremely high ecological relevancy. SFG is being used in several programs to monitor ecosystems. As more experience and data are gathered, its utility should expand.

Research Needs

As with all indicators of stress, research is needed comparing the results from controlled laboratory experiments with changes observed under field conditions. Establishing the background variability and the influence of exogenous factors on SFG endpoints are critical. Of all the stress indicators, SFG is probably the most developed and has been rigorously tested under field conditions. (See Marine Ecology Progress Series, 46:1–278, 1988.) SFG endpoints should be incorporated into routine monitoring programs as soon as possible.

Oxygen:Nitrogen Ratio

As discussed in the section on energy reserves, under conditions of stress and increased energy demand, organisms will mobilize either carbohydrates, lipids, or proteins. The relative importance of the different energy sources depends on the organism and its recent metabolic history. Under conditions of severe stress, organisms increase their protein utilization. An index of relative protein utilization is the ratio between oxygen consumed and nitrogen excreted. If more protein is being utilized as an energy source, nitrogen excretion will increase due to amino acid oxidation and the oxygen:nitrogen (O:N) ratio will decrease. O:N ratios have been measured in crustacea (Conover 1978) and molluscs (Bayne 1975).

History of Use

As with SFG, both extrinsic and intrinsic accessory factors influence the O:N ratio and must be documented before the ratio can be used effectively as an in situ indicator. Seasonal and gametogenic cycles (Widdows 1978), diet (Bayne 1973a,b; Bayne and Scullard 1977), and temperature (Widdows 1978) have been shown to influence the O:N ratio. Widdows (1978) has documented decreases in the O:N ratio in bivalves transplanted to polluted waters. Interpretation of the relative conditions of organism based on specific O:N ratios has been proposed (Widdows 1978). In *Mytilus edulis*, O:N values of 20 or less are indicative of starvation or severe stress, whereas values greater than 50 indicate a healthy organism. Such generalized interpretations are likely highly specific to the study, and thus are not applicable to different species or locations.

Potential Value

The sensitivity of the O:N ratio is currently unknown. It is an easily measured general indicator, but its field utility has not been evaluated. Since it primarily reflects protein metabolism, it is doubtful that it will be as sensitive as other indices. For bivalve molluscs, the ratio directly reflects the reproductive cycle of the organism. This obviously must be considered when attempting to use the ratio as an in situ indicator. This ratio can and should be calculated whenever SFG is investigated. This will provide a comparative data base to assess its utility relative to a more directly applicable endpoint such as SFG.

Research Needs

The O:N ratio should be evaluated whenever SFG measurements are made. This endpoint is easier to measure than SFG, and under some circumstances will provide the same information. It is important to validate the utility of this indicator under field conditions. Testing under field conditions will determine its relative sensitivity in detecting change and will evaluate the background noise or variability associated with the ratio.

ENDOCRINE

To maintain homeostasis, organisms must compensate for the physiological and biochemical alterations incurred by exposure to a chemical stressor. The measurement of hormonal concentrations in the blood, as dictated by synthesis, secretion, metabolism, and clearance, may serve as a measure to assess sublethal effects of chemical contaminants. Such measurements may be used to gauge the impact of contaminants on metabolism, growth, and reproduction. This section addresses the current and potential uses for several hormones as biomarkers of chemical exposure.

In order to use plasma concentrations of hormones effectively as biomarkers, knowledge of production rates, half-life, and clearance rates are required. Profiles of plasma hormones throughout an annual cycle must be established. During that period it will be necessary to establish the influence of age, sex, nutritional and reproductive status, and other circannual rhythms. Also, monitoring plasma hormone levels can establish the ability of endocrine tissues to respond to their appropriate releasing factors.

This section discusses the use and potential of the following hormones as biomarkers: corticosteroids, catecholamines, reproductive steroids, thyroid hormones, insulin, glucagon, and growth hormone.

Corticosteroids/Catecholamines

The corticosteroid hormones are the most extensively studied group of stress-related hormones in fish. In Osteichthyes, this group of hormones is comprised of cortisol, cortisone, and 11-deoxycortisol, with cortisol the predominant form. In birds, the predominant adrenal corticosteroids are corticosterone, 11-deoxycorticosterone, and aldosterone. In fish, release of these hormones from the interrenal tissue is mediated through the hypothalamus (corticotropic releasing factor, CRF) and the rostral pars distalis of the pituitary (adrenocorticotropic hormone, ACTH). The hypothalamic-pituitary-interrenal axis (HPI) is activated by a wide variety of chemical and nonchemical stressors. In fish, corticosteroids induce metabolic activity in various organs and osmoregulatory changes at the gill and kidney. Cortisol produces hyperglycemia due to reduced peripheral glucose utilization and gluconeogenesis. Cortisol also decreases total white blood cell counts and alters differential white blood cell counts (Johansson-Sjobeck et al. 1978). In birds, petroleum has been reported to suppress corticosterone secretion (Holmes 1984), while methyl parathion was reported to elevate plasma corticosterone concentrations (Rattner and Franson 1984). Corticosteroids are measured in plasma by either radioimmunoassay, HPLC, or ELISA.

The HPI axis is rapidly stimulated by a variety of physical and chemical stressors resulting in the release of cortisol into the plasma. The elevation of plasma cortisol depends on the intensity and duration of the stressor. After a short-term stressor is removed, plasma cortisol levels will rapidly return to normal. However, with chronic stress the return of plasma cortisol to normal levels is highly variable and is dependent upon the duration, intensity, and nature of the stressor. Data interpretation is somewhat difficult because of the rapid release, metabolism, and cellular uptake of cortisol. In assessing toxicity, handling stress may mask any effects associated with the toxicant. Also, many biotic factors may affect plasma cortisol concentrations including: diet (Bry 1982), photoperiod (Peter et al. 1978), temperature (Strange 1980), dissolved oxygen concentration (Swift 1981), social stress (Ejike and Schreck 1980), and sex (Donaldson and Dye 1975). There does not appear to be any uniform response to toxicants; exposure to copper and acid caused increases in plasma cortisol (Schreck and Lorz 1978; Brown et al. 1984), cadmium exposure produced no change in plasma cortisol (Thomas 1978), and naphthalene and PCBs caused decreases in plasma cortisol (DiMichelle and Taylor 1978; Sivarajah et al. 1978).

The catecholamines, epinephrine, norepinephrine, and dopamine are synthesized and secreted primarily from chromaffin and neural tissues. Release of catecholamines is under neural control. In fish, catecholamines affect heart and gill physiology by altering blood flow, cause hemodilution in freshwater fish and hemoconcentration in marine fish, induce hypercholesterolemia and hyperglycemia, and inhibit the release of insulin. Epinephrine and norepinephrine concentrations are measured in plasma by either fluorometric or enzyme-linked immunological techniques.

History of Use

Evidence is lacking on the effects of chemical contaminants on plasma catecholamine concentrations. However, plasma catecholamine concentrations rapidly respond to biotic stressors such as capture stress, crowding, elevated temperature (Mazeaud et al. 1977), and hypoxia (Butler et al. 1978). Stressors that alter oxygen uptake will stimulate catecholamine release.

Potential Value

Both the corticosteroids and catecholamines provide information on general stress. Care must be taken in sampling to ensure that any elevation in the plasma concentrations of these hormones is due to environmental perturbations, not handling stress. In animals collected from a contaminated site, these measurements may provide information about the general health of the animals.

Research Needs

Extensive field testing should determine whether these measurements will be useful in predicting the health (level of stress) of feral animals.

Reproductive Steroids

The reproductive steroids of teleost fish can be divided into the androgens (e.g., testosterone and 11-ketotestosterone) produced in the testes and maturing ovaries, and estrogen (estradiol) and progesterone produced in the ovary. The major estrogen in birds is also estradiol, while the major androgens are testosterone and 5-α–dihydrotestosterone. The synthesis and release of these hormones are under pituitary control (via gonadotropins I and II in fish, or LH and FSH in birds) which in turn is regulated by the hypothalamus (via gonadotropin releasing hormone). These steroid hormones regulate sexual differentiation and development of gametes. Estradiol is important for its role in the vitellogenic process.

Alterations in circulating levels of reproductive steroids caused by exposure to chemical contaminants appear to be related directly to steroid synthesis and metabolism in mammals (Thomas 1975) and birds (Rattner et al. 1984), but there is contradictory evidence in fish (Freeman et al. 1980; Thomas 1988). One possible mechanism could relate to the chronic chemical stress-induced elevation of cortisol and the resulting negative feedback on ACTH which regulates interrenal/renal androgen synthesis. Altered concentrations of the sex steroids in plasma are believed to reflect reproductive impairment.

History of Use

Early evidence suggested that some chlorinated hydrocarbons impaired normal cyclical changes in plasma concentrations of estrogen, progesterone, and testosterone in birds (Peakall et al. 1981). Those changes were related to the competition of the chemicals with endogenous steroid hormones at the receptor-binding sites in target tissue rather than the impairment of gonadal steroidogenesis. In birds, it has been demonstrated that DDT, as well as its metabolites and structural analogs, bind to estrogen receptors in the shell gland, reduce circulating levels of androgens and estrogens (due to induction of hepatic mixed-function oxygenases), and inhibit gonadotropin secretion (cf. Rattner et al. 1984). Induction of steroid-metabolizing enzymes and subsequent reduction of plasma androgens were observed during exposure of male Atlantic salmon and winter flounder to the water-soluble fraction of crude oil; however, female winter flounder receiving the same treatment had significantly increased plasma levels of estradiol for a 24-h period (Truscott et al. 1983).

DDE (2,2-bis(p-chlorophenyl)-1,1-dichloroethylene) concentrations of 10 g/g and PCB concentrations of 5 g/g in the blubber of Dall's porpoises were related to significantly reduced testosterone levels in the plasma (Subramanian et al. 1987). Sangalang and Freeman (1974) found that testicular steroidogenesis was delayed and plasma levels of 11-ketotestosterone peaked two weeks later in cadmium-treated (1 g/L, 93 d) three-year-old brook trout than in control fish. The two-week delay resulted in asynchronous spawning, with those males ripening later not being available during the peak spawning period. Concentrations of 11-ketotestosterone and cortisol in plasma of spawning male and female coho salmon from Lake Erie (where fecundity and egg survival were poor) were lower than those from Lake Ontario and Lake Michigan. However, no relationship could be made between exposure concentrations of heavy metals or organochlorine chemicals and plasma concentrations of 11-ketotestosterone. Female English sole from sites heavily contaminated with polynuclear aromatic hydrocarbons were less likely to undergo gonadal recrudescence and had lower plasma estradiol levels than females from uncontaminated sites (Johnson et al. 1988). Mount (1988) observed a reduction in plasma levels of estradiol and vitellogenin in brook trout exposed to low pH (4.5) and high aluminum concentrations (486 g/L), but no impairment of reproduction. Thomas (1988) found that lead, benzo-a-pyrene, and PCBs decreased plasma testosterone levels in Atlantic croaker. In vitro assays showed no effect of those chemicals on steroidogenic capacity of ovarian tissue. In a subsequent study with croaker (Thomas 1989), PCBs were shown to impair ovarian growth and reduce plasma estradiol levels, while cadmium accelerated ovarian growth and increased plasma estradiol concentrations.

Potential Value

Measurement of reproductive hormones in the plasma is indicative of the reproductive status of an individual. A survey of male and female fish at a contaminated site may provide information to predict demographic changes at the population level.

Research Needs

Researchers need to determine which concentrations and classes of chemicals alter circulating levels of reproductive steroids. Those values should be compared with results of chronic tests where reproductive impairment was observed, and field validation of laboratory exposures should be conducted.

Thyroid Hormones

In fish, there are two thyroid hormones: thyroxine (T_4) and triiodothyronine (T_3). Thyroxine is the primary hormone secreted from the thyroid while most T_3 is produced through deiodination in the peripheral tissue. In terrestrial animals, the thyroid also produces reverse T_3 and T_2, which serve primarily as additional means of iodine preservation. The release of these hormones is regulated by the pituitary (via thyroid stimulating hormone, TSH) which in turn is mediated presumably through inhibitory control (negative feedback) by the hypothalamus (via thyrotropin releasing hormone, TRH). The thyroid hormones are involved with intermediary metabolism, thermogenesis in homeotherms, gonadal growth and development, and somatic growth regulation. Thyroxine is a critical element in the metamorphosis of anuran amphibians (Regard et al. 1978), the transition of juvenile salmonids from freshwater to seawater (smoltification) (cf. Hoar 1988), and the metamorphosis of flounder from the pelagic to demersal stage (Miwa et al. 1988).

History of Use

In salmon from the Great Lakes, plasma thyroxine levels are altered as a result of goiter and chlorinated hydrocarbons, in particular PCBs. In subsequent studies where those salmon were fed to rats, the rats also developed thyroid hyperplasia, indicating the presence of goitrogenic chemicals in the flesh of those fish. When salmon were fed organochlorine chemicals (those most commonly observed in feral fish), T_3 and T_4 levels were depressed; however, no effect on thyroid histology was observed (Leatherland and Sonstegard 1978, 1979, 1980). Herring gull colonies from the Great Lakes area also exhibited histological signs of goiter and depressed levels of T_3 and T_4 (Mineau et al. 1984). DDT and PCBs produced dose-dependent hyperthyroidism or hypothyroidism in Japanese quail (Grassle and Biesmann 1982). In smolting coho salmon injected with PCBs, plasma thyroxine concentrations were observed to "surge" or increase later than sham or uninjected controls. That delay in smoltification was associated with decreased seawater survival (Folmar et al. 1982). Similar results were observed in coho salmon smolts exposed to arsenic (Nichols et al. 1984). Plasma T_3 concentrations were elevated in black ducks (*Anas rubripes*) reared on acidified ponds (Rattner et al. 1987). A single oral dose of crude oil caused significant elevations in plasma thyroxine in nestling herring gulls for a two-week period (Peakall et al. 1981). Mice treated with the organophosphate Soman (pinacolyl methylphosphonofluoridate) showed significant, but transient, increases in plasma T_3 and T_4 concentrations, which returned to normal within 24 h (Clement 1985).

Potential Value

Currently, the assay of plasma thyroid hormones can predict general developmental problems in growth (intermediary metabolism), smoltification in salmonids, and metamorphosis in anuran amphibians. Plasma thyroid hormone concentrations may reflect goiter or a predisposition to goiter without labor intensive histopathological evaluations.

Research Needs

Research needs to establish how plasma hormone concentrations respond to different classes of chemicals in the laboratory. Threshold concentration of effects should be developed. Circadian and circannual changes under normal and variable environmental conditions (e.g., temperature, pH, salinity, and rearing density) should be established. If laboratory testing demonstrates thyroid hormones to be responsive either generally or to a specific class of chemical stressors, then the technique must be validated in the field with feral animals collected from contaminated and uncontaminated sites.

Insulin/Glucagon

Insulin is produced in the B-cells of the endocrine pancreas, glucagon in the A-cells. Together, insulin and glucagon regulate intermediary metabolism by facilitating cellular uptake and utilization of plasma glucose (insulin) and stimulating glycolysis and glycogenolysis (glucagon). In both birds and fish, the primary stimulus for the release of insulin is plasma glucose; however, amino acids, other pancreatic hormones, and gastrointestinal hormones are also insulinotropic. The release of insulin is inhibited by hypoglycemia and somatostatin, and results in hypoglycemia, and decreased plasma fatty acids and amino acids. Glucagon promotes the conversion of amino acids into glucose (Chester-Jones et al. 1987; Plisetskaya 1989).

History of Use

Gardner and Yevich (1988) reported a hyperplastic condition of the B-cells of the endocrine pancreas (nesidioblastosis) in winter flounder (*Pseudopleuronectes americanus*) collected from the highly contaminated area of Black Rock Harbor, Connecticut. Measurements of plasma glucose in the exposed fish proved to be an unreliable estimate of insulin secretion. The values were highly variable, probably a result of catecholamine and corticosteroid secretion caused by handling stress. In this case, it would have been ideal to employ a radioimmunoassay (RIA) for plasma insulin to test its response and sensitivity as associated with a chronic exposure to polynuclear aromatic hydrocarbons. Unfortunately, homologous RIAs for insulin have been developed for only a few species (e.g., hagfish and salmonids). Insulin is a highly conserved hormone; however, in some cases it does not demonstrate the expected immunological cross-reactivity. For example, the antibody to salmon insulin shows good cross-reactivity with snails, but not halibut. Rainbow trout exposed to zinc exhibited hyperglycemia and reduced plasma insulin concentrations. After nine days of exposure, plasma glucose and insulin concentrations returned to normal (Wagner and McKeown 1982).

Potential Value

Insulin and glucagon RIAs may reduce the necessity to conduct labor-intensive histopathological evaluations to demonstrate damage to the endocrine pancreas.

Research Needs

Insulin RIAs should be developed for fish species commonly used in laboratory toxicity testing and representative feral fish from different geographical areas. Insulin output in response to clinically induced conditions of hypoglycemia and hyperglycemia should be measured after acute and chronic exposures to a variety of concentrations and classes of chemicals. These techniques require field validation with feral animals.

Growth Hormone

In fish, growth hormone (GH) is synthesized and secreted from the adenohypophysis (proximal pars distalis) of the pituitary presumably under the regulation of the hypothalamus via growth releasing hormone (GRH) which promotes GH release, and somatostatin which inhibits GH release (Chester-Jones et al. 1987). The overall action of GH is to promote growth of cartilage, bone, and soft tissues by stimulating protein synthesis. Growth hormone, like insulin, is anabolic and promotes the transport of amino acids into peripheral cells.

Plasma levels of GH are quite variable under normal circumstances; therefore, random measures of GH are not generally informative. More information can be obtained by injecting experimental animals with GRH or by inducing an abnormally high dietary glucose load to determine the functional state of GH production in the pituitary.

Only recently have homologous RIAs for growth hormone become available for eel (Kishida and Hirano 1988), goldfish (Marchant et al. 1989), and salmonid fishes (Bolton et al. 1986). The established RIA for birds is a heterologous assay based on chicken antisera.

History of Use

In studies with growth-stunted coho salmon, plasma concentrations of GH were significantly elevated above GH levels in normal coho smolts. This finding reinforces the need to consider metabolism and clearance as well as plasma concentrations of hormones (Bolton et al. 1987).

Potential Value

The relationship between GH and growth inhibition must be established in fish exposed to known growth inhibitors such as chlorpyrifos, phenol, or ammonia. If a quantitative relationship exists, GH levels measured during acute toxicity tests may predict chronic growth reduction. Similar measurements must then be made on feral fish from contaminated locations. This information could be used with existing demographic models to predict future population structure.

Research Needs

RIAs need to be developed for other fish species of interest. Consideration should be given to species used both in laboratory toxicity testing and representative feral fish from different geographical areas. Extensive testing is required to establish the ability of the pituitary to respond to somatotrophic or glucose stimulation after acute and chronic exposures to different concentrations and classes of chemicals. Field validation of techniques is also needed with feral animals.

BLOOD CHEMISTRY

Blood chemistry offers the potential for several biomarkers of toxicant stress. Serum enzyme activities, ion levels, and concentrations of glucose and lipids are considered in this section.

Serum Enzymes

Serum enzyme activities have been used extensively to provide simple accurate measures of organ dysfunction in mammals, and recently have received greater attention from aquatic toxicologists. Increased serum enzyme concentrations can result from either of the following:

1. enzyme leakage from a cell with a damaged cell membrane
2. increased enzyme production and leakage from the cell
3. decreased enzyme clearance from the blood

Currently, the specific mechanism by which serum enzyme activities increase is not known. Nonetheless, it is agreed that it can be due to, and diagnostic of, tissue damage (Galen 1975; Chenery et al. 1981). Chemical and nonchemical stressors can lead to release of cellular enzymes (Rattner et al. 1983) possibly due to localized ischemia, depletion of intracellular nucleotides and high energy phosphates, or changes in membrane potential (Cole and Palmer 1979; Freedel et al. 1979; Highman et al. 1965; Wilkinson 1978). In some studies, histopathological changes were noted before the occurrence of increased serum enzyme activities (Dalich et al. 1982).

Certain serum enzymes, by virtue of increased levels in a specific tissue, are indicative of cellular damage or dysfunction in that tissue. For example, in humans, acid phosphatase (ACP) is present at increased concentrations in the mammalian prostate and increased serum activity in males is a specific indicator of prostate cancer (Sullivan et al. 1942). Increased serum activities of specific transaminases are diagnostic of heart and/or liver dysfunction. Elevated serum N-acetyl-β-D-glucosaminidase (NAG) activities (Ackerman et al. 1981) and altered serum NAG isozyme patterns (Tucker et al. 1980) are indicative of a variety of diseases in humans.

History of Use

Serum Lysosomal Enzyme Levels. The lysosomal enzymes are potential biomarkers due to their importance in metabolic and pathologic processes as well as their recent use in toxicological investigations (Sunderman and Horak 1981; Nogawa et al. 1986). Lysosomes function in the cell to catabolize organelles and macromolecules.

Serum and tissue activities of the lysosomal enzymes have received little attention in fish research; however, the available studies indicate their utility in understanding the effects of surfactants, pesticides (Gupta and Dhillon 1983), and metals (Jackim et al. 1970; Versteeg and Giesy 1986) on fish. Versteeg and Giesy (1986) observed an increase in ACP and NAG activities in the serum of bluegill sunfish exposed to cadmium. In the bluegill sunfish, spleen and liver N-acetyl-β-D-glucosaminidase (NAG) activities are approximately equal and have approximately twice the activity of the kidney (Versteeg 1985). Thus, one might expect NAG activity in serum to be a good biomarker for spleen and liver dysfunction. Bouck (1984) found plasma concentrations of the lysosomal enzyme, leucine amino naphthylamidase (LAN), to be useful for quantifying tissue damage in fish.

The mechanism responsible for increased serum levels of lysosomal enzymes is not known but may differ from other serum enzymes. Lysosomes contain increased concentrations of metals and may be active in the degradation of metal binding proteins (Fowler and Nordberg 1978). Metals have also been shown to increase lysosome numbers and reduce lysosomal membrane stability possibly leading to enzyme leakage (Versteeg 1985; Versteeg and Giesy 1986; Leland 1983; Moore and Stebbing 1976). In general, increased serum lysosomal enzymes appear to be indicative of toxicity and may be due to increased enzyme production, decreased lysosomal stability, or tissue damage. Much additional research is needed to understand the mechanisms of lysosomal enzyme appearance in serum; the effect of nonchemical stressors on serum levels; organismal, spatial, and temporal variability; and the correlation of effects on enzyme levels to relevant population-level effects.

Serum Transaminase Activity. Serum transaminases, specifically aspartate aminotransferase (ASAT) and alanine aminotransferase (ALAT), have been widely utilized in mammalian toxicology as a biomarker of specific organ dysfunction (Wroblewski and LaDue 1956). ASAT is a nonspecific cytosolic and mitochondrial enzyme found in a variety of tissues including liver, skeletal muscle, cardiac muscle, and kidney (Verma et al. 1981). ALAT is also a cytosolic enzyme, but it is more tissue-specific and is normally associated with the liver. Both of these enzymes have been measured in fish and invertebrates under stress. Increased serum transaminase activity is usually associated with hepatocyte dysfunction since many compounds are metabolized in the liver, where transaminase activities are high (Gingerich 1982). However, increased activities of the transaminases also occur in other organs, such as the heart (Gaudet et al. 1975; Versteeg 1985). Increased serum activities of the transaminases are good biomarkers of toxic effects at the cellular level; however, transaminase activities in serum do not necessarily correlate with mortality or other adverse effects (Lane and Scura 1970).

Metals have been found to affect the activities of transaminases in fish. Fish exposed to acutely toxic concentrations of cadmium, mercury, or copper had increased transaminase activities (McKim et al. 1970; Roberts et al. 1979; Williams and Wooten 1981; Verma et al. 1984; Versteeg and Giesy 1986). Alternatively, chronic exposure to copper has been reported to decrease serum ASAT activity while chronic cadmium exposure did not affect activities of this transaminase in serum (McKim et al. 1970). Little work has been conducted with serum transaminase activities in invertebrates following metal exposure.

Increases in serum transaminases have also been associated with exposure to organic xenobiotics. Serum ASAT:ALAT ratios in the bluegill sunfish increased after carbon tetrachloride (CCl_4) exposure (Versteeg 1985). The greater increase in serum ASAT compared to serum ALAT suggests: (1) the liver with its low ASAT:ALAT ratio was not the only organ leaking these enzymes, or (2) ASAT is released to a greater degree than ALAT from the liver. The heart, with a relatively high ASAT content, could provide a significant portion of the serum ASAT. In rainbow trout, serum ASAT:ALAT ratios do not change following CCl_4 exposure (Racicot et al. 1975; Statham et al. 1978); however, in the English sole, the serum ratio increased from 0.3 to 5.4 (Casillas et al. 1983). Monocholorobenzene (MCB) injected into rainbow trout caused an increase of ALAT activity in serum, indicating liver dysfunction (Dalich et al. 1982). The response of serum enzymes was approximately as sensitive as behavioral effects but not as sensitive as histopathology.

Serum transaminase levels have also been used effectively as biomarkers of chemical effects in terrestrial organisms. The use of these enzymes as biomarkers in toxicity studies with laboratory rats and mice is widespread (Galen 1975). Utilization of these enzymes with feral species has received less attention.

In a 12-week feeding study of the quail (*Coturnix coturnix japonica*), serum ASAT activities were increased in a dose-dependent manner during exposure to DDE, Aroclor 1254, and malathion (Dieter 1974) while mercury did not alter serum ASAT levels. In this study, use of multiple enzymes (i.e., creatinine phosphokinase, AChE, and LDH) were effective at identifying probable sites of action of each toxicant. Lack of increase in serum transaminases provide useful, but not necessarily exhaustive, information on the lack of effects on the liver. Rattner (1981) used the transaminases and other blood analyses to demonstrate a lack of overt toxicity of crude oil on the adult male mallard (*Anas platyrhynchos*). Neither ASAT nor ALAT levels were increased in serum of ducks fed crude oil for seven days ad lib. In a similar design, serum ASAT and ALAT activities were unaffected after feeding white-footed mice (*Peromyscus leucopus novebora-censis*) and meadow voles (*Microtus pennsylvanicus*) the organophosphate insecticide acephate (Rattner and Hoffman 1984).

Lactate Dehydrogenase. The LDH protein is a tetramer composed of H (heart) and M (muscle) subunits, and may occur in up to five isomeric forms in fish (Markert and Faulhaber 1965). LDH activity in fish occurs at greater concentrations in muscle tissue than in other tissues, making total serum LDH levels a potential biomarker of muscle damage (Oikari et al. 1983; Versteeg 1985). Due to the organ specificity of some LDH isozymes, separation and quantification of isozyme levels in blood can make LDH a specific biomarker of cellular damage in other organs.

Carbon tetrachloride exposure caused a large increase in serum LDH activity. This increase, however, was not due to a specific LDH isozyme (Versteeg 1985), suggesting that the toxicity of CCl_4 is not specific to the liver in the bluegill sunfish. IMOL S-140, a triaryl phosphate oil, produced similar results (Lockhart et al. 1975). Exposed fish had increased serum LDH activities, but the serum LDH was due to release of both muscle and heart-type LDH isozymes. In rainbow trout, Racicot et al. (1975) has demonstrated that the increase in serum LDH after CCl_4 exposure was due primarily to liver LDH isozymes.

Serum LDH was not increased following sublethal exposure of fish to metals (Versteeg and Giesy 1986; Roberts et al. 1979). This may be due to the lack of significant damage to LDH-containing tissues, as indicated by histopathology in metal-exposed fish (Versteeg and Giesy 1986).

Other Serum Enzymes. Activities of a wide variety of enzymes have been used successfully to investigate toxic effects of chemicals. Additional research on these enzymes is encouraged. The present discussion should not be considered a thorough review of the literature, but merely selected examples of appropriate uses for serum enzymes.

Potential Value

Serum enzymes have been demonstrated as useful biomarkers of tissue damage. The relationship between serum enzyme levels and organismal-level effects are well documented. The use of a suite of serum enzymes may prove to be useful for understanding population-level effects.

Research Needs

Serum enzymes have been widely used to assess the effects of xenobiotics in fish. Many of these studies involve short-term exposure to one contaminant in a controlled laboratory setting during which one or a few biotic parameters are assessed. Future research should involve long-term exposures in which both the exposure and recovery phases are extensively sampled. Multiple biotic parameters comprising, but not limited to, serum enzymes should be included to help develop a thorough understanding of the biological status of the organisms. To facilitate the interpretation of altered enzyme profiles in feral populations, additional research is needed to understand: (1) the effects of changes in environmental conditions (e.g., temperature, photoperiod, and diet quality), (2) intraspecies and interspecies differences, (3) the impacts of exposure to multiple compounds, (4) the impact of fluctuating xenobiotic concentrations on scrum enzyme levels, and (5) the long-term consequences of altered serum enzyme profiles. Additional research is also needed on invertebrate hemolymph enzymes.

Ion Levels

Maintenance of constant internal ion concentrations (e.g., sodium, chloride, calcium, and magnesium) is essential, requiring active regulation of water influx and ion efflux in aquatic organisms. Alterations in the ionic balance of environmental organisms can be due to stressor effects on the ion regulatory organs (Gilles and Requeux 1983; Spronk et al. 1971), internal and external sensory receptors involved with detection of changes in osmotic conditions (Inman and Lockwood 1977), the endocrine system (Mazeaud et al. 1977; Harman et al. 1980), metabolism (Greenway 1979), or active transport processes (Inman and Lockwood 1977; Jowett et al. 1981). Thus, ion levels in blood or hemolymph, as measured by osmolality or specific ion concentrations (Na^+, K^+, Cl^-) have potential as sensitive biomarkers of chemical exposure and effects.

History of Use

Acute exposure of freshwater fish to heavy metals usually results in decreased plasma ion levels (Staurnes et al. 1984; Lock et al. 1981). These studies support the use of blood ion levels for the quantification of toxic effects of metals during acute exposures. During long-term exposures of fish to metals, however, osmoregulation is not always impaired. Initial effects may be transient, and ion concentrations can return to normal values despite continued exposure to toxic concentrations (McKim et al. 1970; Skidmore 1970; Watson and Beamish 1980; Hilmy et al. 1982).

Little information exists concerning the effects of organic toxicants on fish osmoregulation. The available information indicates that organic toxicants can increase or decrease plasma Na^+, Cl^-, and osmolality (Grant and Mehrle 1973; Zbanysek and Smith 1984; Srivastava and Mishra 1983; Lidman et al. 1975; McKeown and March 1978). Thus, the utility of ion levels in assessing effects of organic xenobiotics will require additional research before it can be considered a potential biomarker for these compounds.

Although many studies have shown osmoregulatory impairment in invertebrates during toxicant exposure, an equal number have failed to demonstrate alterations in hemolymph ionic composition (Roesijadi et al. 1974, 1976b; Aarset and Zachariassen 1982; Cantelmo et al. 1982; Thurberg et al. 1977; Thomas et al. 1981). In one of the few field studies attempting to relate the ionic composition of hemolymph to pollution gradients, Phelps et al. (1981) were unable to relate osmolyte alterations to the degree of pollution. The varied response noted among different studies represents a primary limiting factor in utilizing hemolymph composition as an indicator of toxicant exposure. Even though organisms are known to be under toxicant-induced stress (as determined by histology, enzyme activity, respiration, growth, or mortality), adaptive mechanisms can enable the organism to maintain a constant internal environment. In general, alterations in osmoregulation do not seem to be sensitive to sublethal toxic exposure in invertebrates.

Potential Value

Effects of stressors on osmoregulation have not been extensively related to effects on survival, growth, or reproduction, so the utility of monitoring the osmoregulatory system for predicting population-level effects is currently low. Many biotic and abiotic factors influence osmoregulation, further complicating the use of osmoregulatory status as a biomarker of population-level effects.

Difficulties with inherent variability, accessory factors, and data interpretation limits the applicability of osmoregulatory alterations as stress indicators in field monitoring studies. However, consideration of these factors in controlled laboratory experiments will enable the researcher to reduce the variability and separate toxicant-induced alterations from normal fluctuations. For this reason, changes in the extracellular ionic composition of an organism has potential use as a hazard assessment tool.

Research Needs

Plasma and hemolymph ion levels may prove to be useful biomarkers, but additional research is necessary. Toxicological research to date has consisted primarily of short-term exposures with the intent of detecting alterations in internal ion pools. Future research should aim at understanding the root cause of changes in ion levels during and following short- and long-term exposures. Only by understanding the reasons for changes in ion levels will we be able to evaluate the utility of ion regulation as a biomarker. Understanding the relationship between xenobiotic effects (e.g., survival, reproduction, and growth) and ion levels is also necessary, as is understanding variability, intraspecies and interspecies difference, and effects of nonchemical stressors.

Serum Glucose and Lipids

The blood is the major system for transporting energy-related biomolecules between storage and utilization sites in fish; the hemolymph is the equivalent system in invertebrates. In acute stress, the endocrine system (i.e., catecholamines and glucocorticoids) generally controls energy mobilization. During chronic stress both endocrine and tissue-level systems (synthesis and degradation) interact to affect energy mobilization. Mobilization of energy stores is required to maintain homeostasis during chemical challenge. Serum and hemolymph glucose and lipid levels are affected by storage tissue release and tissue uptake for utilization. Thus, serum and hemolymph concentration of energy-related biomolecules may be useful biomarkers of toxicant stress. Since energy storage and utilization are so closely tied to diet and activity, a large data base will be required for field application of these techniques to reduce variability associated with accessory factors of nutrition, sex, temperature, activity, reproductive status, and others.

History of Use

Glucose. Acute heavy metal exposure generally causes increased concentrations of glucose in plasma of fish (Watson and McKeown 1976; Shaffi 1980; Chaudhry 1984; Gill and Pant 1983). However, chronic exposure of fish to heavy metals has been reported to either increase (Sastry and Subhadra 1985) or decrease plasma glucose levels (Christensen et al. 1977; Gill and Pant 1983; Sastry and Sunita 1983; Haux and Larsson 1984).

In one study, rainbow trout exposed to titanium showed increased plasma glucose and lactate concentrations when exposed at 13–15°C, but showed no effect on glucose and lactate concentrations when exposed at 7–8°C (Lehtinen et al. 1984). The results of this study indicate some of the problems in interpreting the results of energetic and metabolic indicators of toxicant-induced stress in fish.

Numerous organic compounds cause a rapid increase of glucose concentrations in plasma when fish are exposed to acutely toxic concentrations (McLeay and Brown 1975; DiMichelle and Taylor 1978; Casillas et al. 1983; Thomas et al. 1981; Gluth and Hanke 1984; Soivio et al. 1983; Pant and Singh 1983). The magnitude of the increase can depend on liver glycogen concentration (McLeay 1977). Hypoglycemia has also been reported following acute exposure of fish to fenitrothion (Koundinya and Ramamurthi 1979) and a mixture of aldrin and formithion (Singh and Srivastava 1981).

The response of plasma glucose in fish chronically exposed to organic compounds has not been extensively studied. In the few chronic studies reported, pesticide exposures have led to hyperglycemia (Singh and Singh 1980; Sastry and Siddiqui 1982).

In invertebrates, hyperglycemic responses have been observed in freshwater crabs exposed to DDT (Fingerman et al. 1981), fenitrothion (Bhagyalakshmi et al. 1983b), and BHC (benzene hexachloride) (Sreenivasula et al. 1983); and in blue crabs and polychaete worms exposed to pentachlorophenol (PCP) (Coglianese and Neff 1982; Thomas et al. 1981). These responses have been attributed to the release of hyperglycemic hormones and subsequent mobilization of stored glycogen reserves. However, hyperglycemic responses are not always observed. Thomas et al. (1981) observed increased glucose in the coelomic fluid of *Neanthes virens*, but hypoglycemia generally preceded hyperglycemia during acute PCP exposure. The transitory nature of changes in glucose concentrations in hemolymph may limit its usefulness as an indicator of stress. In addition, the influence of accessory factors on glucose concentrations in hemolymph will impede one's ability to separate toxic effects from background variability. Factors such as sex (Dean and Vernberg 1965), reproduction, seasonal and diurnal cycles, and molt status (Telford 1974) influence glucose levels and need to be considered before using a particular organism.

Lipids. Relatively few studies have investigated the effects of toxicant-induced stress on the lipid content of fish, and the effects of stressors appear equivocal. The free fatty acid concentrations in the plasma can either increase or decrease in response to stress according to species (Mazeaud et al. 1977).

Cholesterol concentrations in serum were increased by exposure of fish to methyl parathion (Rao and Rao 1984b) and decreased during exposure to a variety of chlorinated hydrocarbons (Gluth and Hanke 1984). As there is no consistent trend, it is difficult to interpret chemical effects on cholesterol concentrations. Cholesterol does not appear to be a useful biomarker of effects of toxic chemicals.

Hemolymph lipids have been altered in aquatic invertebrates exposed to toxicants. Hypocholesterolemia has been reported in blue crabs exposed to PCP (Coglianese and Neff 1982), although no differences in the fatty acid profile were observed in similar experiments (Bose and Fujiwara 1978). Alterations in the cholesterol level may be an indirect result of toxicant effects on metabolic enzymes. O'Hara et al. (1985) have shown that polynuclear aromatic hydrocarbons can interfere with the cytochrome P-450 enzyme system such that cholesterol metabolism is impaired.

The greater energy demand caused by the stress of toxicant exposure can result in a mobilization of lipid energy reserves and subsequent decline in the total lipid content of the organism. A decrease in lipid content was noted in amphipods exposed to fuel oil (Lee et al. 1981), while increased lipid catabolism was measured in oysters exposed to naphthalene (Riley and Mix 1981). Capuzzo et al. (1984) found that petroleum hydrocarbons significantly interfere with lipid metabolism resulting in consistently lower triglycerol concentrations in hemolymph of exposed American lobster larvae.

Potential Value

Serum and hemolymph energy-related biomolecules can be used successfully as biomarkers of organismal-level effects. Utilization and interpretation of these biomarkers are optimized when whole-body energy reserves are also evaluated (see the previous section on energetics). These biomarkers can now be used in field situations, but care should be taken in extrapolating effects to the population level. Many different types of lipids exist, and their occurrence and importance are species specific.

Research Needs

Measurement of serum and hemolymph glucose and lipids are useful tools for understanding toxic effects of chemicals in controlled laboratory experiments. Their use as biomarkers in the field will require an understanding of normal ranges and the influences of duration of exposure, energetic status, sex, diet, season, temperature, and many other factors on these ranges. These biomarkers may be most useful when coupled with other measures of energetic status (i.e., whole body calorimetry and key metabolic enzymes) and organism health. Studies are needed comparing effects on serum and hemolymph glucose and lipid levels and effects at the population, community, and ecosystem levels. Additional research on the specific energy-related lipids in environmental organisms is needed to further increase the utility of these biomarkers.

GROWTH RATE

The potential success of biomarkers for ecological effects can be enhanced by selecting those closely related to survival, growth, and reproduction. Growth of an organism is generally used as a sensitive and reliable endpoint in chronic toxicological investigations. Reductions in growth imply that growth rate has decreased during the growth process. Several biomarkers of protein synthesis rate have been developed and may be used as indicators of the growth rate of an animal. These biomarkers tend to be nonspecific because they respond to a wide variety of toxicants. RNA concentration and radiolabeled amino acid incorporation have received considerable attention as biomarkers of growth rate in ecotoxicology. Biomarkers of protein synthesis rate in plants have not been extensively investigated. This is at least partially because of the ease with which traditional methods of measuring growth rate in plants (by changes in biomass with time) can be applied in both laboratory and field.

RNA Concentration

The rate of protein synthesis and the concentration of RNA in rapidly growing cells have shown a positive correlation (Brachet 1960). Since protein is a major component of an organism's dry weight, it has been suggested that RNA concentration be used to make inferences about the growth rate of aquatic microbes (Karl et al. 1981), phytoplankton (Dortch et al. 1983), invertebrates (Lang et al. 1965; Sutcliffe 1970; Dagg and Littlepage 1972; Barnstedt 1983; Wright and Hetzel 1985), and fish (Bulow 1970; Haines 1973; Buckley 1984). RNA concentration is generally expressed relative to protein or DNA content. The ratio of RNA to DNA is generally believed to more accurately reflect growth rate because DNA concentration tends to remain relatively constant among cells (Fiszer-Szafarz and Szafarz 1984).

History of Use

RNA:DNA ratios were initially used in environmental sciences to predict in situ growth rates of zooplankton (Sutcliffe 1970) and fish (Bulow 1970). Research over the past 20 years has been moderately successful in this regard. An important result of this research was the identification of variation which can limit the use of this biomarker. Major sources of variation in RNA concentration in animals are attributed to the age of the organism and changes in environmental conditions such as food supply, temperature, pH, and possibly toxicants (Bulow 1987).

Variation due to environmental conditions can be controlled in the laboratory and variation due to age can be quantified. Controlling variation in this manner has allowed evaluation of toxicant effects on RNA concentration in laboratory studies using both invertebrates and fish. Chronic studies using *Daphnia magna* exposed to several different toxicants indicated that RNA concentration was affected by toxicant exposure during the rapid growth phase of these organisms; however, the effect was not as sensitive as traditional endpoints of growth and reproduction (McKee and Knowles 1986; Knowles and McKee 1987). Laboratory studies with fathead minnows (*Pimephales promelas*) exposed to various toxicants demonstrated that, in most cases, the 28 to 32 d no observed effect concentration (NOEC) could be predicted by effects on RNA concentration after 96 h of exposure (Barron and Adelman 1984). Laboratory studies with salmonids chronically exposed to toxicants have not shown a sufficient increase in sensitivity of RNA concentration relative to traditional endpoints to warrant further laboratory toxicity studies with this group of fish (Cleveland et al. 1986; McKee et al. 1989).

Variation due to environmental conditions cannot be eliminated in field investigations; however, it can be reduced by proper selection of control sites. For instance, several investigations using RNA concentration as an in situ biomarker of toxicant exposure in fish have been conducted. In a lake study using yellow perch exposed to cadmium and zinc pollution, Kearns and Atchison (1979) demonstrated both RNA:DNA ratios and growth rates to be negatively correlated to the pollution. In a study using salmonid fish exposed to carbaryl in situ, however, Wilder and Stanley (1983) identified an increase in RNA:DNA ratios, suspected to be a result of increased consumption of dead insects by the fish. These contrasting studies indicate RNA concentration may be useful in field assessments of growth rate, but allude to the need for identifying particular factors that can influence variability.

Potential Value

RNA concentration is potentially a useful tool in hazard assessment and monitoring in situ. Its usefulness in the laboratory is limited because growth is similar in sensitivity and more easily measured under laboratory conditions. Growth is not as easily measured in the field as in the laboratory, however, increasing the potential for applying biomarkers of growth rate. Moreover, RNA concentration may provide cost-effective alternatives for assessing growth rate compared to age class weight determinations and caged animal techniques.

The major limitation on the use of RNA concentration as a biomarker of growth rate in situ is the variation associated with environmental factors, especially nutrition and temperature. These parameters can affect different species in different ways and, without background information, can make interpretation difficult (Bulow 1987). For example, Goolish et al. (1984) found that acclimation of fish to cold temperatures required an increase in RNA concentration to maintain the growth rate observed at higher temperatures. In this situation, increased RNA concentration may be compensating for reduced efficiency of the protein synthesis process at low temperatures.

Research Needs

The usefulness of this biomarker needs to be validated. One approach would be to analyze RNA concentrations in populations in which growth rates of individuals are known. Proper species and tissue selection is critical. Current recommendations are to use whole animal for most invertebrates and larval fish. For larger fish, white muscle tissue samples are advised (Bulow 1987). Analytical techniques should be refined to reduce interferences and to investigate other, more specific analyses for RNA. Temporal, spatial, and individual variation of biomarkers in situ should be quantified.

Protein Synthesis

Growth is accomplished primarily through protein synthesis (Lied and Rosenlund 1984). Consequently, measurement of protein synthesis rate in certain tissues has been shown to correlate with whole organism growth rate (Adelman 1987). A biomarker of growth based on protein synthesis rate could have wide application in environmental sciences and ecotoxicology. Protein synthesis rate is generally considered to be a nonspecific biomarker since it can be affected by many different toxicants; however, some toxicants can affect protein synthesis directly.

History of Use

Traditionally, the rate of protein synthesis has been determined by measuring the rate at which radiolabeled amino acids are incorporated. In fish, muscle and scales have been the primary tissues utilized to assess protein synthesis rate. Since muscle tissue accounts for about 50% of total fish weight, measuring the protein synthesis rate in muscle may provide a reliable biomarker of growth rate (Lied and Rosenlund 1984). Measurement of protein synthesis in muscle tissue can be determined in vitro in homogenized tissue samples (Lied and Rosenlund 1984). It can also be measured in vivo by introducing the radiolabeled amino acid directly into the bloodstream of cannulated fish (Smith 1981) or by adding it directly to the water (Adelman and Busacker 1982; Fauconneau 1984). Laboratory experiments using these techniques have demonstrated that the calculated protein synthesis rate is related to growth rate and that the biomarker can change as a result of environmental stresses such as food deprivation (Smith 1981).

Uptake of radiolabeled glycine by fish scales in vitro has been used as a relative measure of protein synthetic rate (Adelman 1987). Under certain conditions, this biomarker can relate to growth rate of the fish. Although the relationship between protein synthesis in the scale and whole organism growth rate is more tenuous than muscle protein synthesis rate, the scale model has several advantages. First, the relative ease of performing the assay allows processing a larger number of samples. Second, the technique does not require sacrificing or prolonged restraining of the organism. This method has been shown to be correlated with stressors known to affect fish growth (Adelman 1980).

Viarengo et al. (1980, 1981) have used ^{14}C-labeled leucine incorporation to develop a sublethal stress indicator based on protein synthesis in gill and digestive gland tissue of the marine mussel, *Mytilus galloprovincialis*. These authors correlated copper body burdens with reduced protein synthesis (Viarengo et al. 1980) and applied the technique to organisms collected from polluted and unpolluted environments (Viarengo et al. 1981). A similar assay measures ^{14}C-labeled thymidine uptake by marine invertebrate embryos (Jackim and Nacci 1984). The assay has proven to be a useful predictor of survival and growth in long-term tests.

Potential Value

Protein synthesis rate can be used to make inferences about growth rate of individuals in the laboratory and in the field. The value of protein synthesis rate as a biomarker of growth rate in the laboratory is limited because of the ease of direct growth rate measurements. However, laboratory research can yield information on protein metabolism/turnover rates and the relationship between protein synthesis rate and whole body growth rates.

Estimation of growth rate of fish or other organisms based on protein synthesis rate in field populations has obvious relevance to ecotoxicology. These estimations require clear definition of the relationship between individual growth rate and protein synthesis rate in vivo or in vitro. In the case of muscle tissue protein synthesis rate, the relationship is well-defined, but the methodology for this procedure is complex and not practical for large numbers of samples (Adelman 1987). Measurement of radiolabeled glycine into protein of fish scales does not have these methodological restrictions.

Research Needs

Future research should be directed at clarifying the potential for using scale protein synthesis rate to predict whole organism growth rate. Limitations are similar to other biomarkers of growth rate. Validation research should continue to determine if protein synthesis accurately reflects growth rate of individuals in the field and to identify the influence of environmental factors.

GROSS INDICES

Gross indices are sometimes indicative of toxicant effects. Plant or animal condition as determined by morphology, appearance, and other gross characteristics should not be overlooked in assessing contaminant impact. Skeletal abnormalities also have potential for monitoring organism health and predicting contaminant effects. Measurement of avian eggshell thinning already has wide application as a biomonitoring technique in the field of environmental toxicology.

Condition

A first tier screen to identify potential pollutant exposure and effect can be accomplished through overt and relatively simple measures of condition (Goede 1989). Such measures may serve to identify the most sensitive members of a population or plant community. Various indices have been constructed from different morphological characteristics, ranging from near normality to pathology. Condition indices can serve to guide more definitive evaluations. Such measures are quite general and nonspecific but their low cost, ease, and rapidity make them a valuable tool.

History of Use

Plants. For more than a century, the occurrence of foliar symptoms and the general habit of the plant have been used to identify areas impacted by air pollutants and the nature of airborne toxicants.

A symptom is usually considered to be some change in the normal appearance of the plant, most often its foliage. Generally, these changes involve distortion, discoloration, death, or loss of foliar tissue. The practical significance of foliar symptoms is threefold: (1) they offer a diagnostic means for assessing the nature of the pollutant and degree and time of exposure in the field, (2) by their nature, they represent a functional loss to the plant in photosynthetic capacity, and (3) experimentally, they offer a means of assessing the actions of environmental and biological factors on the exposure-effect relationship for the pollutant.

Animals. Body condition can provide information on potential pollutant impacts. In fish, birds, and mammals, condition is assessed by the hepatosomatic and gonadosomatic index, fat and flesh condition, and overt appearance. In shellfish, it is assessed by measuring the proportion of internal volume occupied by organs. Admittedly such measures are crude and prone to the effects of nonpollutant factors (e.g., season, disease, nutritional plane, and biological rhythms), yet they may serve as a first tier screen indicative of exposure and effect. In addition, condition may provide information on energy reserves and possibly the ability of animals to tolerate toxicant challenges or other environmental stressors.

Potential Value

Plants. Diagnoses evaluate the kind, size, and distribution of lesions on a leaf as well as the pattern of occurrence among leaves on the same plant and different species of plants in the same location. Quantitative measures of effect (on ordinal or arithmetic scales) are based on incidence and severity. Incidence usually refers to the number of leaves per plant or the number of plants per sample with lesions; severity refers to the area of leaf or total amount of foliar tissue of plant that is affected by lesions. Several atlases of symptoms are available for the field recognition of pollutant-induced symptoms (van Haut and Stratmann 1970; Jacobson and Hill 1970; Lacasse and Treshow 1976; Malhotra and Blauel 1980; Manning and Feder 1980). Moreover, these are accompanied with different species ranked or grouped according to tolerance to individual air pollutants.

Animals. Developmental indicators, including contaminant-induced teratological effects in fish (Middaugh et al. 1990) and birds may also serve as an index of exposure and effect. Incidence of anomalies in healthy populations is usually low, and when their frequency increases, contaminant exposure may be suspected. In some instances, a suite of anomalies can be associated with exposure to a particular contaminant. For example, Type I and II teratogenesis is induced by anticholinesterase compounds (Hoffman and Eastin 1981; Hoffman and Sileo 1984) and by selenium and other agricultural drainwater contaminants (Hoffman and Heinz 1988; Hoffman et al. 1988).

Research Needs

No major research thrusts are recommended.

Skeletal Abnormalities

Vertebral and skeletal abnormalities in fish have been attributed to heredity and abnormal embryonic development (Dahlberg 1970), low dissolved oxygen concentration (Blaxter 1969), water temperature variation during development (Brungs 1971), parasitic infection (Hoffman et al. 1962), electric current (Spencer 1967), vitamin C deficiency (Wilson and Poe 1973), and chemical contaminants. Skeletal and vertebral abnormalities have been reported in wild populations of marine fish (Orska 1962; Gill and Fisk 1966; Van de Kamp 1977), and water pollution has often been suspected as the cause (Sneed 1970; Wunder 1975). In another case, vertebral anomalies in pond-cultured channel catfish have been thought to be induced by either deficient nutrition or contaminants (Sneed 1970).

Chemical contaminants have been reported to induce various degrees of vertebral lesions in fish; gross observable lesions such as lordosis, scoliosis, and vertebral damage were induced by exposure to organophosphate pesticides (McCann and Jasper 1972); metals such as zinc, cadmium, and lead (Bengtsson 1975); Kepone (Couch et al. 1977); crude oil (Linden 1976); and toxaphene (Mayer et al. 1977a; Mehrle and Mayer 1975). Vertebral collagen, calcium, and phosphorus were altered in fish exposed to organic contaminants such as PCBs, toxaphene, di(2-ethylhexyl)phthalate, dimethylamine salt of 2,4-dichlorophenoxyacetic acid (Mayer et al. 1977b), triarylphosphate esters (Mayer et al. 1981), Kepone (Mehrle et al. 1981), and methoxychlor. Chronic effects of these toxicants on vertebral composition proved to be an early indication of reduced growth and altered bone development in fish (Mayer et al. 1977b). Furthermore, Hamilton et al. (1981) reported that mechanical properties such as stress, strain, and elasticity of vertebrae in fish exposed to toxaphene were more sensitive than bone composition as indicators of bone structural integrity. In addition, the relationship between bone composition and density was important in assessing the mechanical properties of vertebrae.

Contaminant-induced vertebral lesions can be caused in at least two ways: acute exposures that cause neurotoxic tetanic contractions of skeletal muscle, and chronic exposures that alter bone composition and render the bone more fragile. Regardless of the etiology of the vertebral lesions, several possible adverse effects on essential biological functions have been described, including impaired swimming performance, decreased ability to escape predators, and altered feeding behavior (Kroger and Guthrie 1971). Other adverse effects suggested by Hickey (1972) include decreased territorial defense, decreased ability to compete for sexual partners, and general physiological weakness.

History of Use

Measurement of skeletal deformities in fish has been proposed as a means of monitoring pollution effects in marine environments (Bengtsson 1979; Bengtsson and Bengtsson 1983). Likewise, measurements of biochemical composition and mechanical properties of vertebrae have been shown to be indicators of bone development in fish exposed to contaminants in the laboratory (Hamilton et al. 1981; Mayer et al. 1977b) and in the field (Mehrle et al. 1982). Skeletal abnormalities in fourhorn sculpin (*Myoxocephalus quadricornis*) have been used to monitor the impacts of ore smelter and pulp mill effluents in the Baltic Sea (Bengtsson et al. 1985). Mayer et al. (1988) conducted laboratory and field comparisons of biochemical composition and mechanical properties of vertebrae on the fish.

Effects of organic and inorganic contaminants on bone integrity are similar in that vertebral anomalies are produced, although they may develop through different modes of action. This similarity makes the use of biochemical composition and mechanical properties, as well as vertebral deformities, conducive to assessing the effect of an array of contaminants on fish health. Today, the most useful and direct application is in monitoring the incidence of vertebral anomalies by x-ray examinations (Bengtsson and Bengtsson 1983).

Potential Value

The study of fish vertebrae for skeletal anomalies, biochemical composition, and mechanical properties appears promising for predicting contaminant effects (onset of condition), monitoring fish health in the field, and diagnosing whether problem contaminants are organic or inorganic.

Research Needs

Measuring mechanical properties would be too technically complex and expensive to be used in the field. Research efforts would be better spent on biochemical composition. For predictive purposes, research is needed to determine the amount of deviation from the norm required to result eventually in vertebral anomalies. For monitoring purposes, research is needed to better define differences in vertebral responses to inorganic and organic contamination, and to develop an index of contaminant-induced vertebral anomalies for assessing the degree of response from normal occurrence to various levels of contaminant effects.

Avian Eggshell Thinning and Impaired Reproduction

Eggshell thinning caused by chlorinated hydrocarbons has been observed in several species of birds. The consequences of cracking and crushing of thin-shelled eggs are quite serious, often leading to impairment of reproduction. Historically, this phenomenon may have led to the population decline of some raptorial (e.g., peregrine falcon; *Falco peregrinus*) and fish-eating (e.g., brown pelican; *Pelecanus occidentalis*) species.

History of Use

Ratcliffe first reported thinner egg shells of raptors sampled in 1967 compared to museum specimens from the pre-organochlorine era (Ratcliffe 1967). Since this report, shell thinning has been correlated with organochlorine residue burdens in many avian species (Cooke 1973; Faber and Hickey 1973; King et al. 1978). The principal environmental contaminant affecting shell thinning is DDE, although other DDT metabolites, dieldrin, chloredecone, lindane, and polychlorinated biphenyls, as well as mercury and aluminum have been shown to result in shell thinning (Cooke 1973; Lundholm 1987).

The mechanism by which organochlorines thin eggshells in birds has been studied in some detail (Lundholm 1987). It has been observed in ducks that the rate of calcium translocation from the mucosal cells to the shell gland cavity is reduced by DDE. In vivo DDE administration and in vitro DDE incubation studies revealed dose-dependent inhibition of calcium uptake and calcium-magnesium ATPase activity in homogenates and subcellular fractions of the shell gland mucosa. There is also evidence that DDE has a high affinity for the progesterone receptor, and thereby could interfere with the stimulus-secretion mechanism by inhibiting the binding of progesterone to its cytoplasmic receptor. The mechanism by which some metals thin eggshells is less well studied, but may be related to their affinity for calcium binding sites on calmodulin, which in turn alters calcium-ATPase mediated translocation of calcium (Lundholm and Mathson 1986).

Thickness is determined directly (e.g., average of several measurements at the equator with a micrometer) or indirectly using the thickness index (Ratcliffe 1967):

$$\text{Thickness index} = \frac{\text{Weight of shell (mg)}}{\text{Length of shell (mm)} \times \text{Breadth (mm)}}$$

More recently breaking strength has been compared to thickness indices, and appears to be a more sensitive, and highly quantifiable measure of eggshell quality (Carlisle et al. 1986; Bennett et al. 1988). These techniques are inexpensive, accurate, precise, and apparently quite sensitive indicators of exposure to chlorinated hydrocarbons. In general, measurements are compared to museum specimens or possibly concurrent controls. Nutritional plane and various stressors (elevated ambient temperature) can cause eggshell thinning (Roland et al. 1973; Smith 1974). The stage of incubation in which the sample is collected is another potentially confounding factor; thinning occurs naturally (a source of calcium for the developing embryo) late in incubation.

Potential Value

Measurement of eggshell thinning and its application to biomonitoring are already well recognized tools that are widely used in the field of environmental toxicology.

Research Needs

Only limited research on gross indicators of contaminant exposure and effect is recommended. Eggshell breaking strength shows particular promise as a more sensitive indicator of thinning than shell thickness measurements.

OTHER BIOMARKERS

This final section discusses two other areas that hold potential for nonspecific biomarkers. Lysosomal membrane stability is affected by chemical and nonchemical factors, and may be useful as an integrative biomarker of multiple stressors. Changes in free amino acids are indicative of general stress rather than exposure to any particular class of chemicals.

Lysosomal Membrane Stability

Lysosomes are a morphologically heterogeneous group of membrane-bound subcellular organelles, containing acid hydrolases and ranging in size from 250 Å to 1 μm diameter (Karp 1979). Lysosomal hydrolases are produced in the endoplasmic reticulum and incorporated into membranes by the Golgi complex or Golgi endoplasmic reticulum lysosome (GERL) (Cohn and Fedorko 1969). Lysosomes fuse with membrane-bound vesicles containing intracellular macromolecules or extracellular materials forming secondary lysosomes.

Lysosomal membrane stability is thought to be a general measure of stress. In stable lysosomes, hydrolases are prevented from reacting with substrate by an intact membrane. Theoretically, membrane stability decreases in response to stress as membrane permeability increases. The mechanism causing this alteration in membrane stability is not well understood, but may involve direct effects of chemicals on the membrane or the increased frequency of secondary lysosomes in toxicant-stressed cells. Lysosomal membrane stability has been demonstrated to be a useful measure of environmental stressors in aquatic organisms.

History of Use

Two lysosomal membrane stability techniques, one histochemical and the other cytochemical/biochemical, have been used. Both measure the functional integrity of the lysosomal membrane. In the histochemical procedure, unfixed frozen tissue sections are exposed to low pH in a staining media specific for a lysosomal enzyme. The incubation time required for staining is proportional to the susceptibility of the lysosomal membrane to the pH shock, and is referred to as the latency period (Baccino and Zuretti 1975). The longer the latency period, the more stable the lysosomal membrane is to low pH. This technique has been most successfully applied in describing environmental pollution effects on the marine mussel, *Mytilus edulis*. Lysosomal membranes are destabilized in hepatopancreas cells of marine mollusks injected with anthracene (Moore et al. 1978) or exposed in situ (Moore et al. 1982) and in the laboratory (Widdows et al. 1982) to the water soluble fraction of crude oil. Copper has also been observed to destabilize lysosomal membranes during acute (6 d)

(Viarengo et al. 1981) and chronic (76 d) (Harrison and Berger 1982) exposure of the marine mussel. In addition to the effects of xenobiotics on lysosomal membrane stability, a variety of nonchemical stressors have been observed to affect membrane stability in mussels and in mammals (Moore et al. 1980; Moore 1976; Gabrielescu 1970). Interestingly, injection of cortisol into the mussel was observed to stabilize lysosomal membranes.

The histochemical lysosomal stability technique has several limitations. Quantification of staining is difficult, time consuming, and relies on an expensive microscope-densitometer. The technique is difficult to perform on many organisms, resulting in decreased sample size, reduced replication, and decreased statistical degrees of freedom. As far as we know, this technique has not been applied successfully to fish in assessing lysosomal membrane stability.

The biochemical lysosomal stability technique involves tissue homogenization and differential centrifugation to produce an enriched lysosomal fraction. The amount of enzyme subsequently released by in vitro hypoosmotic shock is then quantified. This technique has been used less frequently than the histochemical procedure to assay lysosomal membrane stability. One drawback of the biochemical technique is that an unknown number of lysosomes are destroyed in the isolation procedure. This may reduce the sensitivity of this assay to toxicant effects as these lysosomes may be the most seriously affected by the stressor.

In toxicity studies with the rainbow trout, *Salmo gairdneri,* acutely toxic concentrations of nitrite (Mensi et al. 1982) and ammonia (Arillo et al. 1981) resulted in lysosomal membrane destabilization in hepatocytes. In the bluegill sunfish, nonlethal concentrations of cadmium, caused a significant destabilization of liver lysosomes (Versteeg and Giesy 1985, 1986). In this study, lysosomal stability was relatively insensitive to factors of size, sex, and stress induced by crowding in fish. Liver lysosomes displayed decreased membrane stability 1 and 7 d after intraperitoneal injection of CCl_4 into bluegill sunfish (Versteeg 1985). However, 3 d after injection, the membrane stability was significantly increased. Stabilization may be due to the formation of small primary lysosomes. Destruction of cells and lysosomes by CCl_4 exposure would cause the average age of lysosomes to be reduced, increasing the percentage of primary lysosomes with a higher average membrane stability. Seven days after CCl_4 injection, membrane stability was again less than controls, indicating continued lysosomal membrane destabilization.

Deceased lysosomal membrane stability as detected by the biochemical procedure has been reported for a variety of nonchemical stressors (Bird 1975; Sidorov et al. 1980).

Potential Value

Lysosomal membrane stability is affected by both chemical and nonchemical factors. Thus, the assay may have utility as an integrative biomarker of multiple stressors. The difficulty in interpreting alterations in lysosomal membrane stability will be in discriminating chemical from nonchemical factors. Currently, fish lysosomal membranes, as assayed by the biochemical technique, appear to be relatively insensitive to accessory factors. However, more research is needed. The successful use of lysosomal membrane stability in field exposures indicates the potential for this assay as an in situ monitor.

Research Needs

Future research should concentrate on improving the drawbacks in the methods for assessing lysosomal membrane stability. Computer imaging should be applied to the histochemical technique to improve quantitation. In addition, histochemical procedures are needed in a wider variety of species, especially fish. Methods to improve the isolation of intact lysosomes should be attempted in the biochemical procedure. We recommend these improvements be investigated prior to further utilization of this biomarker.

Once improved methods to assess toxicant-induced effects on lysosomal membranes are developed, future research should focus on (1) expanding the number of organisms with background and toxicity data bases, and (2) relating effects on lysosomes with effects at the population, community, and ecosystem levels of organization.

Free Amino Acids

The free amino acids (FAA) of aquatic organisms represent a highly dynamic pool of biomolecules responsive to changes in biochemical and physiological processes induced by environmental perturbations. Changes in FAAs represent a general stress indicator and are not indicative of exposure to any particular class of chemicals (Schafer 1961; Jeffries 1972; DuPaul and Webb 1970). The majority of the work on the biochemical and/or physiological regulation of intracellular FAA pools has been conducted with marine or euryhaline invertebrates. During exposure to toxicants, the intracellular FAA pools of certain marine organisms change (Roesijadi et al. 1976a; Kasschau et al. 1980). Very little work has been conducted with freshwater organisms; however, the data available suggests that a similar response may exist (Gardner et al. 1981; Graney and Giesy 1986a, 1986b, 1987, 1988).

The potential mechanisms responsible for changes in the concentration of FAA in chemically or physically stressed aquatic invertebrates can be separated into three categories. The first category encompasses the complex mechanisms controlling osmotic regulation and the role of FAA in that process. Toxicants may influence the FAA pool by directly or indirectly altering the osmoregulatory processes of the organism. The second category involves the effects of anoxia on the FAA pool; this mechanism can be indirectly linked to osmoregulation. The third category covers the effect of stress on protein metabolism. Increased protein degradation and subsequent utilization of the released amino acids for anaplerotic reactions and/or energy production represents an important mechanism for changes in the FAA pool. None of these potential mechanisms is exclusive, and all three may interact to produce the observed FAA alterations. In addition, there is very little conclusive evidence in the literature establishing that any one of these mechanisms causes changes in FAA during toxicant-induced stress. Rather, these proposed mechanisms may be responsible for some of the FAA alterations that have been observed.

History of Use

In marine invertebrates, specific amino acids have been identified as being responsive to certain kinds of stress. Increases in the molar ratio of taurine to glycine were used as a quantitative index of stress in the hard clam, *Macoma inquimator* (Roesijadi and Anderson 1979) and in the mussel, *Mytilus edulis* (Widdows et al. 1981) exposed to petroleum hydrocarbons. However, a decrease in the taurine/glycine ratio was observed in oil-stressed bivalves (Augenfield et al. 1980). Kasschau et al. (1980) found that glycine concentrations in sea anemones (*Bunodosoma cavernata*) were too variable to be an effective indicator of toxicant-induced stress. Other amino acids found in marine invertebrates and shown to be specifically affected by toxicants include glutamate (Kasschau et al. 1980), alanine, aspartate (Riley and Mix 1981; Powell et al. 1982; Roesijadi et al. 1976a), arginine, lysine, and threonine (Augenfield et al. 1980; Roesijadi and Anderson 1979). In general, the inconsistency of organismal responses involving different conditions and exposure to different toxicants makes interpretation of alterations in the concentration of FAA extremely difficult.

Compared with marine organisms, relatively little work has been conducted on toxicant-induced alterations in the FAA pool of freshwater organisms. Changes in the total FAA concentration was measured in freshwater crabs exposed to fenitrothion; however, concentrations of individual amino acids were not reported (Bhagyalakshmi et al. 1983b). Differences in total FAA were also reported for freshwater bivalves collected from polluted and unpolluted environment (Gardner et al. 1981). The increase in total FAA concentrations reported by Gardner et al. (1981) is opposite to the effect observed by Graney and Giesy (1987), where a significant decrease in FAAs occurred in amphipods exposed for 48 h to pentachlorophenol (PCP). Freshwater amphipods (*Gammaus pseudolimaneus*) exposed to sublethal PCP concentrations also had reduced concentrations of FAA (Graney and Giesy 1986b). Conversely, short-term (96 h) and long-term (60 d) exposure of freshwater clams (*Corbicula fluminea*) to sodium dodecyl sulfate (SDS) resulted in a significant increase in the FAA pool (Graney and Giesy 1988). Seasonal variations in the FAA of freshwater amphipods has been demonstrated (Graney and Giesy 1986a).

Potential Value

Presently, there is not enough information on changes in the FAA pool of aquatic organisms for them to be a useful biomarker. Although the data available indicates a potential, considerably more information is needed on the function and metabolism of amino acids in aquatic organisms and the effect of toxicants on these processes. As indicated in the previous discussion, a number of natural stressors can influence the FAA pool and may limit their usefulness as an in situ indicator. In marine organisms, the taurine:glycine ratio shows promise; however, the confounding influence of salinity on this ratio may limit its sensitivity. A large, site-specific data base will likely be required to separate toxicant-related changes from natural variability. Although free amino acids have been shown to be fairly sensitive in laboratory toxicity tests (Graney and Giesy 1986a, 1988), it is not known whether this sensitivity will be realized under variable field conditions.

Another potential limiting factor may be the measurement of FAA. For biomarkers to be most useful, the endpoint should be fairly easy to measure. Individual free amino acids are not easy to measure; however, the total FAA concentration can be measured rather easily using spectrophotometric techniques.

Research Needs

Future research should address measuring the total FAA pool, since much of the available data indicates that the total FAA pool is as sensitive an endpoint as any individual amino acid. The final area of research needed is field confirmation of laboratory-generated responses and subsequent correlation to higher level effects. This data has yet to be generated in a controlled experiment and is necessary to confirm or reject the usefulness of FAA as a biomarker of higher level effects.

REFERENCES

Aarset, A.V., and K.E. Zachariassen. "Effects of Oil Pollution on the Freezing Tolerance and Solute Concentration of the Blue Mussel *Mytilus edulis*," *Mar. Biol.* 72:45–51 (1982).

Ackerman, W., G. Pott, B. Voss, K.M. Muller and U. Gulach. "Serum Concentration of Procollagen-III-peptide in Comparison With the Serum Activity of N-acetyl-glucosaminidase for Diagnosis of the Activity of Liver Fibrosis in Patients With Chronic Active Liver Disease," *Clin. Chim. Acta* 112:365–369 (1981).

Adams, S.M., C.A. Burtis and J.J. Beauchamp. "Integrated and Individual Biochemical Responses of Rainbow Trout (*Salmo gairdneri*) to Varying Directions of Acidification Stress," *Comp. Biochem. Physiol.* 82C:301–310 (1985).

Adelman, I.R. "Uptake of ^{14}C-glycine by Scales as an Index of Fish Growth: Effect of Fish Acclimation Temperature." *Trans. Am. Fish. Soc.* 109:187–194 (1980).

Adelman, I.R. "Uptake of Radioactive Amino Acids as Indices of Current Growth Rate of Fish: A Review," in *Age and Growth of Fish*, R. Summerfelt, and G. Hall, Eds. (Ames, IA: Iowa State University Press, 1987) pp. 65–80.

Adelman, I.R., and G.P. Busacker. "Indicators of Current Growth Rate as Rapid Methods for Toxicity Tests with Fish Larvae," in *Environmental Biology State-of-the-art Seminar*, R.A. Archibald, Ed., EPA Report 60019-82-007 (1982).

Alscher, R., M. Franz and C.W. Jeske. "Sulfur Dioxide and Chloroplast Metabolism," in

Phytochemical Effects of Environmental Compounds, J.A. Saunders, L. Kosak-Channing and E.E. Conn, Eds. (New York: Plenum Publishing 1987), pp. 1–28.

Arillo, A., C. Margiocco, F. Melodia and P. Mensi. "Effects of Ammonia on Liver Lysosomal Functionality in *Salmo gairdneri* Rich," *J. Exp. Zool*, 218:321–326 (1981).

Arillo, A., C. Margiocco, F. Melodia and P. Mensi. "Biochemical Effects on Long Term Exposure to Cr, Cd, Ni on Rainbow Trout (*Salmo gairdneri* Rich.): Influence of Sex and Season," *Chemosphere* 11:47–57 (1982).

Atkinson, D.E. *Cellular Energy Metabolism and Its Regulation*. (New York: Academic Press, Inc., 1987), p. 293.

Augenfield, J.M., J.W. Anderson, D.L. Woodrull and J.L. Webster. "Effects of Prudhoe Bay Crude Oil-Contaminated Sediments on *Protothaca staminea*: Hydrocarbon Content, Condition Index, Free Amino Acid Level." *Mar. Environ. Res.* 4:135–143 (1980).

Baccino, F. M. and M.F. Zuretti. "Structural Equivalents of Latency for Lysosome Hydrolases," *Biochem. J.* 146:97–108 (1975).

Bamstedt, U. "RNA Concentration in Zooplankton: Seasonal Variation in Boreal Species," *Mar. Ecol. Prog. Series* 11:291–297 (1983).

Barclay, M.C., W. Dall and D.M. Smith. "Changes in Lipid and Protein During Starvation and the Moulting Cycle in Tiger Prawn, *Penaeus esculentus* Haswell," *J. Mar. Biol. Ecol.* 68:229–244 (1983).

Barnhart, R.A. "Effects of Certain Variables on Hematological Characteristics of Rainbow Trout," *Trans. Am. Fish. Soc.* 3:411–448 (1969).

Barron, M.G. and I.R. Adelman. "Nucleic Acid, Protein Content and Growth of Larval Fish Sublethally Exposed to Various Toxicants," *Can. J. Fish. Aquat. Sci.* 41:141–150 (1984).

Bayne, B.L. "Physiological Changes in *Mytilis edulis* L. Induced by Temperature and Nutritive Stress," *J. Mar. Biol. Assoc. U.K.* 53:39–58 (1973a).

Bayne, B. "Aspects of the Metabolism of *Mytilus edulis* During Starvation," *Neth. J. Sea Res.* 7:399–410 (1973b).

Bayne, B.L. "Aspects of Physiological Condition in *Mytilus edulis* (L.) with Special Reference to the Effects of Oxygen Tension and Salinity," in *Proceedings of the European Marine Biology Symposium*, H. Barnes, Ed.(Aberdeen: Aberdeen University Press, 1975), pp. 213–238.

Bayne, B.L., D.A. Brown, K. Burns, D.R. Dixon, A. Ivanovici, D.R. Livingstone, D.M. Lowe, M.N. Moore, A.R.D. Stebbing and J. Widdows. "The Effects of Stress and Pollution on Marine Animals" (New York: Praeger Publishers, 1985), p. 384 .

Bayne, B.L., M.N. Moore, J. Widdows, D.R. Livingstone and P. Salkeld. "Measurements of Response of Individuals to Environmental Stress and Pollution: Studies with Bivalve Molluscs," *Phil. Trans. R. Soc. Lond. B.* 286:563–581 (1979).

Bayne, B.L. and R.C. Newell. "Physiological Energetics of Marine Molluscs," in *The Mollusca, Vol 4.*, A. S.M. Saleuddur and K.M. Wilbur, Eds.(London: Academic Press, Inc., 1983).

Bayne, B. L. and C. Scullard. "Rates of Nitrogen Excretion by Species of *Mytilus* (Bivalvia:Mollusca)," *J. Mar. Biol. Assoc. U.K.* 57:355–369 (1972).

Bengtsson, A. and B.-E. Bengtsson. "A Method to Registrate Spinal and Vertebral Anomalies in Fourhorn Sculpin, *Myoxocephalus quadricornis* L. (Pisces)," *Aquilo Ser. Zool.* 22:61–64 (1983).

Bengtsson, B.E. "Vertebral Damage in Fish Induced by Pollutants," in *Sublethal Effects of Toxic Chemicals on Aquatic Animals*, J. H. Koeman, Ed.(New York: Elsevier, 1975), pp. 23–30.

Bengtsson, B.E. "Biological Variables, Especially Skeletal Deformities in Our Fish for Monitoring Marine Pollution," *Philos. Trans. R. Soc. London* 286:457–464 (1979).

Bengtsson, B.-E., A. Bengtsson and M. Himberg. "Fish Deformities and Pollution in Some Swedish Waters," *Ambio* 14:32–35 (1985).

Bennett, J. K., R.K. Ringer, R.S. Bennett, B.A. Williams and P.E. Humphrey. "Comparison of Breaking Strength and Shell Thickness as Evaluators of Eggshell Quality," *Environ. Toxicol. Chem.* 7:351–357 (1988).

Berglind, R., G. Dave and M. Sjobeck. "The Effects of Lead on *s*-Aminolevulinic Acid Dehydratase Activity, Growth, Hemoglobin Content, and Reproduction in *Daphina magna*," *Ecotoxicol. Environ. Safety* 9:216–229 (1985).

Bhagyalakshmi, A., R.S. Reddy and R. Ramamurthi. "Subacute Stress Induced by Sumithion on Certain Biochemical Parameters in *Oziotelphusa senex senex*, the Freshwater Rice-field Crab," *Toxicol. Lett.* 21:127–134 (1984a).

Bhagyalakshmi, A., R.S. Reddy and R. Ramamurthi. "In Vivo Sub-acute Physiological Stress Induced by Sumithion on Some Aspects of Oxidative Metabolism in the Freshwater Crab," *Water, Air Soil Pollut.* 23:257–262 (1984b).

Bhagyalakshmi, A., P.S. Reddy and R. Ramamurthi. "Muscle Nitrogen Metabolism of Freshwater Crab, *Oziotelphusa senex senex*, Fabricuis, During Acute and Chronic Sumithion Intoxication," *Toxicol. Lett.* 17:89–93 (1983a).

Bhagyalakshmi, A., P.S. Reddy and R. Ramamurthi. "Changes in Hemolymph Glucose, Hepatopancreas Glycogen, Total Carbohydrates, Phosphorylase and Amino Transferases of Sumithion-stressed Freshwater Rice-field Crab (*Oziotelphusa senex senex*)," *Toxicol. Lett.* 18:277–284 (1983b).

Bird, J.W.C. "Skeletal Muscle Lysosomes," in *Lysosome in Biology and Pathology, Vol. 4.*, J.T. Dingle and R.T. Dean, Eds.(New York: American Elsevier Publishing, 1975), pp. 75–109.

Bishop, Y.M.M., S.E. Feinberg and P.W. Holland. *Discrete Multivariate Analysis: Theory and Practice.* (Cambridge: MIT Press, 1975), p. 557.

Blackstock, J. "Activities of Some Enzymes Associated With Energy Yielding Metabolism in *Glycera alba* (Mullu) from Three Areas of Lock Eel," in *Physiology and Behavior of Marine Organisms*, D.S. McLusky and A.J. Berry, Eds.(London: Pergamon Press, 1978), pp. 11–20.

Blackstock, J. "A Biochemical Approach to Assessment of Effects of Organic Pollution on the Metabolism of the Non-opportunistic Polychaete, *Glycera alba, Helgolander Meeresunter* 33:546–555 (1980).

Blaxter, J.H.S. "Development: Eggs and Larvae," in *Fish Physiology, Vol. 3*, W.S. Hoar and D.J. Randall, Eds.(New York: Academic Press, Inc., 1969), pp. 177–252.

Bolton, J.P., A. Takahashi, H. Kawauchi, J. Kubota and T. Hirano. "Development and Validation of Salmon Growth Hormone Radioimmunoassay," *Gen. Comp. Endocrinol.* 62:230–238 (1986).

Bolton, J.P., G. Young, R.S. Nishioka, T. Hirano and H.A. Bern. "Plasma Growth Hormone Levels in Normal and Stunted Yearling Coho Salmon (*Oncorhynchus kisutch*)," *J. Exp Zool.* 242-379–382 (1987).

Bose, A.K., and H. Fujiwara. "Fate of Pentachlorophenol in the Blue Crab, *Callinectes sapidus*," in *Pentachlorophenol: Chemistry, Pharmacology and Environmental Toxicology*, K.R. Rao, Ed. (New York: Plenum Press, 1978), p. 402.

Bostrom, S.L., and R.G. Johansson. "Effects of Pentachloro-phenol on Enzymes Involved in Energy Metabolism in the Liver of the Eel," *Comp. Biochem. Physiol.* 41B:359–369 (1972).

Bouck, G.R. "Concentration of Leucine Aminonapthylamidase (LAN) and Soluble Protein in Tissues of Rainbow Trout, *Salmo gairdneri*," *Can. J. Aquat. Sci.* 37:116–120 (1980).

Bouck, G.R. "Physiological Responses of Fish: Problems and Progress Toward Use in Environmental Monitoring," in *Contaminant Effects on Fisheries*, V.W. Cairns, P.V. Hodson and J.O. Nriagu, Eds. (New York: John Wiley & Sons, Inc., 1984), pp. 216–229.

Brachet, J. *The Biological Role of Ribonucleic Acids*, (New York: Elsevier,1960), p. 144.

Brett, J.R. "Implications and Assessments of Environmental Stress," in *Investigations of Fish-power Problems*, P.A. Larkin, Ed. (Vancouver: University of British Columbia Press, 1958), pp. 69–83.

Brown, S.B., J.G. Eales, R.E. Evans and T.J. Hara. "Interrenal, Thyroidal, and Carbohydrate Responses of Rainbow Trout (*Salmo gairdneri*) to Environmental Acidification," *Can. J. Fish. Aquat. Sci.* 41:36–45 (1984).

Brungs, W.A. "Chronic Effects of Constant Elevated Temperature on the Fathead Minnow (*Pimephales promelas Rafineque*)," *Trans. Am. Fish. Soc.* 100:659–644 (1971).

Bry, C. "Daily Variations in Plasma Cortisol Levels of Individual Female Rainbow Trout *Salmo gairdneri*: Evidence for a Post-feeding Peak in Well-Adapted Fish," *Gen. Comp. Endocrinol.* 48:462–468 (1982).

Buckley, L.J. "RNA-DNA Ratio: an Index of Larval Fish Growth in the Sea," *Mar. Biol.* 80:291–298 (1984).

Bull, C.J., and J.E. McInerney. "Behavior of Juvenile Coho Salmon Exposed to Sumithion, an Organophosphate Insecticide," *J. Fish. Res. Bd. Can.* 31:1867–1872 (1974).

Bulow, F.J. "RNA-DNA Ratios as Indicators of Recent Growth Rates of a Fish," *J. Fish. Res. Bd. Can.* 27:2343–2349 (1970).

Bulow, F.J. "RNA-DNA Ratios as Indicators of Growth in Fish: A Review," in *Age and Growth of Fish*, R. Summerfelt and G. Hall, Eds. (Ames, IA: Iowa State University Press, 1987), pp. 45–64.

Butler, P.J., E.W. Taylor, M.F. Capra and W. Davison. "The Effects of Hypoxia on the Levels of Circulating Catecholamines in the Dogfish (*Scyliorhinus canicula*)," *J. Comp. Physiol.* 727:325–330 (1978).

Cairns, V.W., P.V. Hodson and J.D. Nriagu Eds. "Contaminant Effects on Fisheries," *Adv. Environ. Sci. Tech.* 16:1–333 (1984).

Cantelmo, A., L. Mantel, R. Lazell, F. Hospod, E. Flynn, S. Goldberg and M. Katz. "The Effects of Benzene and Dimethylnaphthalene on Physiological Processes in Juveniles of the Blue Crab, *Callinectes sapidus*," in *Physiological Mechanisms of Marine Pollutant Toxicity*, W.B. Vernberg, A. Calabrese, F.P.Thurberg and F.J. Vernberg, Eds. (New York: Academic Press, Inc., 1982), pp. 349–389.

Capuzzo, J.M., B.A. Lancaster and G.C. Sasaki. "The Effects of Petroleum Hydrocarbons on Lipid Metabolism and Energetics of Larval Development and Metamorphosis in the American Lobster (*Homarus americanus* Milne Edwards)," *Mar. Environ. Res.* 14:201–228 (1984).

Carlisle, J.C., D.W. Lamb and P.A. Toll. "Breaking Strength: An Alternative Indicator of Toxic Effects on Avian Eggshell Quality," *Environ. Toxicol. Chem.* 5:887–889 (1986).

Casillas, E., M. Myers and W.E. Ames. "Relationship of Serum Chemistry Values to Liver and Kidney Histopathology in English Sole (*Parophrys vetulus*) after Acute Exposure to Carbon Tetrachloride," *Aquat. Toxicol.* 3:161–78 (1983).

Casillas, E. and L.S. Smith. "Effect of Stress on Blood Coagulation and Hematology in Rainbow Trout (*Salmo gairdneri*)," *J. Fish Biol.* 10:481–491 (1977).

Chapman, A.G., L. Fall and D.E. Atkinson. "Adenylate Energy Charge in *Escherichia coli* During Growth and Starvation," *J. Bacteriol.* 108:1072–1086 (1971).

Chaudhry, H.S. "Nickel Toxicity on Carbohydrate Metabolism of a Freshwater Fish, *Colisa fasciatus*," *Toxicol. Lett.* 20:115–121 (1984).

Chenery, R., M. George and G. Krishna. "The Effect of Ionophore A23187 and Calcium on Carbon Tetrachloride-Induced Toxicity in Cultured Rat Hepatocytes," *Toxicol. Appl. Pharmacol.* 50:241–252 (1981).

Chester-Jones, I., P.M. Ingleton and J.G. Phillips, Eds. *Fundamentals of Comparative Vertebrate Endocrinology* (New York: Plenum Press, 1987), p. 666.

Christensen, G., E. Hunt and J. Fiandt. "The Effect of Methylmecuric Chloride, Cadmium Chloride, and Lead Nitrate on Six Biochemical Factors of the Brook Trout (*Salvelinus fontinalis*)," *Toxicol. Appl. Pharmacol.* 42:523–530 (1977).

Claybrook, D.L. "Nitrogen Metabolism," in *The Biology of Crustacea, Vol. 5, Internal Anatomy and Physiological Regulation*, L.H. Mantel, Ed. (New York: Academic Press, Inc., 1983).

Clement, J.G. "Hormonal Consequences of Organophosphate Poisoning," *Fund. Appl. Pharmacol.* 5:561–577 (1985).

Cleveland, L., E.E. Little, S.J. Hamilton, D.R. Buckler and J.B. Hunn. "Interactive Toxicity of Aluminum and Acidity to Early Life Stages of Brook Trout," *Trans Am. Fish. Soc.* 115:610–620 (1986).

Coglianese, M., and J.M. Neff. "Evaluation of the Ascorbic Acid Status of Two Estuarine Crustaceans: the Blue Crab, *Callinectes sapidus* and the Grass Shrimp, *Palaemonetes pugio*," *Comp. Biochem. Physiol.* 68A:451–455 (1981).

Coglianese, M.P., and J.M. Neff. "Biochemical Responses of the Blue Crab, *Callinectes sapidus* to Pentachlorophenol," in *Physiological Mechanisms of Marine Pollutant Toxicity*, W.B. Vernberg, A. Calabrese, F.P. Thurberg and E.J. Vernberg, Eds. (New York: Academic Press, Inc., 1982).

Cohn, Z.A., and M.E. Fedorko. "The Formation and Fate of Lysosomes," in *Lysosomes in Biology and Pathology. Vol. 1*, J.T. Dingle and H.B. Fell, Eds. (New York: John Wiley & Sons, Inc., 1969), pp. 44–69.

Cole, A.W.G., and T.N. Palmer. "Action of Purine Nucleosides on the Release of Intracellular Enzymes from Rat Lymphocytes," *Clin. Chim. Acta* 92:93–100 (1979).

Conover, R.J. "Transformation of Organic Matter," in *Marine Ecology, Vol. IV, Dynamics*, O. Kinne, Ed. (Chichester: Wiley-Interscience, 1978), pp. 221–500.

Cooke, A.S. "Shell Thinning in Avian Eggs by Environmental Pollutants," *Environ. Poll* 4:85–152 (1973).

Coppage, D.L., and T. Braidech. "River Pollution by Anticholinesterase Agents," *Water Res.* 10:19–24 (1976).

Coppage, D.L., E. Matthews, G.H. Cook and J. Knight. "Brain Acetylcholinesterase Inhibition in Fish as a Diagnosis of Environmental Poisoning by Malathion, 0,0,-dimethyl S-(1,2-Dicarbethy Oxyethyl) Phosphorodithioute," *Pest. Biochem. Physiol.* 5:536–542 (1975).

Couch, J.A., J.T. Winstead and L.R. Goodman. "Kepone-induced Scoliosis and its Histological Consequences in Fish," *Science* 197:585–587 (1977).

Cox, W.S. "Enforcing Insecticide Content Water Quality Standards," *Science* 159:1123–1124 (1968).

Crabtree, B., and E.A. Newsholme. "The Activities of Phosphorylase, Hexokinase, Phosphofructokinase, Lactate Dehydrogenase and the Glycerol-3-phosphate Dehydrogenases in Muscle from Vertebrates and Invertebrates," *Biochem. J.* 126:49–58 (1975).

Cutkomp, L.K., H.H. Yap, D. Desaiah and R.B. Koch. "The Sensitivity of Fish ATPases to Polychlorinated Biphenyls," *Environ. Health Perspect.* 1:165–168 (1972).

Dagg, M.J., and J.L. Littlepage. "Relationships Between Growth Rate and RNA and DNA, Protein and Dry Weight in *Artemia salina* and *Euchaeta elongata*," *Mar. Biol.* 17:162–170 (1972).

Dahlberg, M.D. "Frequencies of Abnormalities in Georgia Estuarine Fishes." *Trans. Am. Fish. Soc.* 99:162–170 (1970).

Daikoku, T., I. Yano and M. Musui. "Lipid and Fatty Acid Compositions and Their Changes in the Different Organs and Tissues of Guppy, *Poecilia reticulata* on Sea Water Adaptation." *Comp. Biochem. Physiol.* 73A:167–174 (1982).

Dalich, G.M., R.E. Larson and W.H. Gingerich. "Acute and Chronic Toxicity Studies with Monochlorobenzene in Rainbow Trout," *Aquat. Toxicol.* 2:1270–142 (1982).

Dean, J.M., and F.J. Vernberg. "Variation in the Blood Glucose Level of Crustacean," *Comp. Biochem. Physiol.* 14:29–34 (1965).

Desaiah, D., and R.B. Koch. "Inhibition of ATPase Activity in Channel Catfish Brain by Kepone and its Reduction Product," *Bull. Environ. Contam. Toxicol.* 13:153–158 (1975).

Dey, A.C., J.W. Kiceniuk, U.P. Williams, R.A. Khan and J.F. Payne. "Long Term Exposure of Marine Fish to Crude Petroleum. I. Studies on Liver Lipids and Fatty Acids in Cod (*Gadus morhua*) and Winter Flounder (*Pseudopleuronectes americanus*)," *Comp. Biochem. Physiol.* 75C:93–101 (1983).

Dickson, G.W., and J.P. Giesy. "The Effects of Starvation on Muscle Phosphoadenylate Concentrations and Adenylate Energy Charge of Surface and Cave Crayfish," *Comp. Biochem Physiol.* 71A:357–361 (1981).

Dickson, G.W., and J.P. Giesy. "Seasonal Variation of Phosphoadenylate Concentrations and Adenylate Energy Charge in Dorsal Tail Muscle of the Crayfish, *Procambarus Acutus, Acutus* (Decapoda: Astacidae)," *Comp. Biochem. Physiol.* 72A:195–299 (1982).

Dieter, M.P. "Plasma Enzyme Activity in *Coturnix cortunix* Quail Fed Graded Doses of DDE, Polychlorinated Biphenyl, Malathion, and Mercuric Chloride," *Toxicol. Appl. Pharmacol.* 27:86–98 (1974).

DiMichelle, L., and M.H. Taylor. "Histopathological and Hysiological Responses of *Fundulus heteroclitus* to Naphthalene Exposure," *J. Fish. Res. Bd. Can.* 35:1060–1066 (1978).

Dively, J.L., J.E. Mudge, W.H. Neff and A. Anthony. "Blood Po_2, Poo_2 and pH Changes in Brook Trout (*Salvelinus fontinalis*) Exposed to Sublethal Levels of Acidity," *Comp. Biochem. Physiol.* 57A:347–351 (1977).

Dixon, D.G. "Some Effects of Chronic Cyanide Poisoning on the Growth, Respiration and Liver Tissue of Rainbow Trout," M.S. Thesis, Concordia University, Montreal, Quebec, Canada, p. 77 (1975).

Dixon, D.G., and J.B. Sprague. "Copper Bioaccumulation and Hepatoprotein Synthesis During Acclimation to Copper by Juvenile Rainbow Trout," *Aquat. Toxicol.* 1:69–81 (1981).

Donaldson, E.M. "The Pituitary-interrenal Axis as an Indicator of Stress in Fish," in *Stress and Fish*, A.D. Pickering, Ed. (London: Academic Press, Inc., 1981), pp. 11–47.

Donaldson, E.M., and H.M. Dye. "Corticosteroid Concentrations in Sockeye Salmon (*Oncorhynchus nerka*) Exposed to Low Concentrations of Copper," *J. Fish. Res. Bd. Can.* 32:533–539 (1975).

Dortch, Q., T.L. Roberts, J.R. Clayton and S.I. Ahmed. "RNA/DNA Ratios and DNA Concentrations as Indicators of Growth Rate and Biomass in Planktonic Marine Organisms," *Mar. Ecol. Prog. Ser.* 13:61–71 (1983).

Dragomirescu, A., L. Raileanu and L. Ababei. "The Effect of Carbetox on Glycolysis and the Activity of Some Enzymes in Carbohydrate Metabolism in the Fish and Rat Liver," *Water Res.* 9:205–209 (1975).

DuPaul, W.D., and K.L. Webb. "The Effect of Temperature on Salinity-induced Changes in the Free Amino Acid Pool of *Mya arenaria*," *Comp. Biochem. Physiol.* 32:785–801 (1970).

Ejike, C., and C.B. Schreck. "Stress and Social Hierarchy Rank in Coho Salmon," *Trans. Am. Fish. Soc.* 109:423–426 (1980).

Faber, A.R., and J.J. Hickey. "Eggshell Thinning, Chlorinated Hydrocarbons and Mercury in Inland Aquatic Bird Eggs 1967 and 1968," *Pestic. Monit. J.* 7:27–36 (1984).

Fauconneau, B. "The Measurement of Whole Body Protein Synthesis in Larval and Juvenile Carp (*Cyprinus carpio*)," *Comp. Biochem. Physiol.* 78B:845–850 (1984).

Fingerman, M., M.M. Hanumate, V.P. Deshpunde and R. Nagabhushan. "Increase in the Total Reducing Substances in the Hemolymph of the Freshwater Crab, *Barytelphusa guerini*, Produced by a Pesticide (DDT) and an Indolealkylamine (Serotonin)," *Experientia* 37:178–179 (1981).

Finley, M.T., M.P. Dieter and L.N. Locke. "α-Amino-Levulinic Acid Dehydratase: Inhibition in Ducks Dosed with Lead Shot," *Environ. Res.* 12:243–249 (1976).

Fiszer-Szafarz, B., and D. Szafarz. "DNA and Protein Content as Cellular Biochemical Parameters. A Discussion with Two Examples: k Acid Phosphatase and Catepsin D in Rat Liver and Hepatoma and Acid Phosphatase in Human Breast Normal Tissue and Adenocarcinoma," *Anal. Biochem.* 148:255–258 (1984).

Fitch, K., and P. Johnson. *Human Life Science* (New York: Holt, Rinehart, and Winston, 1977).

Fletcher, D.J. "Plasma Glucose and Plasma Fatty Acid Levels of *Limanda limanda* (L.) in Relation to Season, Stress, Glucose Loads and Nutritional State," *J. Fish. Biol.* 25:629–648 (1984).

Florkin, M., and B.T. Scheer. *Chemical Zoology, Arthropoda Vol. VI* (New York: Academic Press, Inc., 1970), p. 460.

Folmar, L.C., W.W. Dickhoff, W.S. Zaugg and H.O. Hodgins. "The Effects of Aroclor 1254 and Number Two Diesel Fuel on Smoltification and Seawater Adaptation of Coho Salmon (*Oncorhynchus kisutch*)," *Aquatic Toxicol.* 2:291–299 (1982).

Fowler, B.A. and G.F. Nordberg. "The Renal Toxicity of Cadmium Metallothionein: Morphometric and X-ray Micro-Analytical Studies," *Toxicol. Appl. Pharm.* 46:609–623 (1978).

Frank, J.R., S.D. Sulkin and R.P. Morgan. "Biochemical Changes During Larval Development of the Xanthid Crab *Rhethropanopeus harrisii*. I. Protein, Total Lipid, Alkaline Phosphatase and Glutamic Oxaloacetic Transaminase," *Mar. Biol.* 32:105–111 (1975).

Freedel, R., F. Diederichs and J. Lindena. "Release and Extracellular Turnover of Cellular Enzymes," in Advances in Clinical Enzymology, E. Schmidt, F.W. Schmidt, I. Trautschold and R. Friedel, Eds. (New York: Karger, 1979), pp. 70–105.

Freeman, H.G., J.F. Uthe and G. Sangalang. "The Use of Steroid Hormone Metabolism Studies in Assessing the Sublethal Effects of Marine Pollution," in *Biological Effects of Marine Pollution and the Problem of Monitoring*, A.S. McIntyre and J.B. Pearce, Eds. (Copenhagen, Denmark: Conseil International pour L'Exploration de la Mer, 1980), pp. 16–22.

Gabrielescu, E. "The Lability of Lysosomes During the Response of Neurons to Stress," *Histochem. J.* 2:123–130 (1970).

Galen, R.S. "Multiphasic Screening and Biochemical Profiles: State of the Art, 1975," in *Progress in Clinical Pathology, Volume 6*, M. Stefanini and H.D. Isenberg, Eds.(New York: Grune and Stratton, 1975), pp. 83–110.

Gardner, W.S., H.H. Miller and M.J. Imlay. "Free Amino Acids in Mantle Tissues of the Bivalve

Amblema plicata: Possible Relation to Environmental Stress," *Bull. Environ. Contam. Toxicol.* 26:157–162 (1981).

Gardner, D., and J.P. Riley. "The Component Fatty Acids of the Lipids of Some Species of Marine and Freshwater Molluscs," *J. Mar. Biol Assoc. U.K.* 52:827–838 (1972).

Gardner, G.R., and P.P. Yevich. "Comparative Histopathological Effects of Chemically-Contaminated Sediment on Marine Organisms," *Mar. Environ. Res.* 24:311–316 (1988).

Gaudet, M., J.G. Racicot and C. Leray. "Enzyme Activities of Plasma and Selected Tissues in Rainbow Trout *Salmo gairdneri* Richardson," *J. Fish Biol.* 7:505–512 (1975).

Gibson, R.F., J.L. Ludke and D.E. Ferguson. "Sources of Error in the Use of Fish-brain Acetylcholinesterase as a Monitor for Pollution," *Bull. Environ. Contam. Toxicol.* 4:17–23 (1969).

Giesy, J.P., and P.M. Allred. "Replicability of Aquatic Multispecies Test Systems," in *Multispecies Toxicity Testing*, J. Cairns, Ed. (New York: Pergamon Press, 1985), pp. 187–247.

Giesy, J.P., S.R. Denzer, C.S. Duke and G.W. Dickson. "Phosphoadenylate Concentrations and Energy Charge in Two Freshwater Crustaceans: Responses to Physical and Chemical Stressors," *Verh. Int. Verein. Limnol.* 21:205–220 (1981).

Giesy, J.P., C.S. Duke, R.D. Bingham and G.W. Dickson. "Changes in Phosphoadenylate Concentrations and Adenylate Energy Charge as an Integrated Biochemical Measure of Stress in Invertebrates. The effects of Cadmium on the Freshwater Clam *Corbicula fluminea*," *Toxicol. Environ. Chem.* 6:259–295 (1983).

Giles, G.A. "Electrolyte and Water Balance in Plasma and Urine of Rainbow Trout (*Salmo gairdneri*) During Chronic Exposure to Cadmium," *Can. J. Fish. Aquat. Sci.* 41:1678–1685 (1984).

Gilfillan, E.S., L.C. Jiang, D. Donovan, S. Hanson and D.W. Mayo. "Reduction in Carbon Flux in *Mya arenaria* Caused by a Spill of No. 2 Fuel Oil," *Mar. Biol.* 37:115–123 (1976).

Gilfillan, E.S., D.S. Page, J.C. Foster, D. Vallus, L. Gonzalez, A. Luckerman, J.R. Hotham, E. Pendergast and S. Hebert. "A Comparison of Stress Indicators at the Biochemical, Organismal and Community Level of Organism," *Mar. Environ. Res.* 14:503–504 (1979).

Gilfillan, E.S., and J.H. Vandermeulen. "Alterations in Growth and Physiology in Chemically Oiled Soft-shell Clams, *Mya arenaria*, Chemically Oiled with Bunker C from Chedabucto Bay, Nova Scotia, 1970–1976," *J. Fish. Res. Bd. Can.* 35:630–636 (1978).

Gill, D.C., and D.M. Fisk. "Vertebral Abnormalities in Sockeye, Pink, and Chum Salmon," *Trans. Am. Fish Soc.* 95:177–182 (1966).

Gill, T.S., and J.C. Pant. "Cadmium Toxicity: Inducement of Changes in Blood and Tissue Metabolites in Fish," *Toxicol. Lett.* 18:195–200 (1983).

Gilles, R. "Intermediate Metabolism and Energy Production in Some Invertebrates," *Arch. Int. de Physiol. Biochem.* 78:313–326 (1970).

Gilles, R. "Effects of Osmotic Stresses on the Protein Concentrations and Patterns of *Eriocheir sineses* Blood," *Comp. Biochem. Physiol.* 56A:109–114 (1977).

Gilles, R., and A. Requeux. "Cell Volume Regulation in Crustaceans: Relationship Between Mechanisms for Controlling the Osmolality of Extracellular and Intracellular Fluids," *J. Exp. Zool.* 215:351–362 (1981).

Gilles, R., and A. Requeux. "Interactions of Chemicals and Osmotic Regulation with the Environment," in *The Biology of Crustacea, Vol. 8, Environmental Adaptations*, F.J. Vernberg and W.B. Vernberg, Eds. (New York: Academic Press, Inc., 1983).

Gingerich, W.H. "Hepatic Toxicology of Fishes," in *Aquatic Toxicology, Vol. 1*, L.J. Weber, Ed. (New York: Raven Press, 1982), pp. 55–105.

Gingerich, W.H., and L.J. Weber. "Assessment of Clinical Laboratory Procedures to Evaluate

Liver Intoxication in Fish," Environmental Protection Agency, Duluth, MN, EPA Report, EPA/600/3/79/088 (1979).

Glass, R.L., T.P. Krick, D.M. Sand, C.H. Rahn and H. Schlenk. "Ruranoid Fatty Acids from Fish Lipids," *Lipids* 10:695–702 (1975).

Gluth, G., and W. Hanke. "A Comparison of Physiology Changes in Carp, *Cyprinus carpio*, Induced by Several Pollutants at Sublethal Concentrations. II. The Dependency on the Temperature," *Comp. Biochem. Physiol.* 79C:39–45 (1984).

Goede, R.W. *Fish Health/Condition Assessment Procedures* (Logan, UT: Utah Division of Wildlife Resources, 1989).

Goldstein, I.F. "The Use of Biological Markers in Studies of Health Effects of Pollutants," *Environ. Res.* 25:236–240 (1981).

Gooch, J.W., and F. Matsumura. "Characteristics of the Hepatic Monooxygenase System of the Goldfish (*Carassius auratus*) and its Induction with *o*-Naphthaflavone," *Toxicol. Appl. Pharmacol.* 68:380–391 (1983).

Goolish, E.M., M.G. Barron and I.R. Adelman. "Thermoacclimatory Response of Nucleic Acids and Protein Content of Carp Muscle Tissue: Influence of Growth Rate and Relationship to Glycine Uptake by Scales," *Can. J. Zool.* 62:2164–2170 (1984).

Gould, E. "Low Salinity Stress in the American Lobster, *Homarus americanus*, After Chronic Sublethal Exposure to Cadmium: Biochemical Effects," *Helgolander Meeresunter.* 33:36–46 (1980).

Gould, E. "Monitoring Sea Scallops in the Offshore Waters of New England and the Mid-Atlantic States: Enzyme Activity in Phasic Adductor Muscle," in *Biological Monitoring of Marine Pollutants*, J. Vernberg, A. Calabrese, F. P. Thurberg and W. B. Vernberg, Eds. (New York: Academic Press, Inc., 1981).

Gould, E., and R.A. Greig. "Short-term Low Salinity Response in Lead-exposed Lobsters, *Homarus americanus* (Milne Edwards)," *J. Exp. Mar. Biol. Ecol.* 69:283–295 (1983).

Graney, R.L., and J.P. Giesy. "Seasonal Changes in the Free Amino Acid Pool of the Freshwater Amphipod *Gammarus pseudolimnaeus* Bousfield (Crustacea: Amphipod)," *Comp. Biochem. Physiol.* 85A:535–543 (1986a).

Graney, R.L., and J.P. Giesy. "Effects of Long-term Exposure to Pentachlorophenol on the Free Amino Acid Pool and Energy Reserves of the Freshwater Amphipod *Gammarus pseudolimnaeus* Bousfield (Crustacea, Amphipod)," *Ecotoxicol. Environ. Saf.* 12:233–251 (1986b).

Graney, R.L., and J.P. Giesy. "The Effect of Short-term Exposure to Pentachlorophenol and Osmotic Stress on the Free Amino Acid Pool of the Freshwater Amphipod *Gammarus pseudolimnaeus* Bousfield," *Arch. Environ. Contam. Toxicol.* 16:167–176 (1987).

Graney, R.L., and J.P. Giesy. "Alterations in the Oxygen Consumption, Condition Index and Concentrations of Free Amino Acids in *Corbicula fluminea* (Mollusca: Pelicypoda) Exposed to Sodium Dodecyl Sulfate," *Environ. Toxicol. Chem.* 7:301–315 (1988).

Grant, B.F., and P.M. Mehrle. "Endrin Toxicosis in Rainbow Trout (*Salmo gairdneri*)," *J. Fish. Res. Bd. Can.* 30:31–40 (1973).

Grassle, B., and A. Biesmann. "Effects of DDT, Polychlorinated Biphenyls and Thiouracil on Circulating Thyroid Hormones, Thyroid Histology and Eggshell Quality in Japanese Quail (*Coturnix coturnix japonica*)," *Chem.-Biol. Interact.* 42:371–377 (1982).

Greenway, P. "Freshwater Invertebrates," in *Comparative Physiology of Osmoregulation in*

Animals, G. M. O. Maloig, Ed. (New York: Academic Press, Inc., 1979), pp. 117–173.

Griffiths, C.C., and J.A. King. "Energy Expended on Growth and Gonad Output in the Ribbed Mussel. *Anlacomya ater,*" *Mar. Biol.* 53:217–222 (1979).

Grizzle, J.A., and W.A. Rogers. "Anatomy and Histology of the Channel Catfish," Auburn University Agric. Expt. Station, Auburn (1976).

Grue, C.E., W.J. Fleming, D.G. Busby and E.F. Hill. "Assessing Hazards of Organophosphate Pesticides to Wildlife," *Trans. North Am. Wildl. Nat. Res. Conf.* 48:200–220 (1983).

Guilbault, C.G., R.L. Lozes, W. Moore and S.S. Kuan. "Effect of Pesticides on Cholinesterase from Aquatic Species: Crayfish, Trout and Fiddler Crab," *Environ. Lett.* 3:235–245 (1972).

Gupta, A.K., and S.S. Dhillon. "The Effects of a Few Xenobiotics on Certain Phosphatases in the Plasma of *Clarias batrachus* and *Cirhina mrigala,*" *Toxicol. Lett.* 15:181–186 (1983).

Haines, T.A. "An Evaluation of RNA-DNA Ratio as a Measure of Long-term Growth in Fish Populations," *J. Fish. Res. Bd. Can.* 30:195–199 (1973).

Hamilton, S.J., P.M. Mehrle, F.L. Mayer and J.R. Jones. "Mechanical Properties of Bone in Channel Catfish as Affected by Vitamin C and Toxaphene," *Trans. Am. Fish. Soc.* 110:718–724 (1981).

Hammond, P.B., and R.P. Belile. "Metals," in *Cassarett and Doull's Toxicology: The Basic Science of Poisons,* J. Doull, C.D. Klassen, and M.O. Amdur, Eds. (New York: MacMillan, 1980), pp. 409-467.

Hansen, H.J.M., and S. Abraham. "Influence of Temperature, Environmental Salinity and Fasting on the Patterns of Fatty Acids Synthesized by Gills and Liver of the European Eel (*Anguilla anguilla*)," *Comp. Biochem. Physiol.* 75B:581–587 (1983).

Harman, B.J., D.I. Johnson and L. Greenwald. "Physiological Responses of Lake Erie Freshwater Drum to Capture by Commercial Shore Seine," *Trans. Am. Fish. Soc.* 109:544–551 (1980).

Harrison, F.L., and R. Berger. "Effect of Copper on the Latency of Lysosomal Hexosaminidase on the Digestive Cells of *Mytilus edulis,*" *Mar. Biol.* 68:109–116 (1982).

Harvey, S., J.G. Phillips, A. Rees and T.R. Hall. "Stress and Adrenal Function," *J. Exp. Zool.* 232:633–645 (1984).

Haut, H. van, and II. Stratmann. *Farbtafelatlas uber Schwefeldioxid-Wirkungen and Pflanzen* (Essen: Verlag Girardet, 1970).

Haux, C., and A. Larsson. "Long-term Sublethal Physiological Effects on Rainbow Trout, *Salmo gairdneri,* During Exposure to Cadmium and After Subsequent Recovery," *Aquat. Toxicol.* 5:129–142 (1984).

Haux, C., A. Larsson, G. Lithner and M.-L. Sjobeck. "A Field Study of Physiological Effects on Fish in Lead-contaminated Lakes," *Environ. Toxicol. Chem.* 5:283–288 (1986).

Haya, K., C.E. Johnson and B.A. Waiwood. "Adenylate Energy Charge and ATPase Activity in American Lobster (*Homarus americanus*) from Belledune Harbour," in *Cadmium Pollution of Belledune Harbour,* J.F.Uthe, and V. Zitco, Eds., Tech. Rep. Fish. Aquat. Sci., New Brunswick, Canada (1980), pp. 85–91.

Haya, K., and B.A. Waiwood. "Adenylate Energy Charge and ATPase Activity: Potential Biochemical Indicators of Sublethal Effects Caused by Pollutants in Aquatic Organisms," in *Aquatic Toxicology,* J.O. Nriagu. Ed. (New York: John Wiley & Sons, Inc., 1983), pp. 307–333.

Haya, K., B.A. Waiwood and L. Van Eckhaute. "Energy Metabolism During Several Life Stages of *Salmo salar* upon Exposure to Low pH," in *Workshop on Acid Rain*, R. H. Peterson and H. H. V. Hord. Eds. pp. 57–70 (1983).

Head, E.J.H., and P.A. Gabbott. "Control of NADP⁺ Dependent Isocitrate Dehydrogenase Activity in the Mussel *Mytilus edulis* L.," *Biochem. Soc. Trans.* 7:896–898 (1979).

Heath, A.G. "Changes in Tissue Adenylates and Water Content of Bluegill, *Lepomis machrochirus*, Exposed to Copper," *J. Fish. Biol.* 24:299–309 (1984).

Heath, R.L. "Biochemical Mechanisms of Pollutant Stress," in *Assessment of Crop Loss from Air Pollutants*, W.W. Heck, O.C. Taylor and D.T. Tingey, Eds. (London: Elsevier Applied Science, 1988), pp. 259–286.

Heitz, J.R., L. Lewis, J. Chambers and D. Yarbrough. "The Acute Effects of Empire Mix Crude Oil on Enzymes in Oysters, Shrimp and Mullet," in *Pollution and Physiology of Marine Organisms*, F.J. Vernberg, and W.B. Vernberg, Eds. (New York: Academic Press, Inc., 1974), pp. 311–328.

Hickey, C.R. "Common Abnormalities in Fishes, their Causes and Effects," New York Ocean Sci. Lab. Tech. Rep. 0013 (1972).

Highman, B., P.D. Altland and J. Garbus. "Pathological and Serum-enzyme Changes after Epinephrine in Oil and Adrenergic Blocking Agents," *Arch. Pathol.* 80:332–344 (1965).

Hill, E.F. "Brain Cholinesterase Activity of Apparently Normal Wild Birds," *J. Wildl. Dis.* 24:51–61 (1988).

Hilmy, A.M., M.B. Shabana and M.M. Saied. "Ionic Regulation of the Blood in the Cyprinodont, *Aphanius dispar* rupp, Under the Effect of Experimental Mercury Pollution," *Water Air Soil Pollut.* 18:467–473 (1982).

Hoar, W.S. "The Physiology of Smolting Salmonids," in *Fish Physiology, Vol. 11, Part B*, W.S. Hoar, and D.J. Randall, Eds. (New York: Academic Press, Inc., 1988), pp. 275–343.

Hodson, P.V., B.R. Blunt, D.J. Spry and K. Austen. "Evaluation of Erythrocyte-amino Levulinic Acid Dehydratase Activity as a Short-term Indicator in Fish of a Harmful Exposure to Lead," *J. Fish. Res. Bd. Can.* 34:501–508 (1977).

Hodson, P.V., B.R. Blunt and D.M. Whittle. "Biochemical Monitoring of Fish Blood as an Indicator of Biologically Available Lead," *Thal. Jugos.* 1:389–396 (1980).

Hodson, P.V., B.R. Blunt and D.M. Whittle. "Monitoring Lead Exposure of Fish," in *Contaminant Effects on Fisheries*, V.W. Cairns, P.V. Hodson and J.O. Nriagu, Eds. (Toronto: John Wiley & Sons, 1984), pp. 87–97.

Hoffman, D.J., and W.C. Eastin, Jr. "Effects of Malathion, Diazinon, and Parathion on Mallard Embryo Development and Cholinesterase Activity," *Environ. Res.* 26:472–485 (1981).

Hoffman, D.J., and G.H. Heinz. "Embryotoxic and Teratogenic Effects of Selenium in the Diet of Mallards," *J. Toxicol. Environ. Health* 24:477–490 (1988).

Hoffman, D.J., H.M. Ohlendorf and T.W. Aldrich. "Selenium Teratogenesis in Natural Populations of Aquatic Birds in Central California," *Arch. Environ. Contam. Toxicol.* 17:519–525 (1988).

Hoffman, D.J., and L. Sileo. "Neurotoxic and Teratogenic Effects of an Organophosphorus Insecticide (Phenyl Phosphonothioic Acid-O-ethyl-O[4-nitrophenyl] ester) on Mallard Development," *Toxicol. Appl. Pharmacol.* 73:284–294 (1984).

Hoffman, G.L., E.C. Dunbar and A. Bradford. "Whirling Disease of Trouts Caused by *Myxosoma cerebralis* in the United States," U.S. Fish Wild. Serv. Spec. Sci. Rept. Fish No. 427 (1962).

Holland, D.L. "Lipid Reserves and Energy Metabolism in the Larvae of Benthic Marine Invertebrates," in *Biochemical and Biophysical Perspectives in Marine Biology, Vol. 4,* D.C. Malins, and J.R. Sargent, Eds. (New York: Academic Press, Inc., 1978), pp. 85–123.

Holland, H.T., D.L. Coppage and P.A. Butler. "Use of Fish Brain Acetylcholinesterase to Monitor Pollution by Organophosphorus Pesticides," *Bull. Environ. Contam. Toxicol.* 2:156–162 (1967).

Holmes, W.N. "Petroleum Pollutants in the Marine Environment and Their Possible Effects on Seabirds," in *Reviews in Environmental Toxicology I,* E. Hodgson, Ed. (Amsterdam: Elsevier, 1984), pp. 251–317.

Inman, C.B.E. and A.P.M. Lockwood. "Some Effects of Methylmercury and Lindane on Sodium Regulation in the Amphipod *Gammaus duebeni* During Changes in the Salinity of its Medium," *Comp. Biochem. Physiol.* 58C:67–75 (1977).

Ivanovici, A.M. "Adenylate Energy Charge. Potential Value as a Tool for Rapid Determination of Toxicity Effects," in *Proc. 5th Ann. Aquatic Toxicity Workshop,* pp. 241–255. Fish. Mar. Serv. Tech. Rep., Hamilton, Ontario, Canada, p. 682 (1979).

Ivanovici, A.M., and W.J. Wiebe. "For Working Definition of "Stress": A Review and Critique," in *Stress and Natural Ecosystems,* G.W. Barrett, and R. Rosenberg, Eds. (New York: John Wiley & Sons, Inc., 1982), pp. 13–27.

Jackim, E., J.M. Hamlin and S. Sonis. "Effects of Metal Poisoning on Five Liver Enzymes in the Killifish (*Fundulus heteroclitus*)," *J. Fish. Res. Bd. Can.* 27:383–390 (1970).

Jackim, E., and D. Nacci. "A Rapid Aquatic Toxicity Assay Utilizing Labelled Thymidine Incorporation in Sea Urchin Embryos," *Environ. Toxicol. Chem.* 3:631–636 (1984).

Jacobson, J.S., and A.C. Hill, Eds. *Recognition of Air Pollution Injury to Vegetation: A Pictorial Atlas.* (Pittsburgh, PA: Air Pollution Control Association, 1970).

Jeffries, H.P. "A Stress Syndrome in the Hard Clam, *Mercenaria mercenaria," J. Invert. Pathol.* 20:242–251 (1972).

Johansson-Sjobeck, M.-L., G. Dave, A. Larsson, K. Lewander and U. Lidman. "Hematological Effects of Cortisol in the European Eel, *Anguilla anguilla* L.," *Comp. Biochem. Physiol.* 60A:165–168 (1978).

Johansson-Sjobeck, M.-L., and A. Larsson. "Effects of Inorganic Lead on Delta-aminolevulinic Acid Dehydratase Activity in the Rainbow Trout, *Salmo gairdneri," Arch. Environ. Contam. Toxicol.* 8:419–431 (1979).

Johnson, L.L., E. Casillas, T.K. Collier, B.B. McCain and U. Varanasi. "Contaminant Effects on Ovarian Development in English Sole (*Parophrys vetulus*) from Puget Sound Washington," *Can. J. Fish. Aquat. Sci.* 45:2133–2146 (1988).

Jowett, P.E., M.M. Rhead and B.L. Bayne. "In vivo Changes in the Activity of Gill ATPase and Haemolymph Ions of *Carcinus maenas* Exposed to p,p'-DDT and Reduced Salinities," *Comp. Biochem. Physiol.* 69C:399–402 (1981).

Karl, D.M., C.D. Winn and D.C. Long. "RNA Synthesis as a Measure of Microbial Growth in Aquatic Environments. I. Evaluation, Verification and Optimization of Methods," *Mar. Biol.* 64:1–12 (1981).

Karp, G. *Cell Biology* (New York: McGraw-Hill Book Co., 1979), p. 846.

Kasschau, M.R., M.M. Skaggs and E.C.M. Chen. "Accumulation of Glutamate in Sea Anemones Exposed to Heavy Metals and Organic Amines," *Bull. Environ. Contam. Toxicol.* 25:873–878 (1980).

Kearns, P.K., and G.J. Atchison. "Effects of Trace Metals on Growth of Yellow Perch (*Perca flavescens*) as Measured by RNA-DNA Ratios," *Env. Biol. Fishes.* 4:383–387 (1979).

King, K.A., E.L. Flickinger and H.H. Hildebrand. "Shell Thinning and Pesticides Residues in Texas Aquatic Bird Eggs 1970," *Pestic. Monit. J.* 12:16–21 (1978).

Kishida, M., and T. Hirano. "Development of Radio-Immunoassay for Eel Growth Hormone," *Nippon. Suisan Gakkaishi* 54:1321–1327 (1988).

Klaverkamp, J.F., M. Duangsawadsi, W.A. MacDonald and H.S. Majewski. "An Evaluation of Fenitrothion Toxicity in Four Life Stages of Rainbow Trout, *Salmo gairdneri,*" in *Aquatic Toxicology and Hazard Evaluation,* F.L. Mayer, and J.L. Hamelink, Eds. ASTM STP 634, American Society for Testing Materials, Philadelphia, PA (1977), pp. 231–240.

Knowles, C.J. "Microbial Metabolic Regulations by Adenine Nucleotide Pools," in *Microbial Energetics,* B.A. Haddock, and W.A. Hamilton, Eds., Twenty-Seventh Symposium of the Society for General Microbiology, Imperial College, London (Cambridge, England: Cambridge University Press, 1977).

Knowles, C.O., and M.J. McKee. "Protein and Nucleic Acid Content in *Daphnia magna* During Chronic Exposure to Cadmium," *Ecotox. Environ. Safety* 13:290–300 (1987).

Koundinya, P.R., and R. Ramamurthi. "Effect of Organophosphate Pesticide Sumithion (Fenitrothion) on Some Aspects of Carbohydrate Metabolism in a Freshwater Fish, *Sarotherodon mossambicus* (Peters)," *Experimentia* 35:1632–1633 (1979).

Kroger, R.L., and J.F. Guthrie. "Incidence of Crooked Vertebral Columns in Juvenile Atlantic Menhaden, *Brevoortia tyrannus,*" *Chesapeake Sci.* 12(4):276–278 (1971).

Krueger, H.M., J.B. Saddler, G.A. Chapman, I.J. Tinsley and R.R. Lowry. "Bioenergetics, Exercise and Fatty Acids of Fish," *Am. Zool.* 8:119–129 (1968).

Lacasse, N.L., and M. Treshow, Eds. *Diagnosing Vegetation Injury Caused by Air Pollution,* Air Pollution Training Institute, EPA Contract 68-020/344. Applied Science Associates (1976).

Lane, C.E., and E.D. Scura. "Effects of Dieldrin on Glutamic Oxaloacetic Transaminase in *Poecilia latipinna,*" *J. Fish. Res. Bd. Can.* 27:1869–1871 (1970).

Lang, C.A., H.Y. Lau and D.J. Jefferson. "Protein and Nucleic Acid Changes During Growth and Aging in the Mosquito," *J. Biochem.* 95:372–377 (1965).

Larsson, A., B. Bengtsson and C. Haux. "Disturbed Ion Balance in Flounder, *Platichthys flesus* L., Exposed to Sublethal Levels of Cadmium," *Aquat. Toxicol.* 1:19–35 (1981).

Larsson, A., C. Haux and M. Sjobeck. "Fish Physiology and Metal Pollution: Results and Experiences from Laboratory and Field Studies," *Ecotoxicol. Environ. Saf.* 9:250–282 (1985).

Leatherland, J.F., and R.A. Sonstegard. "Lowering of Serum Thyroxine and Triiodothyronine Levels in Yearling Coho Salmon, *Oncorhynchus kisutch,* by Dietary Mirex and PCBs," *J. Fish. Res. Bd. Can.* 34:1285–1289 (1978).

Leatherland, J.F., and R.A. Sonstegard. "Effect of Dietary Mirex and PCB (Aroclor 1254) on Thyroid Activity and Lipid Reserves in Rainbow Trout, *Salmo gairdneri,*" *J. Fish Dis.* 2:43–48 (1979).

Leatherland, J.F., and R.A. Sonstegard. "Effect of Dietary Mirex and PCBs in Combination with Food Deprivation and Testosterone Administration on Thyroid Activity and Bioaccumulation of Organochlorines in Rainbow Trout, *Salmo gairdneri* Richardson," *J. Fish Dis.* 3:115–124 (1980).

Lee, W.Y., S.A. Macko and J.A.C. Nicol. "Changes in Nesting Behavior and Lipid Content of a Marine Amphipod (*Amphithoe valida*) to the Toxicity of a No. 2 Fuel Oil," *Water Air Soil Pollut.* 15:185–195 (1981).

Lehtinen, K., A. Larsson and G. Klingstedt. "Physiological Disturbances in Rainbow Trout, *Salmo gairdneri* (R.) Exposed at Two Temperatures to Effluents from a Titanium Dioxide Industry," *Aquatic Toxicol.* 5:155–166 (1984).

Leland, H.V. "Ultrastructural Changes in the Hepatocytes of Juvenile Rainbow Trout and Mature Brown Trout Exposed to Copper or Zinc," *Environ. Toxicol. Chem.* 2:353–368 (1983).

Lidman, U., G. Dave and M.L. Johansson-Sjobeck. "Metabolic Effects of PCB (Polychlorinated biphenyls) Rich," in *Sublethal Effects of Toxic Chemicals on Aquatic Animals*, J.H. Koeman, and J.J.T.W.A. Strik, Eds.(New York: Elsevier, 1975), pp. 207–212.

Lied, E., and G. Rosenlund. "The Influence of the Ratio of Protein Energy to Total Energy in the Feed on the Activity of Protein Synthesis In Vitro, the Level of Ribosomal RNA and the RNA-DNA Ratio in White Trunk Muscle of Atlantic Cod (*Gadus moshua*)," *Comp. Biochem. Physiol.* 77A:489–494 (1984).

Linden, O. "The Influence of Crude Oil and Mixtures of Crude Oil Dispersants on the Ontogenic Development of Baltic Herring, *Clupea harengus membras* L.," *Ambio* 5:136–140 (1976).

Livingstone, D.R. "Induction of Enzymes as a Mechanism for the Seasonal Control of Metabolism in Marine Invertebrates: Glucose-6-phosphate Dehydrogenase from the Mantle and Hepatopancreas of the Common Mussel *Mytilus edulis* L.," *Comp. Biochem. Physiol.* 69B:147-156 (1981).

Livingstone, D.R. "General Biochemical Indices of Sublethal Stress," *Mar. Pollut. Bull.* 13:261–263 (1982).

Livingstone, D.R. "Biochemical Measures," in *The Effects of Stress and Pollution on Marine Animals*, B.L. Bayne, D.A. Brown, K. Burns, D.R. Dixon, A. Ianovici, D.R. Livingstone, D.M. Lowe, M.N. Moore, A.R.D. Stebbing and J. Widdows, Eds. (New York: Praeger Publishers, 1985), pp. 81–132.

Lock, R.A.C., P.M.J.M. Cruijsen and A.P. van Overbeeke. "Effects of Mercuric Chloride and Methylmercuric Chloride on the Osmoregulatory Function of the Gills in Rainbow Trout, *Salmo gairdneri* Richardson," *Comp. Biochem. Physiol.* 68C:151–159 (1981).

Lockhart, W.L., and D.A. Metner, in *Contaminant Effects on Fisheries*, V. W. Cairns, P.V. Hodson, and J.O. Nriagu, Eds. (New York: John Wiley & Sons, Inc., 1985), pp. 73–85.

Lockhart, W.L., R. Wagemann, J.W. Clayton, B. Graham and D. Murray. "Chronic Toxicity of a Synthetic Tri-aryl Phosphate Oil to Fish," *Environ. Physiol. Biochem.* 5:361–369 (1975).

Lorz, H.W., and B.P. McPherson. "Effects of Copper or Zinc in Freshwater on the Adaptation to Seawater and ATPase Activity, and the Effects of Copper on Migratory Disposition of Coho Salmon (*Oncorhynchus kisutch*)," *J. Fish Res. Bd. Can.* 33:2023–2030 (1976).

Love, R.M. *The Chemical Biology of Fishes* (London: Academic Press, Inc., 1970), p. 547.

Lowe-Jinde, L., and A.J. Niimi. *Arch. Environ. Contam. Toxicol.* 13:759–764 (1984).

Ludke, J.L., E.F. Hill and M.P. Dieter. "Cholinesterase (ChE) Responses and Related Mortality Among Birds Fed ChE Inhibitors," *Arch. Environ. Contam. Toxicol.* 3:1–21 (1975).

Lugo, A.E. "Stress and Ecosystems," in *Energy and Environmental Stress in Aquatic Systems*, J.H. Thorp and J.W. Gibbons, Eds. Tech. Inf. Cent., U.S. Dept. of Energy, Washington, D.C. (1978), pp. 62–101.

Lundholm, C.E., and K. Mathson "Effect of Some Metal Compounds on the Ca^{2+} Binding and Ca^{2+}-Mg^{2+}-ATPase Activity of Eggshell Gland Mucosa Homogenate from the Domestic Fowl," *Acta. Pharmacol. Toxicol.* 59:410–415 (1986).

Lundholm, E. "Thinning of Eggshells in Birds by DDE: Mode of Action on the Eggshell Gland," *Comp. Biochem. Physiol.* 88C:1–22 (1987).

Malhotra, S.S., and R.A. Blauel. "Diagnosis of Air Pollutant and Natural Stress Symptoms on Forest Vegetation in Western Canada," Northern Forest Research Center, Canadian Forestry Service, Information Report NOR-X-228 (1980), p. 84.

Manning, W.J., and W.A. Feder. *Biomonitoring Air Pollutants with Plants* (London: Applied Science Publishers, 1980), p. 142.

Marchant, T.A., J.G. Dulka and R.E. Peter. "Relationship Between Serum Growth Hormone Levels and the Brain and Pituitary Content of Immunoreactive Somatostatin in the Goldfish, *Carrasius auratus* L.," *Gen. Comp. Endocrinol.* 73:458–568 (1989).

Markert, C.L., and I. Faulhaber. "Lactate Dehydrogenase Isozyme Patterns of Fish," *J. Exp. Zool.* 159:319–332 (1965).

Mason, J.W. "A Historical View of the Stress Field. Part I," *J. Human Stress* 1:6-12 (1975a).

Mason, J.W. "A Historical View of the Stress Field. Part II," *J. Human Stress* 1:22–36 (1975b).

Matton, R., and Q.N. Lattam. "Effect of the Organophosphate Dylox on Rainbow Trout Larvae," *J. Fish. Res. Bd. Can.* 26:2193–2200 (1969).

Mayer, F.L., Jr. "Clinical Tests in Aquatic Toxicology: A Paradox?" *Environ. Toxicol. Chem.* 2:139–140 (1983).

Mayer, F.L., W.J. Adams, M.T. Finley, P.R. Michael, P.M. Mehrle and V.W. Saeger. "Phosphate Ester Hydraulic Fluids: An Aquatic Environmental Assessment of Pydrauls 50E and 115E," in *Aquatic Toxicology and Hazard Assessment*, D.R. Branson and K.L. Dickson, Eds. ASTM STP 737 Philadelphia: American Society for Testing Materials (1981), pp.103–123.

Mayer, F.L., B.E. Bengtsson, S.J. Hamilton and A. Bengtsson. "Effects of Pulp Mill and Ore Smelter Effluents on Vertebrae of Fourhorn Sculpin: Laboratory and Field Comparisons," in *Aquatic Toxicology and Hazard Assessment*, W.J. Adams, G.A. Chapman and W.G. Landis, Eds., ASTM STP 971 Philadelphia: American Society for Testing Materials (1988), pp. 406–419.

Mayer, F.L., Jr., K.S. Mayer and M.R. Ellersieck. "Relation of Survival to Other Endpoints in Chronic Toxicity Tests with Fishes," *Environ. Toxicol. Chem.* 5:737–748 (1986).

Mayer, F. L., Jr., P. M. Mehrle, Jr. and W. P. Dwyer. "Toxaphene: Chronic Toxicity to Fathead Minnows and Channel Catfish," Ecol. Res. Ser. No. EPA-600/3-77-069. U.S. Environmental Protection Agency, Duluth, MN (1977a).

Mayer, F.L., P.M. Mehrle and R.A. Schoettger. "Collagen Metabolism in Fish Exposed to Organic Chemicals," in *Recent Advances in Fish Toxicology—A Symposium*, R.A. Tubb, Ed., Ecol. Res. Ser. No. EPA-600/7-77-085, U.S. Environmental Protection Agency, Corvallis, OR (1977b), pp. 31–54.

Mazeaud, M., F. Mazeaud and E.M. Donaldson. "Primary and Secondary Effects of Stress in Fish: Some New Data with a General Review," *Trans. Am. Fish. Soc.* 106:201–212 (1977).

McCann, J.A., and R.L. Jasper. "Vertebral Damage to Blue-gills Exposed to Acute Levels of Pesticides," *Trans. Am. Fish. Soc.* 101:317–322 (1972).

McCune, D.C., L.H. Weinstein and J.F. Mancini. "Effects of Hydrogen Fluoride on the Acid-soluble Nucleotide Metabolism of Plants," *Contrib. Boyce Thompson Inst.* 24:213–226 (1970).

McKee, M.J., and C.O. Knowles. "Effects of Fenvalerate on Biochemical Parameters, Survival, and Reproduction of *Daphnia magna*," *Ecotoxicol. Environ. Safety* 12:70–84 (1986).

McKee, M.J., C.O. Knowles and D.R. Buckler. "Effects of Aluminum on the Biochemical Composition of Atlantic Salmon," *Arch. Environ. Contam. Toxicol.* 18:243–248 (1989).

McKeown, B.A., and G.L. March. "The Acute Effect of Bunker C Oil and Oil Dispersant on: I. Serum Glucose, Serum Sodium and Gill Morphology in Both Freshwater and Seawater Acclimated Rainbow Trout (*Salmo gairdneri*)," *Water Res.* 12:157–163 (1978).

McKim, J.M., G.M. Christensen and E.P. Hunt. "Changes in the Blood of Brook Trout (*Salvelinus fontinalis*) after Short-term and Long-term Exposure to Copper," *J. Fish Res. Bd. Can.* 27:1883–1889 (1970).

McLeay, D.J. "Development of a Blood Sugar Bioassay for Rapidly Measuring Stressful Levels of Pulpmill Effluent to Salmonid Fish," *J. Fish Res. Bd. Can.* 34:477–485 (1977).

McLeay, D.J., and D.A. Brown. "Effects of Acute Exposure to Bleached Draft Pulpmill Effluent on Carbohydrate Metabolism of Juvenile Coho Salmon (*Oncorhynchus kisutch*) During Rest and Exercise," *J. Fish Res. Bd. Can.* 32:753–760 (1975).

McWhinnie, M.A., R.J. Kirchenberg, R.J. Urbanski and J.E. Schwarz. "Crustecdysone Mediated Changes in Crayfish" *Am. Zool.* 12:357–372 (1972).

Mehrle, P.M., Jr., T.A. Haines, S.J. Hamilton, J.L. Ludke, F.L. Mayer, Jr. and M.A. Ribick. "Relationship Between Body Contaminants and Bone Development in East Coast Striped Bass," *Trans. Am. Fish. Soc.* 111:231–241 (1982).

Mehrle, P.M., and F.L. Mayer. "Toxaphene Effects on Growth and Bone Composition of Fathead Minnows, *Pimpehales promelas*," *J. Fish. Res. Bd. Can.* 32:593–598 (1975).

Mehrle, P.M., and F.L. Mayer. "Clinical Tests in Aquatic Toxicology: State of the Art," *Environ. Health Perspect.* 34:139–143(1980).

Mehrle, P.M., and F.L. Mayer. "Biochemistry/Physiology," in G.M. Rand, and S.R. Petrocelli, Eds. (New York: Hemisphere Publ. Co., 1985), pp. 264–282.

Mehrle, P.M., F.L. Mayer and D.R. Buckler. "Kepone and Mirex: Effects on Bone Development and Swim Bladder Composition in Fathead Minnows," *Trans. Am. Fish. Soc.* 110:636–641 (1981).

Mensi, P., A. Arillo, C. Margiocco and G. Schenone. "Lysosomal Damage Under Nitrate Intoxication in Rainbow Trout (*Salmo gairdneri* Rich.)," *Comp. Biochem. Physiol.* 73C:161–165 (1982).

Middaugh, D.P., J.W. Fournie and M.J. Hemmer. "Vertebral Abnormalities in Juvenile Inland Silversides *Menidia beryllina* Exposed to Terbufos During Embryogenesis," *Dis. Aquat. Org.* 9:109–116 (1990).

Miller, D.S. "Heavy Metal Inhibition of P-Aminohippurate Transport in Flounder Renal Tissue: Sites of HgCl$_2$ Action," *J. Pharm. Exp. Ther.* 219:428–434 (1981).

Miller, N.G.A., M.W. Hill and M.W. Smith. "Positional and Species Analyses of Membrane Phospholipids Extracted from Goldfish Adapted to Different Environmental Temperatures," *Biochem. Biophys. Acta.* 455:644–654 (1976).

Mineau, P., G.A. Fox, R.J. Norstrom, D.V. Weseloh, D.J. Hallett and J.A. Ellenton. "Using the Herring Gull to Monitor Levels and Effects of Organochlorine Contamination in the Canadian Great Lakes," in *Toxic Contaminants in the Great Lakes. Advances in Environmental Science and Technology, Vol. 14*, J.O. Nriagu, and M.S. Simmons, Eds. (New York: Wiley Interscience, 1984), pp. 311–314.

Mitz, S.V., and J.P. Giesy. "Sewage Effluent Biomonitoring. I. Survival, Growth, and Histopathological Effects in Channel Catfish," *Ecotoxicol. Environ. Saf.* 10:22–39 (1985a).

Mitz, S.V., and J.P. Giesy. "Sewage Effluent Biomonitoring. II. Biochemical Indicators of Ammonia Exposure in Channel Catfish," *Ecotoxicol. Environ. Saf.* 10:40–52 (1985b).

Miwa, S., M. Tagawa, Y. Inui and T. Hirano. "Thyroxine Surge in Metamorphosing Flounder Larvae," *Gen. Comp. Endocrinol.* 70:158–163 (1988).

Montague, M.D., and E.A. Dawes. "The Survival of *Peptococcus prevotii* in Relation to the Adenylate Energy Charge," *J. Gen. Microbiol.* 80:291–299 (1975).

Moore, M.N. "Cytochemical Demonstration of Latency of Lysosomal Hydrolases in Digestive Cells of the Common Mussel, *Mytilus edulis*, and Changes Induced by Thermal Stress," *Cell Tissue Res.* 175:279–287 (1976).

Moore, M.N., R.K. Koehn and B.L. Bayne. "Leucine Aminopeptidase (Aminopeptidase-I), *N*-acetyl-_-hexosaminidase and Lysosomes in the Mussel, *Mytilus edulis* L., in Response to Salinity Changes," *J. Exp. Zool.* 214:239–249 (1980).

Moore, M.N., D.M. Lowe and P.M. Fieth. "Lysosomal Responses to Experimentally Injected Anthracene in the Digestive Cell of *Mytilus edulis*," *Mar. Biol.* 48:297–302 (1978).

Moore, M.N., R.K. Pipes and S.V. Farrar. "Lysosomal and Microsomal Responses to Environmental Factors in *Littoria littorea* from Sullom Voe," *Mar. Pollut. Bull.* 13:340–345 (1982).

Moore, M.N., and A.R.D. Stebbing. "The Quantitative Cytochemical Effects of Three Metal Ions on a Lysosomal Hydrolase of a Hydroid," *J. Mar. Biol. Assoc. U.K.* 56:995–1005 (1976).

Morata, P., M.J. Faus, M. Perez-Palomo and F. Sanchez-Medina. "Effect of Stress on Liver and Muscle Glycogen Phosphorylase in Rainbow Trout (*Salmo gairdneri*)," *Comp. Biochem Physiol.* 72B:421–425 (1982).

Morgan, R.P., II, R.F. Fleming, V.J. Rasin, Jr. and D.R. Heinle. "Sublethal Effects of Baltimore Harbor Water on the White Perch, *Morone americana*, and Hogchoker, *Trinectes maculatus*," *Chesapeake Sci.* 14:17–27 (1973).

Morris, R.J. "Seasonal and Environmental Effects in the Lipid Composition of *Neomyses integer*," *J. Mar. Biol. Assoc. U.K.* 51:21–31 (1971).

Mount, D.R. "Physiological and Toxicological Effects of Long-term Exposure to Acid, Aluminum and Low Calcium on Adult Brook Trout (*Salvelinus fontinalis*) and Rainbow Trout (*Salmo gairdneri*)," PhD Thesis, University of Wyoming, Laramie, WY, p. 157 (1988).

Munck, A., P.M. Guyre and N.J. Holbrook. "Physiological Functions of Glucocorticoids in Stress and Their Relation to Pharmacological Actions," *Endocrinol. Rev.* 5:25–44 (1984).

Murty, A.S., and A.P. Devi. "The Effect of Endosulfan and its Isomers on Tissue Protein, Glycogen, and Lipids in the Fish (*Channa punctatus*)," *Pest. Biochem. Physiol.* 17:280–286 (1982).

Nelson, D.H. "Pharmacology of the Anticosteroids and Adrenocorticotrophic Hormone," in *Essentials of Pharmacology*, J.A. Beveam, and J.H. Thompson, Eds. (Philadelphia: Harper & Row, 1983), pp. 469–476.

Newell, R.I.E., and B.L. Bayne. "Seasonal Changes in the Physiology, Reproductive Condition and Carbohydrate Content of the Cockle *Cardium* (-Cerastoderma) *edule* (Bivalvia: Cardiidae)," *Mar. Biol.* 56:11–19 (1980).

Nichols, J.W., G.A. Wedemeyer, F.L. Mayer, W.W. Dickhoff, S.V. Gregory, W.T. Yasutake and L.S. Smith. "Effects of Freshwater Exposure to Arsenic Trioxide on the Parr-smolt Transformation of Coho Salmon (*Oncorhynchus kisutch*)," *Environ. Toxicol. Chem.* 3:143–149 (1984).

Nicholson, H.P. "Pesticide Pollution Control," *Science* 158:871–876 (1967).

Nogawa, K., Y. Yamada, T. Kido, R. Honda, M. Ishizaki, I. Tswitsmi and E. Kobayaski. "Significance of Elevated Urinary N-acetyl-β-D-glucosaminidase Activity in Chronic Cadmium Poisoning," *Sci. Total Environ.* 53:173–178 (1986).

Nolan, C.V., and E.J. Duke. *Aquat. Toxicol.* 4:153–164 (1983).

O'Brien, R.D. *Insecticides—Action and Metabolism.* (New York: Academic Press, Inc., 1967).

O'Connor, J.D., and L.I. Gilbert. "Alterations in Lipid Metabolism Associated with Premolt Activity in a Land Crab and Freshwater Crayfish," *Comp. Biochem. Physiol.* 29:889–904 (1969).

O'Hara, S.C.M., A.C. Neal, E.D.S. Corner and A.L. Pulsford. "Interrelationships of Cholesterol and Hydrocarbon Metabolism in the Shore Crab, *Carcinus*," *J. Mar. Biol Assoc. U.K.* 65:113–131 (1985).

Oikari, A., B.-E. Lonn, M. Castren, T. Nakari, B. Snickars-Nikinmaa, H. Bister and E. Virtanen. "Toxicological Effects of Dehyroabietic Acid (DHAA) on the Trout, *Salmo gairdneri* Richardson, in Freshwater," *Water Res.* 17:81–89 (1983).

Orska, J. "Anomalies in the Vertebral Columns of the Pike (*Esox lucius* L.)," *Acta Biol. Cracov.* 5:327–345 (1962).

Pant, J.C., and T. Singh. "Inducement of Metabolic Dysfunction by Carbamate and Organophosphorus Compounds in a Fish, *Puntius conchonius*," *Pestic. Biochem. Physiol.* 20:294–298 (1983).

Patton, J.F. "Indocyanine Green: A Test of Hepatic Function and a Measure of Plasma Volume in the Duck," *Comp. Biochem. Physiol.* 60A:21–24 (1978).

Peakall, D.B., J. Tremblay, W.B. Kinter and D.S. Miller. "Endocrine Dysfunction in Seabirds Caused by Ingested Oil," *Environ. Res.* 24:6–14 (1981).

Pell, E., and E. Brennan. "Changes in Respiration, Photosynthesis, Adenosine 5′-triphosphate, and Total Adenylate Content of Ozonated Pinto Bean Foliage as They Relate to Symptom Expression," *Plant Physiol.* 51:378–381 (1973).

Peter, R.E., A. Hontela, A.F. Cook and C.R. Paulenau. "Daily Cycles in Serum Cortisol Levels in the Goldfish: Effects of Photoperiod, Temperature and Sexual Condition," *Can. J. Zool.* 56:2443–2448 (1978).

Phelps, D.K., W. Galloway, F.P. Thurberg, E. Gould and M.A. Dawson. "Comparison of Several Physiological Monitoring Techniques as Applied to the Blue Mussel, *Mytilus edulis*, along a Gradient of Pollutant Stress in Narragansett Bay, Rhode Island," in *Biological Monitoring of Marine Pollutants*, J. Vernberg, A. Calabrese, F.P. Thurberg and W.B. Vernberg, Eds. (New York: Academic Press, Inc., 1981).

Phillips, A.M., F.E. Lovelace, H.A. Podoliak, D.R. Brockway and G.C. Balzer. "The Nutrition of Trout," *Fish. Res. Bull.* 19:56 (1956)

Pickering, A.D. *Stress and Fish* (New York:Academic Press, Inc., 1981), p. 367.

Plisetskaya, E.M. "Physiology of the Fish Endocrine Pancreas," *Fish Physiol. Biochem.* 7:39–48 (1989).

Pohorecky, L.A., and R.J. Wurtman. "Adrenocortical Control of Epinephrine Synthesis," *Pharmacol. Rev.* 23:1–35 (1971).

Post, G., and R.A. Leisure. "Sublethal Effect of Malathion to Three Salmonid Species," *Bull. Environ. Contam. Toxicol.* 12:312–319 (1974).

Powell, E.N., M. Kasschau, E. Chen, M. Koenig and J. Pecon. "Changes in the Free Amino Acid Pool During Environmental Stress in the Gill Tissue of the Oyster, *Crassostrea virginica*," *Comp. Biochem. Physiol.* 71A:591–598 (1982).

Racicot, J.G., M. Gaudet and C. Leray. "Blood and Liver Enzymes in Rainbow Trout (*Salmo gairdneri* Rich.) with Emphasis on their Diagnostic Use: Study of CCl_4 Toxicity and a Case of Aeromonas Infection," *J. Fish Biol.* 7:825–835 (1975).

Rao, K.S.P., and K.V.R. Rao. "Tissue Specific Alterations of Amino-transferases and Total ATPases in the Fish (*Tilapia mossambica*) Under Methyl Parathion Impact," *Toxicol. Lett.* 20:53–57 (1984a).

Rao, K.S.P., and K.V.R. Rao. "Changes in the Tissue Lipid Profiles of Fish (*Oreochromis mossambicus*) During Methyl Parathion Toxicity—A Time Course Study," *Toxicol. Lett.* 21:147–153 (1984b).

Ratcliffe, D.A. "Decrease in Eggshell Weight in Certain Birds of Prey," *Nature (London)* 215:208–210 (1967).

Rattner, B.A. "Tolerance of Adult Mallards to Subacute Ingestion of Crude Petroleum Oil," *Toxicol. Lett.* 8:337–342 (1981).

Rattner, B.A., V.P. Eroschenko, G.A. Fox, D.M. Fry and J. Gorsline. "Avian Endocrine Responses to Environmental Pollutants," *J. Exp. Zool.* 232:683–689 (1984).

Rattner, B.A., and J.C. Franson. "Methyl Parathion and Fenvalerate Toxicity in American Kestrels: Acute Physiological Responses and Effects of Cold," *Can. J. Physiol. Pharmacol.* 62:787–792 (1984).

Rattner, B.A., G.M. Haramis, D.S. Chu, C.M. Bunck and C.G. Scanes. "Growth and Physiological Condition of Black Ducks Reared on Acidified Wetlands," *Can. J. Zool.* 65:2953–2958 (1987).

Rattner, B.A., and D.J. Hoffman. "Comparative Toxicity of Acephate in Laboratory Mice, White-footed Mice, and Meadow Voles," *Arch. Environ. Contam. Toxicol.* 13:483–491 (1984).

Rattner, B.A., D.J. Hoffman, and C.M. Marn. "Use of Mixed-function Oxygenases to Monitor Contaminant Exposure in Wildlife," *Environ. Toxicol. Chem.* 8:1093–1102 (1989).

Rattner, B.A., S.D. Michael and P.D. Altland. "Age-related Responses to Mild Restraint in the Rat," *J. Appl. Physiol. Respir. Environ. Exerc. Physiol.* 55(5):1408–1412 (1983).

Regard, E., A. Taurog and T. Nakashima. "Plasma Thyroxine and Triiodothyronine Levels in Spontaneously Metamorphosing *Rana castesbiana* Tadpoles and in Adult Anuran Amphibians," *Endocrinology* 102:674–684 (1978).

Reinert, R.E., and D.W. Hobreiter. "Adenylate Energy Charge as a Measure of Stress in Fish," in *Contaminant Effects on Fisheries*, V.W. Cairns, P.V. Hodson and J.O. Nriagu, Eds. (New York: John Wiley & Sons, Inc., 1984), pp. 151–161.

Riley, R.T., and M.C. Mix. "The Effects of Naphthalene on Glucose Metabolism in the European Flat Oyster *Ostrea edulis*," *Comp. Biochem. Physiol.* 70C:13–20 (1981).

Roberts, K.S., A. Cryer, J. Kay, J.F. De, L.G. Solbe, J.R. Warfe and W.R. Simpson. "The Effects of Exposure to Sublethal Concentrations of Cadmium on Enzyme Activities and

Accumulation of the Metal in Tissues and Organs of Rainbow and Brown Trout (*Salmo gairdneri*, Richardson and *Salmo trutta fario* L.)," *Comp. Biochem. Physiol.* 62C:135–140 (1979).

Roesijadi, G., and J.W. Anderson. "Condition Index and Free Amino Acid Content of *Macoma inquinata* Exposed to Oil Contaminated Marine Sediments," in *Marine Pollution: Functional Responses*, W.B. Vernberg, A. Calabrese, F.P. Thurberg and F.J. Vernberg, Eds. (New York: Academic Press, Inc., 1979), pp. 69–83.

Roesijadi, G., J.W. Anderson and C.S. Giam. "Osmoregulation of the Grass Shrimp *Palaemonetes pugio* Exposed to Polychlorinated Biphenyls (PCBs).II. Effect on Free Amino Acids of Muscle Tissue," *Mar. Biol.* 38:357–363 (1976).

Roesijadi, G., J.W. Anderson, S.R. Petrocelli and C.S. Giam. "Osmoregulation of the Grass Shrimp *Palaemonetes pugio* Exposed to Polychlorinated Biphenyls (PCBs).I. Effect on Chloride and Osmotic Concentrations and Chloride and Water-exchange Kinetics," *Mar. Biol.* 38:343–355 (1976).

Roesijadi, G., S.R. Petrocelli, J.W. Anderson, B.J. Presley and R. Sims. "Survival and Chloride Ion Regulation of the Porcelain Crab *Petrolisthes armatus* Exposed to Mercury," *Mar. Biol.* 27:213–217 (1974).

Roland, D.A., Dr., D.R. Sloan, H.R. Wilson and R.H. Harms. "Influence of Dietary Calcium Deficiency on Yolk and Serum Calcium, Yolk and Organ Weights and Other Selected Production Criteria of the Pullet," *Poult. Sci.* 52:2220–2225 (1973).

Rush, S.B., and B.L. Umminger. "Elimination of Stress-induced Changes in Carbohydrate Metabolism of Goldfish (*Carassius auratus*) by Training," *Comp. Biochem. Physiol.* 60A:69–73 (1978).

Sangalang, G.B., and H.C. Freeman. "Effects of Sublethal Cadmium on Maturation and Testosterone and 11-Ketotestosterone Production in vivo in the Brook Trout," *Biol. Repro.* 11:429–435 (1974).

Sastry, K.V., and A.A. Siddiqui. "Chronic Toxic Effects of the Carbamate Pesticide Sevin on Carbohydrate Metabolism in a Freshwater Snakehead Fish, *Channa punctatus*," *Toxicol. Lett.* 14:123–130 (1982).

Sastry, K.V., and K. Subhadra. "In vivo Effects of Cadmium on Some Enzyme Activities in Tissues of the Freshwater Catfish, *Heteropneustes fossilis*," *Environ. Res.* 36:32–45 (1985).

Sastry, K.V., and K. Sunita. "Enzymological and Biochemical Changes Produced by Chronic Chromium Exposure in a Teleost Fish, *Channa punctatus*," *Toxicol. Lett.* 16:9–15 (1983).

Saunders, R.L., E.B. Henderson, P.R. Harmon, C.E. Johnston and K. Davidson. "Physiological Effects of Low pH on the Smolting Process in Atlantic Salmon," in *Workshop on Acid Rain*, R.H. Peterson, H.H.V. Hord, Eds. p. 49 (1983).

Schafer, R.D. "Effects of Pollution on the Free Amino Acid Content of Two Marine Invertebrates," *Pac. Sci.* 15:49–55 (1961).

Schcuhammer, A.M. "Erythrocyte α-aminolevulinic Acid Dehydratase in Birds. I. The Effects of Lead and Other Metals in vitro," *Toxicology* 45:155–163 (1987).

Scholz, F., H.R. Gregorius and D. Rudin, Eds. *Genetic Effect of Air Pollutants in Forest Tree Populations* (Berlin: Springer-Verlag, 1989), p. 201.

Schreck, C.B., and H.W. Lorz. "Stress Response of Coho Salmon (*Oncorhynchus kisutch*) Elicited by Cadmium and Copper and Potential Use of Cortisol as an Indicator of Stress," *J. Fish. Res. Bd. Can.* 35:1124–1129 (1978).

Selye, H. *Stress* (Montreal, Ontario: Acta, Inc., 1950a).

Selye, H. "Stress and the General Adaptation Syndrome," *Br. Med. J.* 1:1384–1392 (1950b).

Selye, H. *The Stress of Life* (New York:McGraw-Hill Book Co., Inc., 1956).

Selye, H. *Stress in Health and Disease* (Boston, MA: Butterworth Publishers, 1976).

Shaffi, S,A. "Zinc Intoxication in Some Freshwater Fishes. I. Variations in Tissue Energy Reserves," *Ann. Limnol.* 16:91–97 (1980).

Sibly, R,M., and P. Calow. *Physiological Ecology of Animals: An Evolutionary Approach* (London:Blackwell Scientific Publications, 1986).

Sidorov, V.S., R.U. Vysotskaya and Y.V. Kostylev. "Lysosomal Enzyme Activity in Adult Females of *Salmo salar* During Pre-spawning Maturation," *J. Icthyology* 20(4):111–116 (1980).

Siegel, H.S. "Physiological Stress in Birds," *BioScience* 30:529–534 (1980).

Simon, L.M., J. Nemcsok and L. Boross. "Studies on the Effect of Paraquat on Glycogen Mobilization in Liver of Common Carp (*Cyprinus carpio* L.)," *Comp. Biochem. Physiol.* 75C:167–169 (1983).

Singh, H., and T.P. Singh. "Effects of Two Pesticides on Testicular ^{32}P Uptake, Gonadotrophic Potency, Lipid and Cholesterol Content of Testes, Liver and Blood Serum During Spawning Phase in *Heteropneustes fossilis* (Bloch)," *Endokrin. Band.* 76:228–296 (1980).

Singh, N., and A.K. Srivastava. "Effect of a Paired Mixture of Aldrin and Formithion on Carbohydrate Metabolism in a Fish, *Heteropneustes fossilis*," *Pestic. Biochem. Physiol.* 15:257–261 (1981).

Sivarajah, K., C.S. Franklin and W.P. Williams. "The Effects of Polychlorinated Biphenyls on Plasma Steroid Levels and Hepatic Microsomal Enzymes in Fish," *J. Fish Biol.* 13:401–409 (1978).

Skidmore, J.F. "Respiration and Osmoregulation in Rainbow Trout with Gills Damaged by Zn Sulfate," *J. Exp. Biol.* 52:481–494 (1970).

Skjoldal, H.R., and T. Bakke. "Anaerobic Metabolism of the Scavenging Isopod *Cirolana borealis* Lilljeborg: Adenine Nucleotides," in *Physiology and Behavior of Marine Organisms*, D.S. McLusky, and A.J. Berry, Eds. (Oxford, England: Pergamon Press, 1978), pp. 67–74.

Smith, A.J. "Changes in the Average Weight and Shell Thickness of Eggs Produced by Hens Exposed to High Environmental Temperature—a Review," *Trop. Anim. Health Prod.* 6:237–244 (1974).

Smith, M.A.K. "Estimation of Growth Potential by Measurement of Tissue Protein Synthetic Rates in Feeding and Fasting Rainbow Trout, *Salmo gairdneri* Richardson," *J. Fish. Biol.* 19:213–220 (1981).

Sneed, K.E. "Warmwater Fish Cultural Laboratories," in Progress in Sport Fisheries Wildlife," U.S. Fish Wildl. Serv. Resour. Publ. 106 (1970), pp. 189–215.

Soivio, A., S. Lindgren and A. Oikari. "Seasonal Changes in Certain Metabolic and Haematologic Responses of *Salmo gairdneri* Acutely Exposed to Dehydroabietic Acid (DHAA)," *Comp. Biochem. Physiol.* 75C:281–284 (1983).

Spencer, S.L. "Internal Injuries of Largemouth Bass and Bluegills Caused by Electricity," *Prog. Fish Cult.* 29:168–169 (1967).

Spronk, N., F.G. Brinkman, R.J. VanHoek and D.L. Knook. "Copper in *Lymnaea stagnales* L. II. Effect on the Kidney and Body Fluids," *Comp. Biochem. Physiol.* 38A:309–316 (1971).

Sreenivasula, R., P. Bhagyalakshmi and R. Ramamurthi. "In vivo Acute Physiological Stress Induced by BHC in Hemolymph Biochemistry of *Oziotelphusa senex senex*, the Indian Rice Field Crab," *Toxicol. Lett.* 18:35–38 (1983).

Srivastava, A.K., and A.B. Gupta. "The Effect of Sodium Salt of 2,4-D on Carbohydrate Metabolism in the Indian Catfish *Heteropneustes fossilis* (Bloch)," *Acta. Hydrobiol.* 23:259–268 (1981).

Srivastava, A.K., and J. Mishra. 1983. Effects of Fenthion on the Blood and Tissue Chemistry of a Teleost Fish (*Heteropneustes fossilis*)," *J. Comp. Path.* 93:27–31 (1983).

Statham, C.N., W.A. Croft and J.J. Lech. "Uptake Distribution, and Effects of Carbon Tetrachloride in Rainbow Trout (*Salmo gairdneri*)," *Toxicol. Appl. Pharmacol.* 45:131–140 (1978).

Staurnes, M., T. Sigholt and O.B. Reite. "Reduced Carbonic Anhydrase and Na-K-ATPase Activity in Gills of Salmonids Exposed to Aluminum-containing Acid Water," *Experientia* 40:226–227 (1984).

Stickle, W.B., and T.D. Sabourin. "Effects of Salinity on the Respiration and Heart Rate of the Common Mussel, *Mytilis edulis*, and the Black Chiton, *Katherina tunicata* (Wood)," *J. Exp. Mar. Biol. Ecol.* 41:252–268 (1979).

Strange, R.J. "Acclimation Temperature Influences Cortisol and Glucose Concentrations in Stressed Channel Catfish," *Trans. Am. Fish. Soc.* 109:298–303 (1980).

Subramanian, A., S. Tanabe, R. Tatsukawa, S. Saito and N. Miyazaki. "Reduction in the Testosterone Levels by PCBs and DDE in Dall's Porpoises of Northwestern Northern Pacific," *Mar. Pollut. Bull.* 18(12):643–646 (1987).

Sullivan, T.J., E.B. Gutman and A.B. Gutman. "Theory and Application of the Serum "Acid" Phosphatase Determination in Metastasizing Prostate Carcinoma; Early Effects of Castration," *J. Urology* 48:426–458 (1942).

Sunderman, F.W., and E. Horak. "Biochemical Indices of Nephrotoxicity, Exemplified by Studies of Nickel Nephropathy," in *Organ-Directed Toxicity Chemical Indices and Mechanisms*, S.S. Brown, and D.S. Davis, Eds. Proceedings of the Symposium on Chemical Indices and Mechanisms of Organ-Directed Toxicity, Barcelona, Spain. (New York: Pergamon Press, 1981), pp. 55–67.

Sutcliffe, W.H., Jr. "Relationship Between Growth Rate and Ribonucleic Acid Concentration in Some Invertebrates," *J. Fish. Res. Bd. Can.* 27:606–609 (1970).

Suter, G.W., A.E. Rosen, E. Linden and D.F. Parkhurst. "Endpoints for Responses of Fish to Chronic Toxic Exposure," *Environ. Toxicol. Chem.* 6:793–809 (1987).

Swift, D.J. "Changes in Selected Blood Component Concentrations of Rainbow Trout *Salmo gairdneri* Richardson, Exposed to Hypoxia or Sublethal Concentrations of Phenol or Ammonia," *J. Fish Biol.* 19:45–61 (1981).

Symons, P.E.K. "Behavior of Young Atlantic Salmon Exposed to or Force-fed Fenitrothion, an Organophosphate Insecticide," *J. Fish. Res. Bd. Can.* 30:651–655 (1973).

Telford, M. "Blood Glucose in Crayfish. I. Variations Associated with Molting," *Comp. Biochem. Physiol.* 47A:461–468 (1974).

Thomas, J.A. "Effects of Pesticides on Reproduction," *Molecular Mechanisms of Gonadal Hormone Action* 1:205–223 (1975).

Thomas, P. "Effect of Cadmium Exposure on Plasma Cortisol Levels and Carbohydrate Metabolism in Mullet *(Mugil cephalus)*," *J. Endocrinol.* 94 (Suppl.) (1982), p. 35.

Thomas, P. "Reproductive Endocrine Function in Female Atlantic Croaker Exposed to Pollutants," *Mar. Environ. Res.* 24:179–183 (1988).

Thomas, P. "Effects of Aroclor 1254 and Cadmium on Reproductive Endocrine Function and Ovarian Growth in Atlantic Croaker," *Mar. Environ. Res.* 28:499–504 (1989).

Thomas, P., H.W. Wofford and J.M. Neff. "Biochemical Stress Responses of Striped Mullet *(Mugil cephalus* L.) to Fluorene Analogs, *Aquat. Toxicol.* 1:329–342 (1981).

Thurberg, F.P., A. Calabrese, E. Gould, R.A. Greig, M.A. Dawson and R.K. Tucker. "Response of the Lobster, *Homarus americanus,* to Sublethal Levels of Cadmium and Mercury," in *Physiological Responses of Marine Biota to Pollutants,* F.J. Vernberg, A. Calabrese, F.P. Thurberg and W.B. Vernberg, Eds. (New York: Academic Press, Inc., 1977), pp. 185–197.

Truscott, B., J.M. Walsh, M.P. Burton, J.F. Payne and D.R. Idler. "Effect of Acute Exposure to Crude Petroleum on Some Reproductive Hormones in Salmon and Flounder," *Comp. Biochem. Physiol.* 75C:121–130 (1983).

Tucker, R.K. "Effect of in vivo Cadmium Exposure on ATPases in Gill of the Lobster, *Homarus americanus,*" *Bull. Environ. Contam. Toxicol.* 23:33–35 (1979).

Tucker, S.M., R.J. Price and G.R. Price. "Characteristics of Human *N*-acetyl-_-D-glucosaminidase Isoenzymes as an Indicator of Tissue Damage in Disease," *Clin. Chim. Acta* 102:29–40 (1980).

Turner, C.D., and J.T. Bagnara. *General Endocrinology.* (Philadelphia: W.B. Saunders Co., 1976).

Van de Kamp, G. "Vertebral Deformities of Herring Around the British Isles and Their Usefulness for a Pollution Monitoring Programme," *Int. Counc. Explor. Sea Coop. Res. Rep. Ser. B* 5 (1977).

Verma, S.R., S. Rani, I.P. Tonk and R.C. Dalela. "Pesticide-induced Dysfunction in Carbohydrate Metabolism in Three Freshwater Fishes," *Environ. Res.* 32:127–133 (1983).

Verma, S.R., M. Saxena and T.P. Tonk. "The Influence of Idet 20 on the Biochemical Composition and Enzymes in the Liver of *Clarias batrachus,*" *Environ. Pollut.* 33:245–255 (1984).

Verma, S.R., I.P. Tink, A.K. Gupta and R.C. Dalela. "Role of Ascorbic Acid in the Toxicity of Pesticides in a Freshwater Teleost," *Water Air Soil Pollut.* 16:107–114 (1981).

Versteeg, D.J. "Lysosomal Membrane Stability, Histopathology, and Serum Enzyme Activities as Sublethal Bioindicators of Xenobiotic Exposure in the Bluegill Sunfish *(Lepomis macrochirus* Rafinesque)," PhD Thesis, Michigan State University, East Lansing, MI, p. 153 (1985).

Versteeg, D.J., and J.P. Giesy. "Lysosomal Enzyme Release in the Bluegill Sunfish *(Lepomis macrochirus* Rafinesque) Exposed to Cadmium," *Arch. Environ. Contam. Toxicol.* 14:631–664 (1985).

Versteeg, D.J., and J.P. Giesy. "The Histological and Biochemical Effects of Cadmium Exposure in the Bluegill Sunfish (*Lepomis macrochirus*)," *Ecotoxicol. Environ. Saf.* 11:31–43 (1986).

Versteeg, D.J., R.L. Graney and J.P. Giesy. "Field Utilization of Clinical Measures for the Assessment of Xenobiotic Stress in Aquatic Organisms," in *Aquatic Toxicology and Hazard Assessment, ASTM STP 971*, W.J. Adams, G.A. Chapman and W.G. Landis, Eds. (Philadelphia: American Society for Testing Materials, 1988), pp. 289–306.

Viarengo, A., M. Pertica, G. Mancinelli, R. Capelli and M. Orunesu. "Effects of Copper on the Uptake of Amino Acids, on Protein Synthesis and on ATP Content in Different Tissues of *Mytilus galloprovincialis* L.," *Mar. Environ. Res.* 4:145–152 (1980).

Viarengo, A., M. Pertica, G. Mancinelli, S. Palmero, G. Zanicchi and M. Oranesu. "Evaluation of General and Specific Stress Indices in Mussels Collected from Populations Subjected to Different Levels of Heavy Metal Pollution," *Mar. Environ. Res.* 6:235–243 (1981).

Voogt, P.A. "Lipids: Their Distribution and Metabolism," in *The Mollusca, Vol. 1. Metabolic Biochemistry and Molecular Biomechanics*, P.W. Hochachka, Ed. (London: Academic Press, Inc., 1983), p. 510.

Wagner, G.F., and B.A. McKeown. "Changes in Plasma Insulin and Carbohydrate Metabolism of Zinc-stressed Rainbow Trout, *Salmo gairdneri*," *Can. J. Zool.* 60:2079–2084 (1982).

Warren, G.E., and G.E. Davis. "Laboratory Studies on the Feeding, Bioenergetics and Growth of Fish," in *The Biological Basis of Freshwater Fish Production*, S.D. Gerhuy, Ed. (Oxford, England: Blackwell Scientific Publications, 1967), pp. 175–214.

Watson, T.A., and F.W.H. Beamish. "Effect of Zinc on Branchial ATPase Activity in vivo in Rainbow Trout, *Salmo gairdneri*," *Comp. Biochem. Physiol.* 66C:77–82 (1980).

Watson, T.A., and B.A. McKeown. "The Effect of Sublethal Concentrations of Zinc on Growth and Plasma Glucose Levels in Rainbow Trout, *Salmo gairdneri* (Richardson)," *J. Wildl. Dis.* 12:263–270 (1976).

Wedemeyer, G.A., and D.J. McLeay. "Methods for Determining the Tolerance of Fishes to Environmental Stressors," in *Stress and Fish*, A.D. Pickering, Ed. (New York: Academic Press, Inc., 1981), pp. 247–275.

Wedemeyer, G.A., D.J. McLeay and C.P. Goodyear. "Assessing the Tolerance of Fish and Fish Populations to Environmental Stress: the Problems and Methods of Monitoring," in *Contaminant Effects on Fisheries*, V.W. Cairns, P.V. Hodson and J.O. Nriagu, Eds. (New York: John Wiley & Sons, Inc., 1984), pp. 164–186.

Wedemeyer, G.A., and N.C. Nelson. "Statistical Methods for Estimating Normal Blood Chemistry Ranges and Variance in Rainbow Trout (*Salmo gairdneri*) Shasta Strain," *J. Fish. Res. Bd. Can.* 32:551–554 (1975).

Wedemeyer, G.A., and W.T. Yasutake. "Clinical Methods for the Assessment of the Effects of Environmental Stress on Fish Health," U.S. Fish and Wildlife Service Tech. Paper 89 (1977).

Weiss, C.M. "The Determination of Cholinesterase in the Brain Tissue of Three Species of Fresh Fish and its Inactivation in vivo," *Ecology* 39:194–199 (1958).

Weiss, C.M. "Physiological Effect of Organic Phosphorous Insecticides on Several Species of Fish," *Trans. Am. Fish. Soc.* 90:143–152 (1961).

Weiss, C.M. "Use of Fish to Detect Organic Insecticides in Water," *J. Water Pollut. Control Fed.* 37:647–658 (1965).

Weiss, C.M., and J.H. Gakstatter. "Detection of Pesticides in Water by Biochemical Assay," *J. Water Pollut. Control Fed.* 37:647–658 (1964).

Wellburn, A.R. "The Influence of Atmospheric Pollutants and Their Cellular Products upon Photophosphorylation and Related Events," *Gaseous Air Pollut. Plant Metab.* 1982:233–221 (1984).

Wellburn, A.R. "Why are Atmospheric Oxides of Nitrogen Usually Phytotoxic and not Alternative Fertilizers?" *New Phytol.* 115:395–429 (1990).

Wellburn, A.R., T.M. Capron, H.S. Chen and D.C. Horsman. "Biochemical Effects of Atmospheric Pollutants on Plants," *Sem. Ser.-Soc. Exp. Biol.* 1:105–114 (1976).

Wellburn, A.R., C. Higginson, D. Robinson and C. Walmsley. "Biochemical Explanations of More than Additive Inhibitory Effects of Low Atmospheric Levels of Sulphur Dioxide Plus Nitrogen Dioxide Upon Plants." *New Phytol.* 88:223–237 (1981).

Westlake, G.E., A.D. Martin, P.I. Stanley and C.H. Walker. "Control Enzyme Levels in the Plasma, Brain and Liver from Wild Birds and Mammals in Britian," *Comp. Biochem. Physiol.* 76C:15–24 (1983).

Widdows, J. "Combined Effects of Body Size, Food Concentration, and Season on the Physiology of *Mytilus edulis*," *J. Mar. Biol. Assoc. U.K.* 58:109–124 (1978).

Widdows, J., T. Bakke, B.L. Bayne, P. Donkin, D.R. Livingston, D.M. Lowe, M.N. Moore, S.V. Evans and S.L. Moore. "Responses of *Mytilus edulis* on Exposure to Water-accommodated Fraction of North Sea Oil," *Mar. Biol.* 67:15–31 (1982).

Widdows, J., and B.L. Bayne. "Temperature Acclimation of *Mytilus edulis* with Reference to its Energy Budget," *J. Mar. Biol. Assoc. U.K.* 51:827–843 (1971).

Widdows, J., B.L. Bayne, P. Donkin, D.R. Livingstone, D.M. Lowe, M.N. Moore and P.N. Salkeld. "Measurement of the Responses of Mussels to Environmental Stress and Pollution in Sullom Vol: A Baseline Study," *Proc. R. Soc. Edin. B.* 80:323–338 (1981).

Wilder, I.B., and J.G. Stanley. "RNA-DNA Ratio as an Index to Growth in Salmonid Fishes in the Laboratory and in Streams Contaminated by Carbaryl," *J. Fish. Biol.* 22:165–172 (1983).

Wildish, D.J., and N.A. Lister. "Biological Effects of Fenitrothion in the Diet of Brook Trout," *Bull. Environ. Contam. Toxicol.* 10:333–339 (1973).

Wilkinson, J.H. "Plasma Enzyme Activities," in *Enzyme Pathophysiology*, D. M. Goldberg, and J.H. Wilkinson, Eds. (New York: Karger, 1978), pp. 1–7.

Williams, A.K., and R.C. Sova. "Acetylcholinesterase Levels in Brains of Fishes from Polluted Water," *Bull. Environ. Contam. Toxicol.* 1:198–204 (1966).

Williams, H.A., and R. Wooten. "Some Effects of Therapeutic Levels of Formalin and Copper Sulphate on Blood Parameters in Rainbow Trout," *Aquaculture* 24:341–353 (1981).

Wilson, R.T., and W.E. Poe. "Impaired Collagen Formation in the Scorbutic Channel Catfish," *J. Nutr.* 103:1359–1364 (1973).

Winberg, G.G. *Methods for the Estimation of Production of Aquatic Animals* (London, England: Academic Press, Inc., 1971), p. 175.

Woltering, D.M. "The Growth Response in Fish Chronic and Early Life Stage Toxicity Tests: A Critical Review," *Aquat. Toxicol.* 5:1–21 (1984).

Wright, D.A., and E.W. Hetzel. "Use of RNA:DNA Ratios as an Indicator of Nutritional Stress in the American Oyster *Crassostrea virginica*," *Mar. Ecol. Prog. Ser.* 25:199–206 (1985).

Wroblewski, F., and J.S.V. LaDue. "Glutamic Pyruvic Transaminase in Cardiac and Hepatic Disease," *P.S.E.B.M.* 91:569–571 (1956).

Wunder, W. "Ver Kruppelete Felchen uas der Biodensee," *Zool. Anz.* 194:279–292 (1975).

Zammitt, V.A., and E.A. Newsholme. "The Maximum Activities of Hexokinase, Phosphorylase, Phosphofructokinase, Glycerol Phosphate Dehydrogenase, Lactate Dehydrogenase, Octopine Dehydrogenase, Phosphoenolpyruvate Carboxykinase, Nucleoside Diphosphatekinase, Glutamate-oxaloacetate Transaminase and Arginine Kinase in Relation to Carbohydrate Utilization in Muscles of Marine Invertebrates," *Biochem. J.* 160:447–462 (1976).

Zbanysek, F., and L.S. Smith. "The Effect of Water-soluble Aromatic Hydrocarbons on Some Hematological Parameters of Rainbow Trout, *Salmo gairdneri* Richardson, During Acute Exposure," *J. Fish. Biol.* 24:545–552 (1984).

CHAPTER 2

Metabolic Products
as Biomarkers

Mark J. Melancon, Ruth Alscher, William Benson, George Kruzynski, Richard
F. Lee, Harish C. Sikka, and Robert B. Spies

INTRODUCTION

This chapter focuses on the possible uses for metabolites of xenobiotic chemicals
and endogenous metabolites as biomarkers of exposure to xenobiotic chemicals. The
discussion on metabolites of xenobiotic chemicals considers animals only, while that
on endogenous metabolites includes both animals and plants.

Metabolites of Xenobiotic Chemicals

The exposure of an organism to persistent xenobiotic chemicals such as
polychlorinated biphenyls (PCBs) can be assessed by direct measurement of body
burdens of these chemicals. However, the exposure to certain commonly occurring
environmental contaminants, including polynuclear aromatic hydrocarbons (PAHs)
and aromatic amines, cannot be assessed by analyzing directly for these chemicals,
because they are rapidly converted to a variety of metabolites by the organism (Thakker
et al. 1985; Thorgeirsson et al. 1983). In such instances measurement of metabolites
may provide evidence of exposure to these chemicals. The metabolites of xenobiotic
chemicals may accumulate to high levels in certain tissues or body fluids or bind to

specific tissue macromolecules in a manner that facilitates detection of exposure and indicates potential harm to the organism. These topic are addressed in this and other chapters in this book.

Endogenous Metabolites

As used in this chapter, the term *endogenous metabolites* means products of "normal" metabolic pathways that may be affected qualitatively or quantitatively by exposure of an organism to chemical (environmental contaminant) or other stress.

The levels of individual endogenous metabolites and the ratios of related endogenous metabolites are modified by a number of xenobiotic chemicals, in both plants and animals. In some cases the modification is due to the fact that the endogenous metabolites are involved with the metabolism or detoxification of xenobiotic chemicals and/or their metabolites. Examples of such endogenous metabolites are the oxidized and reduced forms of glutathione and the constituents of the glucuronic acid pathway. In other cases, the modification is due to the effects of some xenobiotic chemicals and/ or their metabolites that interfere with normal metabolic pathways, leading to the production of excess levels of "normal" metabolites or to the production of metabolites not normally seen. Examples include uroporphyrin and other highly carboxylated porphyrins.

Approach

The remainder of this chapter is divided into two main sections, metabolites of xenobiotic chemicals and endogenous metabolites. Metabolites of xenobiotic chemicals in various species are covered together because there are many similarities in the metabolism and effects of xenobiotic chemicals in different species. A discussion of needs and applications covers metabolites of specific chemicals in particular species. The endogenous metabolites are covered individually, and in some cases implications for both plants and animals are discussed.

METABOLITES OF XENOBIOTIC CHEMICALS

Background

The metabolism of xenobiotics usually involves two types of enzymatic reactions: phase I and phase II reactions. During phase I reactions, polar groups are introduced into the xenobiotic molecule through oxidative, hydrolytic, or reductive processes, making it a suitable substrate for phase II reactions. The first step in the metabolism of most lipophilic chemicals is oxidation by the cytochrome P450-dependent monooxygenase system localized in the endoplasmic reticulum of the liver and other tissues. Typically, this oxidation results in the introduction of a hydroxyl group into the chemical. The metabolites formed by phase I reactions may undergo conjugation with various polar endogenous substrates such as glucuronic acid, sulfate, glutathione, amino acids, and others. The conjugation reactions, called phase II reactions, tend to form highly water-soluble products that are readily excreted via bile, kidney, or gill. Conjugated metabolites may also be hydrolyzed and the free metabolite reabsorbed in the intestinal tract (enterohepatic circulation). Phase I reactions can also lead to formation of reactive metabolites that may bind covalently to tissue macromolecules, resulting in subcellular damage.

The metabolic reactions discussed above may change a toxic xenobiotic chemical to either a less toxic form or to a form that is more toxic than the parent chemical. It is well known that some PAHs undergo biotransformation by the cytochrome P450 system to form mutagenic/carcinogenic metabolites. The metabolism of PAHs also involves detoxification reactions (Thakker et al. 1985). Halomethanes and halogenated aromatic hydrocarbons are often hepatotoxic after being metabolized by cytochrome P450 (Anders and Pohl 1985). Thus depending on the xenobiotic chemical, its metabolism may lead to a form that is more readily monitored and thus can be used as a biomarker of exposure to the parent xenobiotic chemical.

A variety of observations have provided direction for the use of metabolites of xenobiotic chemicals in biomonitoring.

Creaven et al. (1965) used fluorescence measurements to detect the presence of hydroxybiphenyl from the metabolism of biphenyl in liver preparations of 11 species, including rainbow trout (*Salmo gairdneri*), thus establishing that metabolites of PAHs can be formed and be present in fish tissues. Later, using in vivo exposure to ^{14}C-labeled PAHs, Lee et al. (1972) isolated and tentatively identified hydroxynaphthalene and benzo(a)pyrene (BaP) tetrols in the lipid extracts of several marine fish species. Statham et al. (1976) used nine different xenobiotic compounds to show that concentrations (including metabolites) in the bile of exposed rainbow trout could be up to 10,000 times that of the exposure water. Additional studies (Melancon and Lech 1979) showed a 142,000:1 ratio for the bioconcentration of 2-methylnaphthalene and its metabolites in common carp (*Cyprinus carpio*) bile as compared to the water concentration of 2-methylnaphthalene. These studies demonstrated the importance of biliary excretion of xenobiotic compounds, and that metabolites of xenobiotics in bile can be used to determine the exposure of fish to dilute concentrations of such compounds in water. Eventually, the determination of the metabolites of xenobiotic chemicals in bile was successful in linking exposure to effect in feral populations of fish. For example, PAH metabolite concentrations in the bile of English sole (*Parophrys vetulus*) from Puget Sound were associated with an increased occurrence of idiopathic liver lesions in this species (Krahn et al. 1986).

A similar progression in the measurement of metabolites bound to subcellular macromolecules also occurred in the 1970s and 1980s. In 1979, Ahokas et al. (1979) tentatively identified the metabolites cis- and trans-7,8-diol-9,10 epoxides of BaP bound to DNA after incubation of BaP with lake trout (*Salmo trutta lacustris*) microsomes, which indicated that fish metabolically activate BaP via the same mechanism as mammals. Binding of BaP metabolites to DNA was also found after incubation of BaP with hepatic microsomes from starry flounder (*Platichthys stellatus*) (Varanasi and Gmur 1980), coho salmon (*Oncorhynchus kisutch*) (Varanasi and Gmur

1980), and English sole (Varanasi et al. 1981, 1982), and after exposing English sole to ^{3}H-BaP (Varanasi et al. 1982).

Types of Metabolites of Xenobiotic Chemicals

The metabolism of a wide variety of xenobiotic chemicals has been studied both in vivo and in vitro. Many of these metabolism studies used laboratory animals, generally mammals, and numerous studies have also been performed with fish, birds, and a variety of other land and marine organisms. The classes of xenobiotics studied include pesticides, petroleum and petroleum-derived chemicals, halogenated hydrocarbons, and plasticizers. Biomonitoring via metabolites of xenobiotic chemicals requires knowledge of the extent of metabolism and the types of metabolites of a particular xenobiotic chemical produced by an organism. However, it is beyond the scope of this chapter to review the metabolism of every xenobiotic chemical. The xenobiotic chemicals included for examples demonstrate both phase I and phase II metabolism and cover a wide range of chemical types.

The presence of stable metabolites/degradation products of a number of chlorinated hydrocarbons in organisms and in the environment has been known for some time (Stickel 1973; Brooks 1974). For example, when 2,2,bis(p-chlorophenyl)-1,1,1-trichloroethane (DDT) is to be measured in environmental samples, DDT and 2,2,bis(p-chlorophenyl)-1,1-dichloroethylene (DDE), and sometimes 2,2,bis(p-chlorophenyl)-1,1-dichloroethane(DDD) are generally measured, with the total expressed as ΣDDT. Analysis is done for the epoxidation product of aldrin, dieldrin, and for both oxychlordane and chlordane. The concept of determining metabolites of xenobiotic chemicals in environmental samples has been in practice for decades.

The metabolism of PAHs has been studied because PAHs are common environmental contaminants and some of their metabolites are potential mutagens. The nature and proportions of metabolites of the PAH BaP have been studied extensively. The information provided in this chapter is for wildlife species rather than laboratory species, such as white rats and mice, and will frequently be provided for aquatic species because of the extensive research done on aquatic environments.

Diamond and Clark (1970) reported the metabolism of ^{3}H-BaP to water-soluble forms by cell cultures from rainbow trout, bluegill sunfish (*Lepomis microchirus*), and leopard frog (*Rana pipiens*). Lee et al. (1972) studied BaP metabolism by the speckled sanddab (*Citharichthys stigmaeus*), and found the 7,8-dihydrodiol was formed in greater proportions than the 9,10-dihydrodiol. This suggested that speckled sanddab exposed to BaP may risk developing carcinomas from high concentrations of the 7,8-dihydrodiol, a precursor of the 7,8-diol-9,10-epoxide of BaP that is the ultimate carcinogen formed from BaP (see review by Conney 1982).

Varanasi and Gmur (1981) found that a larger proportion of glucuronide metabolites than sulfate metabolites were formed from both BaP and naphthalene in English sole. The percentages of naphthalene metabolites in the bile 168 h after the start of exposure to contaminated sediments were glucuronides, 88%; sulfate/glucosides, 3%; mercapturic acids, 2%; 1,2-dihydrodiols, 0.3%; and unclassified, 7%. No unconjugated 1- and 2-naphthols were detected.

In a study of phenanthrene metabolism in the flounder (*Platichthys flesus*) Solbakken and Palmork (1981) found that in the bile, stomach, intestine, and urine, the 1,2-dihydro-1,2-hydroxy metabolite was preferentially formed (greater than 90%) over the 3,4- and 9,10-isomers as well as over the five monohydroxy isomers identified.

Similarly, Krahn et al. (1987) reported the predominance of 1,2-dihydro-1,2-dihydroxyphenanthrene in the bile of English sole, although the phenanthrene dihydrodiol was dehydrated during gas chromatography of the nonsilanized sample. They also reported the presence of the following other PAH metabolites in bile samples of English sole from contaminated areas of Puget Sound: four dibenzofuranols, five fluorenols, three fluoranthenols, one pyrenol, and two hydroxyBaPs.

In another study, Stein et al. (1987) reported that after 56 d of exposure to sediments containing both BaP and PCBs, the proportions of metabolites formed from radiolabeled BaP in the liver of English sole were as follows: glutathione-derived conjugates, 70%; metabolites bound to macromolecules, 20%; unconjugated metabolites, 3.2%; and glucuronides and sulfates, 1.9%.

The nitrogen heterocyclic compound quinoline is rapidly metabolized in rainbow trout (Bean et al. 1985). After 48- h exposure to ^{14}C-quinoline, plus 24 h in clean water, greater than 60% of the body burden of quinoline and metabolites in juvenile fish was found in the gall bladder with decreasing amounts in the muscle, gut, liver, gill, and kidney. After base catalysis, hydroxyquinolines and quinothiols were identified in the bile samples. The bile contained mainly hydroxyquinoline glucuronide and the other tissues contained mainly the quinothiols.

Although PCBs are not metabolized as quickly as PAHs, Melancon and Lech (1976) isolated an acid-soluble metabolite of 2,5,2',5'-tetrachlorobiphenyl from bile of rainbow trout exposed to ^{14}C-2,5,2',5'-tetrachlorobiphenyl. After treatment with β-glucuronidase the metabolite had the same chromatographic characteristic as 4-OH-tetrachlorobiphenyl. Solbakken et al. (1984) were unable to identify any metabolites of 2,4,5,2',4 ',5',-hexachlorobiphenyl after administering a 0.2 mg/kg dose to flounder by gavage. Similarly, the chlorinated pesticide DDT is very resistant to metabolism in fish, but Pritchard et al. (1973) found DDE and other metabolites constituted a small fraction (less than 2%) of the total compound burden in all tissues in winter flounder (*Pseudopleuronectus americanus*) given large doses of DDT in the laboratory. More than 20% of the DDT in urine was in the form of metabolites. Earlier reports of high concentrations of DDT and PCB metabolites in fish near the large municipal sewage outfalls in southern California (Brown et al. 1982, 1986, 1987; Gossett et al. 1984) are now in doubt (Gossett 1988).

Chlorinated phenols and their conjugation products also have been studied in fish bile. Glickman et al. (1977) examined the distribution and metabolism of ^{14}C-pentachlorophenol (PCP) and ^{14}C-pentachloroanisole (PCA). The bile of rainbow trout exposed to PCP contained high concentrations of pentachlorophenyl-glucuronide; bile of those exposed to PCA also contained pentachlorophenyl-glucuronide, indicating that demethylation of the PCA takes place in vivo. While Glickman's rainbow trout study found only the glucuronide conjugate of PCP, the liver of goldfish (*Carassius auratus*) forms this conjugate as well as pentachlorophenylsulphate (Kobayashi 1978). The latter compound is eliminated through the gills and kidney and is not found in the bile (Kobayashi 1979). In more recent studies with complex effluents, Oikari (1986) found that over 99% of the chlorinated phenols (2,4,6-trichlorophenol, 3,4,5,6-tetrachlorophenol, 4,5,6-trichloroguaiacol, 3,4,5,6-tetrachloroguaiacol and PCP) detected in the bile of roach (*Rutilis rutilis*) and perch (*Perca fluviatilis*) downstream from a Finnish pulp papermill outfall occurred as base-hydrolyzable conjugates. A similar finding was obtained in analyses of the chlorinated phenols from bile of rainbow trout exposed to pulp papermill effluent in the laboratory (Oikari and Anas 1985). Upon treatment of the conjugates with β-glucuronidase and a sulphatase, Oikari and Anas (1985) determined that 68 to 90% of the conjugates were glucuronides and the remainder sulfates.

In addition to PAHs, pesticides, and chlorinated hydrocarbons, the occurrence of resin acids and their conjugation products in the bile of rainbow trout near pulp mills has been studied (Oikari et al. 1984; Oikari and Kunnamo-Ojala 1987). Concentrations of pimaric, sandaracopimaric, isopimaric, palustric+levopimaric, abietic, neoabietic, and dehydroabietic acids of 11 to 76 g/ml were determined in the bile of rainbow trout exposed to a 1% dilution of untreated pulp mill effluent for 30 d (Oikari et al. 1984). Over 99% of these compounds occurred as conjugated metabolites that were released by base hydrolysis.

Reactive Metabolites of Xenobiotic Chemicals

A number of xenobiotic chemicals undergo metabolism to highly reactive electrophilic intermediates that can bind covalently to cellular macromolecules (e.g., proteins, DNA, and RNA). This covalent binding is considered the initiating event for many toxicological processes such as mutagenicity, carcinogenicity, and cellular necrosis (Koss et al. 1980; Miller and Miller 1985; Bartolone et al. 1989).

A toxicologically relevant measure of exposure to carcinogenic chemicals is the *biologically effective dose*; i.e., the amount of the activated agent that has reacted with critical cellular macromolecules (Perara 1987). Assays for the concentrations of carcinogens in tissues, body fluids, or excreta may not provide an accurate assessment of the biologically effective dose. However, a method that quantitates the in vivo reaction products of the activated carcinogen might be useful in estimating the biologically effective dose of the carcinogen. Two promising markers of the biologically effective dose are carcinogen-DNA adducts and their surrogate carcinogen-protein adducts. Because covalent binding to DNA is considered a critical early event in the process of chemical carcinogenesis (Miller and Miller 1985), the most relevant approach would be to measure DNA adducts formed by the reaction of carcinogenic metabolites with cellular DNA. The formation of PAH-DNA adducts has been reported in a variety of biological systems (Philip and Sims 1979). The amounts of carcinogen-DNA adducts formed by the activation of carcinogens in biological systems vary considerably. This variation in the amount of DNA adducts formed depends on many factors, including total exposure to the carcinogen, absorption and transport, the metabolic balance between activation and detoxication reactions, and the capacity of the cells to repair DNA adducts.

Formation of DNA adducts in fish exposed to PAHs in the laboratory and in the field has recently been reported. Several fish species including English sole (Varanasi et al. 1989a), brown bullhead (Sikka et al. 1990), and bluegills (Shugart et al. 1987) can metabolize BaP to reactive metabolites that covalently bind to liver DNA in vivo. The predominant DNA adduct formed in the livers of the three fish species was *anti* BaP-7,8-diol-9,10-epoxide-deoxyguanosine, suggesting that these fish metabolically activate BaP by the same mechanism as the mammalian systems susceptible to the carcinogenic effect of the hydrocarbon. Dunn et al. (1987) detected aromatic carcinogen-DNA adducts in the livers of brown bullheads (*Ictalurus nebulosus*) collected from the Buffalo and Detroit Rivers, known to be contaminated with high levels of PAHs. DNA adducts were also found in the livers of English sole collected from Duwamish Waterway and Eagle Harbor, Puget Sound, Washington, and winter flounder sampled from Boston Harbor, Massachusetts (Varanasi et al. 1989b).

Several methods for detecting carcinogen-DNA adducts in exposed organisms are available, including high performance liquid chromatography (HPLC), synchronous fluorescence spectrophotometry, detection by monoclonal or polyclonal antibodies, and ^{32}P-postlabeling assay. These techniques vary in their sensitivities to detecting DNA adducts. For more details, see the chapter on DNA alterations.

The formation of carcinogen-DNA adducts has been investigated in animals and in cultured human tissues and cells (Weston et al. 1985; Harris et al. 1982). Studies show that the amounts of carcinogen-DNA adducts formed by the activation of carcinogens in biological systems vary considerably. This variation depends on many factors, including total exposure to the carcinogen, absorption and transport, the metabolic balance between activation and detoxification reactions, and the capacity of the cells to repair DNA adducts.

Because activated carcinogens form protein adducts, protein alkylation has been proposed as a monitor for exposure to chemical carcinogens (Calleman et al. 1978). In principle, any protein could be used for monitoring protein adducts; however, hemoglobin is favored as the target protein because of its abundance and ready availability in erythrocytes, and long lifetime (120 d). A broad spectrum of direct- and indirect-acting carcinogens and mutagens bind to hemoglobin in vivo (Pereira and Chang 1981). The stability of hemoglobin adducts throughout the life span of the protein (Segerback et al. 1978; Tannenbaum et al. 1983) results in the accumulation of hemoglobin-bound metabolites following repeated exposure to the parent carcinogen, which may provide an integrated history of exposure over several months. The extent of adduct formation to hemoglobin has been shown to be proportional to the erythrocyte dose and the exposure duration (Segerback et al. 1978). In animal carcinogen studies, a strong correlation has been noted between protein binding and DNA binding (Pereira et al. 1981; Ehrenberg et al. 1983; Murthy et al. 1984), suggesting that carcinogen-protein adducts are useful as readily accessible surrogates for DNA adducts.

Another protein to which reactive metabolites of xenobiotic chemicals may bind is glutathione-S-transferase, a major protein of the hepatic cytosol (Jakoby and Keen 1977). In addition to its enzymatic role of conjugating active electrophiles to glutathione, this protein can bind to various metabolites of xenobiotic compounds (Aniya et al. 1988). With respect to its binding properties, glutathione-S-transferase may have a role similar to albumin in vertebrate plasma. There are other specific cytosolic proteins that also bind to a variety of metabolites. For example, glutathione-S-transferases were not important binding proteins for metabolites of acetaminophen but a 58,000 dalton cytosolic protein accounted for most of the binding instead (Birge et al. 1989; Nishanian et al. 1989). In addition, a sulfur-containing metabolite of a PCB (a methyl sulfone) was found to bind to a specific 16,000 dalton cytosolic protein from rat and mouse lungs (Brandt et al. 1985; Lund et al. 1985). Blue crabs (*Callinectes sapides*) fed ^{14}C-BaP, had significant radioactivity in the hepatopancreas cytosol. This radioactivity included conjugates and metabolites bound to cytosolic lipoproteins and glutathione S-transferase (Lee 1989). Similarly, up to 35% of the metabolites of tributyltin in crabs were bound to mitochondrial and cytosolic proteins of the hepatopancreas (Lee 1989).

Xenobiotic metabolism by microsomal cytochrome P450 systems can result in active metabolites that covalently bind to microsomal proteins (Shimada 1976; Miller et al. 1978; van Ommen et al. 1986; Juedes et al. 1987; Buben et al. 1988; Narasimhan et al. 1988; Weller et al. 1988). The compounds studied include active metabolites of hexachloroethane, bromobenzene, hexachlorobenzene, PCBs, and hexachlorophene. Among these compounds, less binding of these metabolites to nuclear and mitochondrial fractions was observed (Miller et al. 1978).

Thus, there is evidence that metabolites of xenobiotic chemicals bind to organelle and cytosolic proteins in both vertebrate and invertebrate animals. While the majority of the studies on xenobiotic chemicals oxidized to metabolites have analyzed forms resulting from two-electron oxidations, the formation of free radicals from one-electron oxidation is also possible. Free radicals are extremely reactive species that are well-known agents of biological damage (Pryor 1982). The work of Malins et al. (1983) with free radicals in microsomes of English sole from contaminated areas of Puget Sound remains the only study of its type with fish. That study provided evidence that the concentrations of free radicals in the microsomes of English sole with idiopathic liver lesions are higher than in microsomes from sole without such lesions.

Tissue Distribution of Metabolites of Xenobiotic Chemicals

In Lee et al.'s (1972) early experiments using three species of fish, the ^{14}C-labeled parent and metabolites of naphthalene and BaP were not analyzed separately, but the dynamic distribution of the radioactivity in the tissues was studied. The results suggested that uptake by the gills was followed by accumulation and metabolism in the liver, transfer to the bile, and elimination. The urine also appeared to be an important route for elimination, thus implicating the kidney in metabolite excretion.

High concentrations of metabolites of xenobiotic compounds in fish bile was first documented by Lech and co-workers (Lech 1973; Statham et al. 1975, 1976), who found high concentrations of xenobiotic contaminants in the bile of exposed fish. The importance of bile has since been documented in numerous other studies. In general, studies have confirmed the accumulation of organic xenobiotic compounds through the gut and gills, and their conveyance to the lipid-rich tissues, especially the liver, where the compounds are metabolized and excreted. Although it has not been studied as extensively, the kidney may also play a key role by excreting some water-soluble metabolites in the urine.

The distribution and occurrence of naphthalene and several of its metabolites in the liver, blood, and brain of rainbow trout were studied by Collier et al. (1980). They found conjugated derivatives of naphthalene in the blood and liver but very little in the brain; however, a nonconjugated derivative, 1,2-dihydro-1,2-dihydroxy naphthalene, was found in the brain. Varanasi et al. (1979) also studied the distribution of naphthalene and its metabolites in ten tissues and fluids of juvenile starry flounder at 24, 48, and 168 h after an administered dose of the radiolableled compound in oil and of rock sole (*Lepidopsetta bilineata*) at 24, 48, 168, and 1008 h. The data indicated that, as expected, the proportions of metabolites increased with time after exposure in most tissues. Clearance of the parent compound and metabolites from most tissues, except the gall bladder, also was indicated. After 168 h, concentrations of metabolites in tissues and fluids were in the following order: bile > intestine > kidney > liver > stomach > gills > brain > blood > muscle, skin for starry flounder, and bile > intestine > liver > kidney > stomach > blood > gills > brain > skin > muscle for rock sole. In addition, the concentration ratio of naphthalene metabolites in bile as compared to that of the next highest tissue was approximately 250 for starry flounder and 500 for rock sole. By 1008 h in rock sole the concentration of naphthalene metabolites was only 0.01% of that at 168 h.

The distribution of phenanthrene metabolites 48 h after administration of a dose by gavage to the flounder, showed that the greatest amounts of metabolites were in the bile (mainly as 1,2-dihydro-1,2-dihydroxyphenanthrene), with much lower concentrations in the urine, stomach, and intestine (Solbakken and Palmork 1981). Bean et al. (1985) found that rainbow trout, after a 48 h exposure to quinoline, had the greatest concentration of the compound and its metabolites in the gall bladder, with amounts in the remaining tissues as follows: muscle > gut > eyes > liver > gill > kidney.

Gingerich (1986) studied the kinetics of radiolabeled rotenone in rainbow trout. More than 80% of the activity in bile and urine was present as a highly polar metabolite (not hydrolyzable by β-glucuronidase or sulfatase). Some 40% of the rotenone in the liver, kidney, and muscle was associated with the mitochondrial fraction and 80% of this rotenone was not metabolized. The high concentrations of radioactivity in the bile implicated hepatobiliary excretion as the main route; in addition, an estimated 5% of the dose was eliminated in 4 h by the gills and less than 2% was eliminated by the urine in 48 h.

After exposure of rainbow trout to treated papermill effluent, the concentrations of three chlorinated phenols in the gall bladder were about twice the concentrations in plasma (Oikari and Anas 1985). However, concentrations of the conjugates of these phenols in bile were from 50 to 400 times the concentrations in plasma.

Varanasi et al. (1983) studied the distribution of ^{3}H-BaP and its metabolites in sexually mature male and female English sole. The dose was given by gavage, and BaP and metabolite concentrations were measured in bile, blood, gonads, and liver. After 168 h the highest concentrations of BaP and metabolites were found in the bile, followed in order by the liver, blood, and gonads. The proportions of metabolites and parent compound differed among tissues. Liver and bile had high proportions of aqueous-soluble metabolites (greater than 90%), with smaller proportions of organic-soluble metabolites and parent compound. The blood and gonads had approximately 50 to 60% of the totals as aqueous-soluble metabolites, with the remainder as parent compound and organic-soluble metabolites.

Determination of Metabolites of Xenobiotic Chemicals in Field-Exposed Organisms

Only a few studies have analyzed the concentrations of metabolites in tissues of wild fish populations. Work in Puget Sound, Washington, has shed light on the toxicology of PAHs taken up from contaminated sediments by English sole (Malins et al. 1984). While routine chemical analyses for parent hydrocarbons in fish tissue indicated only trace amounts, specialized analytical techniques were developed to estimate concentrations of fluorescent aromatic compounds (metabolites of PAHs) in bile (Krahn et al. 1984). English sole bile from polluted waterways in Puget Sound was analyzed by an HPLC-fluorescence technique that estimates the concentrations of metabolites displaying fluorescence characteristics of naphthalene, phenanthrene, and BaP (Krahn et al. 1984). English sole bile from polluted waterways had fluorescent-equivalent concentrations that were 9 to 14 times higher at these three wavelength pairs than in fish from reference areas within Puget Sound. Krahn et al. (1984) also found that within a polluted site, sole with liver lesions had higher equivalents of metabolites fluorescing at BaP wavelengths than did sole without lesions. A later study of 11 sites in Puget Sound found that there were significant correlations among the mean equivalents of bile metabolites and the frequencies of occurrence of neoplasms, foci of cellular alteration, megalocytic hepatosis, and total hepatic lesions (Krahn et al. 1986).

The first report of PAH-like metabolites adducted to DNA in fish caught in the wild was published recently (Dunn et al. 1987). This study used the ^{32}P-postlabeling technique (Reddy et al. 1984) to show that brown bullheads from areas containing high concentrations of sediment-bound PAHs had indications of adducts with chromatographic properties similar to those of PAH adducts, while control fish raised in aquaria did not. Estimated adduct concentrations were 70.1 nmol/mol DNA in fish from the Buffalo River and 52 to 56 nmol/mol DNA in fish from the Detroit River. A similar finding was obtained using English sole from contaminated and uncontaminated sites in Puget Sound. Autoradiographs of thin-layer chromatograms of ^{32}P-postlabeled hepatic DNA digests showed several diagonal radioactive zones (characteristic of PAH adducts) that were not present in hepatic DNA preparations of fish from uncontaminated areas of Puget Sound (Varanasi et al. 1989b).

Another environmental fate study examined pulp mill effluents, discharges characterized by extreme temporal variability in composition and toxicity (Oikari et al. 1985). Since a single bleached kraft-process mill may discharge 50 tons of chlorinated organic substances per day, much attention has been focused recently on the environmental fate of these xenobiotic chemicals. Among those chemicals identified are chlorophenols (Oikari and Kunnamo-Ojala 1987), chlorinated dibenzodioxins and dibenzofurans, and resin acids (natural products present in wood that are solubilized during the pulping process and discharged to waste). Kruzynski (1979) studied the uptake and sublethal effects of one of the most persistent resin acids, dehydroabietic. Subsequent work in Finland by Oikari et al. (1984) demonstrated both in vitro and in vivo inhibition of UDP-glucuronyltransferase, a major phase II conjugating enzyme, by dehydroabietic acid. The research demonstrated that most (95%) of the dehydroabietic acid and several other resin acids were excreted via the hepatobiliary route as glucuronide conjugates. Routine analysis of hexane bile extracts missed virtually all evidence of previous low-level dehydroabietic acid exposure. Only hydrolysis (chemical or β-glucuronidase enzyme treatment) revealed the actual situation (Oikari et al. 1984).

Further studies (Oikari et al. 1985; Oikari and Holmbom 1986) confirmed the utility of using analyses of hydrolyzed bile to trace previous exposure of fish to chlorophenols and resin acids in controlled laboratory and field experiments. The distribution and accumulation by feral fish of xenobiotic compounds released from pulp mills has been studied extensively in Finland. In these studies, roach and perch were collected in southern Lake Saimaa at various distances from an effluent outfall (Oikari 1986). The bile from one to several pooled fish samples at each site was analyzed for chlorinated phenols and resin acids, both as free compounds and as conjugates (after base hydrolysis). There was a general decrease in the concentration of all xenobiotic compounds (more than 95% conjugates) in roach bile as the distance from the outfall increased. The concentrations of resin acids in perch bile followed a similar pattern, except peak concentrations of chlorinated phenols occurred 4 km rather than 1 km from the source. The number of samples were fewer than three for most stations, so it is not known if any special interpretation of this latter distribution is warranted.

A later study examined rainbow trout placed in cages for 10–15 d 2 km upstream and 1, 4, 6, and 11 km downstream from the mill outfall (Oikari and Kunnamo-Ojala 1987). Free and conjugated chlorinated phenols were detected in plasma of fish collected up to 6 km from the outfall, while free and conjugated resin acids were detected in plasma samples from fish up to 4 km from the source. Detectable amounts of most individual resin acids and chlorinated phenols in bile were found in fish at 11 km from the outfall, the furthest distance sampled. These gradients in bile concentrations reflected the gradients of these compounds measured in water samples, indicating a dose-response relationship. Coupling the bioconcentrating ability of fish with that of the hepatobiliary system has provided a powerful monitoring tool for fate and distribution studies of complex effluents such as chlorine-treated kraft pulp mill effluent. Because the relative proportions of toxicant in bile vs plasma varied depending on exposure, chronic long-term input can be differentiated from periodic spills (Oikari et al. 1984).

Van Kreijl et al. (1986) reported another example of the utility of metabolite monitoring. The bile of rainbow trout placed in Rhine River water was mutagenic, again indicating bioconcentration of xenobiotic chemicals from the water. Chemical fractionation of the bile indicated that the mutagenic compounds may have been nitroaromatic compounds or aromatic amines. Treatment of bile with β-glucuronidase increased mutagenic activity. In this case, conjugation with glucuronide seems to have inactivated these mutagens before biliary excretion. Thus an understanding of metabolite biochemistry coupled with the concentrating ability of the hepatobiliary system provided a useful tool for ecotoxicological studies.

The pair PCP/PCA is an example of a situation where knowledge of the types of metabolites that might be formed, the relative toxicities of the different forms, and the appropriate analytical procedures are important to correctly monitor for exposure. PCP, a ubiquitous contaminant, has a half life of 23.7 h in rainbow trout, while the methylated derivative PCA has a half life of 23.4 d (Glickman et al. 1977). If PCP were to be methylated by bacteria, a completely different toxicological problem would ensue. It is unfortunate, therefore, that many residue analysis protocols in use by regulatory agencies derivatize with diazomethane prior to gas chromatographic determination, thereby losing the ability to separate PCP from PCA.

A number of relatively stable metabolites of organochlorine pesticides are also found in tissues of a wide variety of organisms. These metabolites include DDE, DDD, dieldrin, and oxychlordane. Brooks (1974) presented an extensive review of the metabolism of chlorinated pesticides. The number of published reports on the presence of organochlorine pesticide metabolites in wildlife is too extensive to present here. During the last 25 years there have been hundreds, if not thousands, of such reports. For examples, see journals such as *Pesticide Monitoring Journal, Archives of Environmental Contamination and Toxicology*, and *Bulletin of Environmental Contamination and Toxicology*.

Considerations in Using Metabolites of Xenobiotic Chemicals as Biomarkers

Before a biomarker can be used effectively, a number of important factors to be considered include:

- the environmental question being addressed
- the nature of the chemicals of interest
- the species appropriate to the situation
- the types of metabolites anticipated
- the nature of the biological sample (e.g., tissue, fluid, or excreta) to be used
- possible modifying factors specific to the situation (e.g., time of year, temperature, reproductive status)
- sampling strategy
- limits of detection for the analytical procedures
- quality assurance/quality control considerations
- cost

The design of environmental studies that include monitoring of metabolites of xenobiotics should take into account the specific questions being asked about the presence and/or effects of one or more pollutants. Biomonitoring by determining parent chemical and metabolite profiles in specific organs or tissues has thus far been used largely for delineating effluent discharge zones or the extent of contamination. With this goal in mind, a variety of indigenous species can be selected, including sessile species (e.g., clams and oysters) or ones with life histories that limit migration (e.g., species that are behaviorally restricted to specific kelp beds). This approach maximizes the analytical power of the biomonitoring technique.

Alternatively, migratory species can be sampled at the appropriate time if the residence period in the zone of influence is known. For example, analyses of migratory bird or fish shortly after they arrived in a new area would not give an assessment of that area. A commonly used technique is caging routinely-used species, such as salmonids, at various distances from a suspected source of contamination. If bioavailability from sediments is a concern, more realistic exposures use enclosures that include the sediment interface. For many PAHs and chlorinated organics, bioturbation or consumption of organisms residing in contaminated sediments controls bioavailability. Distinctions between the aqueous and sediment routes can be made by comparing residues in organisms suspended above the bottom with those in organisms residing on the bottom.

The purpose of the study must be defined before tissues and techniques can be selected. Advantages can be gained by using the bioconcentrating ability of the vertebrate hepatobiliary system or invertebrate hepatopancreas analogs in invertebrates as powerful amplifiers of exposure. For example, if the purpose of the study includes concerns about the health of the organism being monitored, preliminary screening can identify the compound classes likely to be present. If, on the other hand, there are concerns about human consumption of residues, select appropriate tissues of commercially important species, such as the edible portions of shellfish or finfish. In the latter case, analysis for metabolites in organs that are not consumed may be of secondary interest. Some exceptions do occur. For example, in 1988, concerns about chlorinated organic compounds, including tetrachlorodibenzodioxin and tetrachlorodibenzofuran, in crab hepatopancreas prompted the Canada Department of Fisheries and Oceans to close the commercial crab fishery in Howe Sound, Vancouver, British Columbia (personal communication, M. Nassichuk, Department of Fisheries and Oceans, Vancouver). These initial closures were expanded in 1989 and 1990 and now cover receiving waters around most British Columbia coastal pulp mills; an area estimated at 670 km^2 (Colodey et al. 1990).

Concerns about organic chemical contamination led to investigations of tissue residues in eulachons (*Thaleichthys pacificus*) moving up the Fraser River during spawning migrations. Rogers et al. (1990) demonstrated uptake of PCBs and chloroguaiacols by eulachons as they moved upstream to spawn. Euchalons have a very high lipid content (approximately 15%), and thus should be excellent integrators of lipophilic xenobiotic compounds. Their oil is rendered to "eulachon grease," valued as a food item by Indians in British Columbia. In this case, a detailed study of contaminants and their metabolites in "euchalon grease" would not only serve as a biomarker of fish exposure, but also provide valuable information on potential health hazards to the native populations. Likewise, because Upper Fraser River Indian bands also consume chinook (*Oncorhynchus tshawytscha*) and sockeye (*O. nerka*) salmon livers, hearts, and roe as part of their diet (personal communication, P. Quaw, Chief, Fort George Band) results of analysis of only muscle tissue would be inadequate to determine risk from contaminants.

Fish tissues shown to accumulate xenobiotic metabolites include gall bladder, liver, kidney, digestive tract, muscle, and nerve tissue. Invertebrate tissues that accumulate metabolites include digestive gland/hepatopancreas, muscle, and nervous tissue. Bile, blood, and urine are fluids that primarily contain conjugates. Metabolites bound to hemoglobin may also be of use as a biomarker.

Information is needed on the nature of metabolites formed by the selected species. For example, invertebrates form glucose conjugates of xenobiotic metabolites while vertebrates form glucuronide conjugates. A variety of factors such as reproductive state, temperature, and dietary status can affect metabolite production. For example, during feeding by vertebrates the bile and its associated metabolites are discharged from the gall bladder. These metabolites may be excreted via the feces or may undergo enterohepatic recycling, with increased residence time in the organism and possible additional metabolism. During egg production in female vertebrates, a number of biochemical changes can affect metabolite production. These changes include steroid-synthesizing cytochrome P450 isozymes that can affect rates of xenobiotic metabolism and types of metabolites produced. Also, increased lipid synthesis and mobilization needed for egg production have been shown to cause contaminant mobilization from depot lipids (Guiney et al. 1979), and may therefore facilitate metabolite production. Eggs that sequester lipophilic contaminants may also sequester lipophilic metabolites of these contaminants.

If metabolite measurements are to be undertaken in wild populations of organisms, sampling efforts must be designed so that temporal and spatial differences can be tested and actual environmental differences detected. Sufficient organism numbers and sample replicates at each site and time must be analyzed, depending on individual variability and the precision required, in order to answer the questions asked. Also, factors such as variability in feeding times (as in the case of biliary metabolites), sex, maturity, and environmental temperatures must be considered and may dictate a stratified sampling program. A statistician should be involved in the planning process to ensure that an adequate sampling design is developed.

Analytical techniques and protocols also must be designed to assure accuracy and precision in metabolite analysis. Replicate samples, field blanks, clean techniques, laboratory blanks, system blanks, sample spikes, instrument internal standards, and interlaboratory comparisons are all standard practice in environmental analytical chemistry. Careful consideration should be given to applying these practices to metabolite analysis. Analysts should develop standards and reference materials when widespread monitoring for particular compounds in organisms is envisioned. Furthermore, in designing a program of rational monitoring, the analysis costs must be considered. Techniques that measure important metabolites simply and at low cost are obviously desirable in monitoring programs.

Development Work Needed

Analytical Techniques

The feasibility of using xenobiotic metabolite formation as a biomarker of exposure depends on the sensitivity of the analytical methods employed for detecting and quantifying metabolites. The presence of such metabolites may be assessed by measuring free and conjugated metabolites in tissues, body fluids, or excreta, and by detecting and quantifying reactive metabolites formed by activation reactions.

Most of the analytical procedures used for measuring free and conjugated metabolites involve chromatographic techniques, including gas chromatography and HPLC (with or without enzymatic hydrolysis). Many of these procedures are limited by the lengthy preparation time required before the sample is analyzed. Thus, there is a need to develop procedures including immunoassay for sensitive, rapid measurement of metabolites, particularly conjugated metabolites.

A number of sensitive methods for detecting DNA adducts in organisms exposed to genotoxic chemicals are currently available. Because an organism in the environment may be exposed to a variety of genotoxic chemicals simultaneously, methods are needed to detect specific adducts in complex mixtures of carcinogen-DNA adducts in biological samples. Both the physicochemical and immunological assays offer potential in that area. (For more information, see Chapter 3 on DNA alterations.) Current methods of measuring xenobiotic-protein adducts include gas chromatography/mass spectroscopy and ion exchange amino acid analysis. One of the drawbacks of analyzing protein adducts is expensive equipment requirements. Much research is needed, including the development of less laborious techniques, before the measurment of protein adducts can be used routinely as a biomarker of exposure. A particularly promising approach that warrants extensive research is the use of highly sensitive immunological techniques to detect the levels of such adducts.

Information on Other Species

Information presently available enables us to monitor via metabolites for exposure to certain contaminants in a number of fish species and, in some cases, to relate levels of such metabolites to harmful effects. In order to evaluate exposure and effects of contaminants in systems other than aquatic ones, further information is needed on target species, whether birds, amphibians, mammals, or invertebrates. Studying a target's natural history (e.g., feeding habits and range) and metabolism of contaminants enables researchers to evaluate the exposure potential and the nature of expected metabolites.

More Linkage to Effects—Batteries of Tests

Much of the biomonitoring that currently uses metabolites of xenobiotics is based on the fact that, in many cases, analysis for metabolites of xenobiotic compounds in biological material is more effective for evaluating the presence of the xenobiotics than analysis for the parent xenobiotic chemicals themselves. Maximizing the use of xenobiotic metabolites as biomarkers requires relating the presence of particular metabolites or classes of metabolites to specific harmful effects and further developing the dose-response or dose-rate correlations.

SUMMARY

Table 1 summarizes the results of an evaluation of unbound metabolites, DNA adducts and protein adducts of metabolites involving xenobiotic chemicals. Chlorinated hydrocarbon and PAH metabolites in tissues and PAH, chlorinated phenol, and resin acid metabolites in bile can presently be used as biomarkers in environmental biomonitoring. DNA adducts of xenobiotic chemical metabolites is a relatively new metabolite biomarker that appears to be close to useable.

The species-specific information needed to expand the use of xenobiotic chemical metabolites as biomarkers beyond fish and the aquatic environment should be acquired and a better understanding is needed to relate the presence of specific metabolites of xenobiotics in organisms to toxic effects.

ENDOGENOUS METABOLITES

Introduction

Plant metabolites of potential interest as environmental stress biomarkers include glutathione, phytochelatins, polyamines, and stress ethylene. In animals, comparable metabolites include metallothioneins, porphyrins, reproductive hormones, vitellogenin, and polyamines. Metabolites such as glucaric acid and thioethers, produced as the result of conjugation and detoxification of xenobiotic compounds, are also promising as biomarkers of exposure and effects.

The concentration of a chemical or metabolite in blood, urine, saliva, sweat, expired air, hair, and adipose tissue has been measured in a number of monitoring programs. For example, measuring urinary concentrations of pesticide metabolites has been used successfully to monitor workers occupationally exposed to agricultural chemicals. In animals, many xenobiotic compounds are biotransformed to conjugates of glutathione, glycine, sulfate, acetic acid, or glucuronic acid; often this biotransformation is accompanied by induction of transformation pathways (Notten and Henderson 1977). In plants, comparable conjugation reactions occur with glutathione acting as the conjugate (Lamoureux and Rusness 1989).

Accumulated evidence suggests there are major mechanisms by which oxidants damage photosynthetically active plant cells. Molecular oxidants such as the diphenylether herbicides, paraquat, or the air pollutants ozone or sulfur dioxide, may form toxic free radical products in photosynthetically active plant cells (Asada 1980; Schmidt and Kunert 1986; Heath 1987; Alscher and Amthor 1988). The toxic radicals are removed enzymatically and chemically through the mobilization of antioxidant reserves. Mechanisms of resistance and chemical constituents that scavenge free radicals have been identified. These mechanisms protect photosynthetically active plant cells against oxygen toxicity. Glutathione plays an important role in these resistance mechanisms.

Endogenous Metabolites as Biomarkers

Metallothionein and Phytochelatins

Metallothioneins are intracellular, low molecular-weight (6000 to 7000 daltons), cysteine-rich proteins that are ubiquitous in animals and capable of binding heavy metals. Although not all the precise functions of metallothioneins are known, their major function appears to be the regulation of intracellular metabolism of two essential metals, zinc and copper. Metallothionein is also believed to play a role in acclimation-induced tolerance to heavy metals (Benson and Birge 1985). Other than cases of occupational exposure in humans, there is little evidence that measuring metallothionein provides more information concerning exposure and environmental effects than does measuring tissue metal concentrations. In addition, it has been demonstrated that physiological conditions (Olsson et al. 1987; Overnell et al. 1987) and environmental stressors (Baer and Benson 1987; Sabourin et al. 1985; Benson et al. 1988; Thomas 1989) influence tissue concentrations of metallothionein and the amount of metal bound to metallothionein. A better understanding of the normal function and patterns of metallothionein biosynthesis, as well as the effects of other environmental factors, is required before metallothioneins can be used as reliable biomarkers in evaluating

Table 1. Attributes of Metabolic Biomarkers.

Attribute	Metabolites of Xenobiotic Chemicals					Endogenous Metabolites					
	Unbound Metabolites	DNA Adducts	Protein Adducts	Vitellogenin	Ethylene	Polyamines (plants)	Phyto-chelatins	GSH	Porphyrins	Lipid Peroxidase	Reproductive Hormones
Exposure marker	++++	++++	++++	-	+	++	++++	++	++	+	±
Effects marker	-	in progress	-	+	+	+(?)	acute ++	++	+	+	+
Biological specificity	±	+	+	+++	-	-	+(?)	-	+	+	+++
Chemical specificity	++++	++	+	-	-	-	+++	-	++	-	-
General indicator	-	-	-	+	++	+	-	an +, pl +++	-	+++	+
Relative sensitivity	++	±	±	-	+++	+	+	mod an ±, pl +	+	±	±
Certainty	+++	+	+	-	++	+	+	an ±, pl +	+	-	±
Time to endpoint	hrs-ds ds-wks	hrs-ds ds-wks	hrs-ds hrs-wks-mos	?	mins	mins	mins-hrs	mins	wks-mos	mins-hrs	?
Permanence	mod	mod	?	var	mins	-	?	mod	?	-	ne
Reliability				+++	+	+	±	++	+	+	var
Linkage to higher effects	±	-	-	-	?	?	?	an ?, pl -	+	+	+
Field applicability	+++	++	+	++	?	?	++	an -, pl +	+	+	ne
Field validation	++	+	-	-	ne	ne	?	an -, pl +	+	?	no
Precision of method	+	+	+	++	++	+	+++	++++	+++	++	+
Cost/ease of method	mod-exp	mod-exp	exp	low	low	mod	mod	low-mod	low	mod	low
Status of utility	current 1–5 yrs	>5 yrs	1–5 yrs	1–5 yrs	1–5 yrs	1–5 yrs	1–5 yrs	current 1–5 yrs	1–5 yrs		
Research needed	more spp, analytical techniques	immun methods, relate to effects	immun methods, relate to effects	dose-resp, molecular techniques	define source of ethylene	dose-resp	field val, chronic experience	field val, time course dose-resp	other spp	field val, dose-resp	field val, dose-resp

Key to abbreviations:

an = animal
pl = plant
val = validation

resp = response
spp = species
hrs = hours

ds = days
wks = weeks
mos = months

mins = minutes
ne = not established
exp = expensive

mod = moderate
var = varies
immun = immunological

exposure and environmental effects of heavy metals (Benson et al. 1988; Thomas 1990). (For additional information, see Chapter 6 on enzymes and proteins.)

In plants, the detoxification of heavy metals involves a class of nonprotein polypeptides derived from glutathione that are called phytochelatins (Steffens 1990). Phytochelatins fulfill the function of metal sequestration in plants analogous to that of metallothioneins in animals. Phytochelatins may facilitate the transport of potentially toxic metals across the tonoplast membrane into the vacuole (Steffens 1990). The phytochelatins have great promise as biomarkers of acute heavy metal stress in plant cells because no other stimulus is known to induce their synthesis. The capacity to synthesize phytochelatins has been found in all plant species tested, or about 0.01% of all known plant species (Grill et al. 1986, 1987; Grill 1989). Phytochelatin biosynthesis is sensitive to inhibitors of glutathione formation, providing direct evidence that these polypeptides are synthesized from glutathione moieties (Steffens et al. 1986).

Considerable progress has been made in using phytochelatins as biomarkers in environmental studies. Mutoh and Hayashi (1988) showed unequivocally that phytochelatins are essential for cadmium tolerance. It now appears that phytochelatins can be used directly as biomarkers of acute heavy metal stress, although their role in chronic heavy metal stress is not as well defined. A thorough understanding of the possible transport role of these compounds (cytoplasm to vacuole) would enable a more precise assessment of the protection process.

Glucaric Acid

Many chemicals are biotransformed to glucuronic acid conjugates; often this biotransformation is accompanied by enhancement of total glucuronidation. Stimulation of glucuronidation results in increased glucaric acid excretion, primarily through decreased production of glycogen, possibly with a direct induction of pathway enzymes (Notten and Henderson 1975). Although certain disease states (e.g., liver and kidney disorders) are known to influence glucaric acid excretion, available evidence indicates that enhanced urinary excretion of glucaric acid may be useful as a nonspecific exposure biomarker to certain compounds. These compounds include primary amines produced by reduction of azides and nitrites and oxidation of alkylamines, as well as alkyl and aryl alcohols produced from oxidation of aromatic hydrocarbons, ethers, and alkyl and aryl hydrocarbons, or from hydrolysis of esters (Brewster 1988). Glucaric acid has not yet been used as a biomarker of xenobiotic exposure or effects in ecological studies.

Glutathione and Thioethers

The tripeptide glutathione is widely distributed in plant cells (Rennenberg 1987). It functions in sulfur transport and in the detoxification of xenobiotic compounds, such as herbicides (Matringe and Scalla 1988) and the air pollutants ozone and sulfur dioxide (Heath 1987; Alscher and Amthor 1988). Glutathione is also implicated in plant adaptation to environmental stresses such as drought and temperature extremes (Esterbauer and Grill 1978; de Kok and Oosterhuis 1983; Burke et al. 1985; Wise and Naylor 1987). In some cases, glutathione has been specifically implicated in resistance to stress (Burke et al. 1985; Alscher et al. 1987).

In animals, as in plants, electrophilic agents (a category that includes most genotoxic compounds) may be inactivated by reaction with glutathione or other sulfhydryl compounds (Henderson et al. 1984). These conjugates often appear in urine as mercapturic acids or other thioether (R-S-R) products. To date, the use of thioether concentrations as biomarkers has only been qualitative, because metabolite identities are unknown, losses are variable during extraction, and dose-response relationships have not been determined. Clearly, developing more selective assays of sulfur-containing compounds in urine is needed for use in ecological biomarker applications (Henderson et al. 1984; van Doorn et al. 1981).

T.A. Smith (1985) and I.K. Smith et al. (1984, 1985) have provided important information concerning the role of glutathione in the stress responses of plants. In a catalase-deficient mutant of barley (*Hordeum vulgare* L.) under photorespiring conditions and in wild-type barley treated with aminotriazole (an inhibitor of catalase), glutathione accumulated to concentrations three times higher than those occurring in nonphotorespiring plants. Glutathione afforded some protection to the photosynthetic apparatus in the mutant under these conditions. Perhaps this protection is related to the larger fraction of the glutathione produced in the mutant that was initially in the form of GSSG (the oxidized form of glutathione). When hybrid poplar plants were exposed to ozone for 3 h, photosynthesis was inhibited by about 60% (see Gupta et al., manuscript submitted). Prior to any detectable change in photosynthesis (15 min after initiation of the exposure), GSSG began to accumulate. Within 20 h following the fumigation period, photosynthesis recovered partially, GSSG levels dropped dramatically, and GSH (the reduced form of glutathione) levels increased about 2.5-fold. These data are consistent with the concept that a diagnostic feature of a plant experiencing or not yet recovered from oxidative stress is the accumulation of detectable amounts of GSSG. If GSSG accumulates, thiol-containing enzymes can be inactivated through the formation of mixed disulfides (Halliwell and Gutteridge 1985). GSSG has also been shown to inhibit protein synthesis (Kan et al. 1988; Dhindsa 1987).

For glutathione to be protective it must be in its reduced state. This is normally the case under "unstressed" conditions, and the enzyme glutathione reductase (GR) maintains the glutathione pool in its "useful" or reduced state. For this reason GR would be expected and has been experimentally proved to be a crucial factor in resistance to oxidative stress.

Transcriptional and translational events are also implicated in the responses of glutathione metabolism to oxidative stress. Tanaka et al. (1988) demonstrated, by Western blot (a technique that combines gel electrophoresis and a specific antibody), an increased GR production in response to ozone stress in spinach. Schmidt and Kunert (1987) compared the response of two *Escherichia coli* strains, differing with respect to their GR complement, to an agent of lipid peroxidation. The presence of the enzyme clearly conferred a resistance to oxidative stress. Glutathione reductase has also been shown to be positively associated with the resistance of higher plants to various oxidative stresses such as paraquat and diphenylether pesticides, heat, and drought stress (Burke et al. 1985; Schmidt and Kunert 1986).

The reduced form, but not the oxidized form, of glutathione was recently shown to act as an elicitor of the transcription of mRNA coding for the enzymes of the phytoalexin (PAL) pathway (Wingate et al. 1988). The requirement for GSH in this stimulation is a most intriguing result in light of the many responses to oxidative stress known to involve induction mechanisms. The increase in glutathione that often follows the onset of the stress stimulus not only directly protects the cell against the generation of free radicals, but may activate the whole panoply of stress-resistance responses.

The responses of glutathione metabolism to environmental stress are well documented. Historically, increases in reduced glutathione have been detected as a consequence of plant exposure to oxidative stress. A biomarker approach, however, has not been developed.

Oxidized glutathione inhibits protein synthesis in animal systems through an interaction with one of the initiation factors for translation (Kan et al. 1988). GSSG appears to have the same effect in a plant system, as demonstrated by Dhindsa (1987). An accumulation of GSSG may therefore, result in deleterious effects on developmental responses or on those stress resistance processes involving induction mechanisms.

Glutathione forms conjugates with several pesticides groups in higher plants (Rennenberg 1987). The reaction is catalyzed in each case by the enzyme glutathione-S-transferase (GST). N-malonylcytseine derivatives are major end-products of herbicide conjugation/detoxification in plant cells and may eventually be deposited in the vacuole.

The damaging effects of the herbicide atrazine on corn are ameliorated by a class of compounds called "antidotes" or "safeners," which induce a novel form of GST at the transcriptional level and induce the constitutive form of the enzyme. (See reviews by Shah et al. 1986 and Rennenberg 1987.)

Both the glutathione pesticide conjugates and the GST isozymes are potential stress biomarkers in higher plants. The responses of glutathione metabolism to stress are also beginning to be used as long-term environmental stress biomarkers. However, the significance of the relative sizes of the oxidized and reduced pools of the antioxidant as well as the activity of GR must always be evaluated through independent, controlled experiments over the same time frame for this system to be applicable to any particular plant system. For instance, in a study of ozone exposure effects on seasonal changes in antioxidants in red spruce, GSSG was found to accumulate in ozone-treated seedlings during autumn when the plant acquires cold tolerance (Hausladen et al. 1990). In a parallel study, Alscher and Hausladen (1991) demonstrated an accumulation of GSH and GSSG during autumn in the needles of high elevation spruce (Whiteface Mountain, New York, 1200 m) with decline symptoms. These results indicate that GSSG accumulation has potential as a biomarker of relatively acute oxidative stress.

Glutathione metabolism of each plant species must be characterized in detail to make informed evaluations of stress-mediated changes. Characterization should include an understanding of metabolic and molecular events that occur as a consequence of oxidative stress at the cellular and organellar levels. The role of GR and the significance of its response at the genetic level warrants further work because GR is a potential biomarker of oxidative stress in plants.

In a non-enzymatic context, glutathione and ascorbate both have a potential antioxidant role in conjunction with α-tocopherol. Since α-tocopherol is the major free radical scavenger of thylakoid membranes, this finding is particularly relevant for studying chloroplast metabolism. α-tocopherol itself was shown to increase as a consequence of exposure to ozone in a long-term study of the effects of the pollutant on red spruce seedlings (Hausladen et al. 1990).

Porphyrins

It has been known for some time that chemicals may disturb porphyrin metabolism in mammals and birds. Chemicals that can have this effect include chlorinated aromatics, such as PCBs, and heavy metals, such as lead. In chemically induced porphyrias, these chemicals or their metabolites modify the activity of one or more of the enzymes involved in heme biosynthesis, altering the size and/or composition of the porphyrin pool (Goldstein et al. 1973; Strik 1979). Porphyrin accumulation patterns in tissues and excreta may be used to predict the sites of chemical action within the heme biosynthesis pathway (Marks 1985). Conversely, when a chemical is known to have a specific effect on heme biosynthesis, abnormalities of porphyrin metabolism may provide a method for assessing exposure (Elder and Urquhart 1987).

Lead inhibits the activity of heme synthetase, the enzyme that incorporates iron into protoporphyrin IX to form heme. As a result, protoporphyrin accumulates in the peripheral blood where it can be measured by a simple fluorescence technique. Using a hematofluorometer, Roscoe et al. (1979) reported increased levels of protoporphyrin in a single drop of untreated blood following administration of lead shot to mallard ducks (*Anas platyrhynchos*). Following the administration of 1 to 18 #4 lead shot, the blood concentrations of protoporphyrin IX were related to clinical signs of lead poisoning. Protoporphyrin IX concentrations less than 40 g/dL corresponded to no evidence of toxicity while concentrations over 800 g/dL corresponded to death. Intermediate values corresponding to sublethal effects in pen-kept ducks could well result in death in a wild population. In other lead exposure studies in both pen-kept and wild black ducks (*Anas rubripes*) and mallards, toxicity and lethality corresponded to much lower blood protoporphyrin IX concentrations (Pain and Rattner 1988; Rattner et al. 1989). Studies in a number of other bird species have shown similar results (Beyer et al. 1986).

The toxicity of a class of halogenated aromatic hydrocarbons (HAHs) in the etiology of porphyria involves the activation of the xenobiotic compounds by cytochrome P450. The porphyrins affected are not the same as those involved in lead toxicity. The activation of the HAHs requires metabolism by the type of cytochrome P450 that is induced by 3-methylcholanthrene-type inducers. The halogenated organic compounds that cause porphyria all appear to be capable of inducing their own metabolism. The major porphyrin that accumulates in liver tissue in response to these chemicals is uroporphyrin, but other highly carboxylated porphyrins are also found. Analysis involves acid extraction of porphyrins from the liver or other tissue (Kennedy et al. 1986; Reddy et al. 1987) and determination of individual protoporphyrins by their fluorescence. Uroporphyrin can be determined directly on the acid extract by its specific fluorescence. The spectrum of protoporphyrins present can be determined by HPLC with fluorescence detection (Kennedy et al. 1986; Reddy et al. 1987). Thus, the analyses of these compounds is relatively simple and rapid.

In a study of herring gulls (*Larus argentatus*) from the Great Lakes (Fox et al. 1988) showed that gulls from contaminated areas have considerably higher concentrations of highly carboxylated porphyrins in liver tissue than gulls from "clean" (control) areas. In the contaminated areas studied, the frequency of concentrations greater than ten times the median control values varied from 22 to 100%.

Malondialdehyde as an Indicator of Lipid Peroxidation

Lipid peroxidation is a free-radical process in which the degradation of polyunsaturated fatty acids in tissues to hydroperoxy- , keto- , epoxy- , and hydroxy-derivatives occurs via an autocatalytic reaction sequence. Lipid peroxidation occurs as a consequence of environmental stressors such as extremes of temperature, light, or exposure to xenobiotic compounds (Halliwell and Gutteridge 1985). Elevated concentrations of malondialdehyde, a breakdown product of lipid endoperoxides, are an expression of lipid peroxidation (Pompella et al. 1987; Sunderman 1987). An advantage to the use of malondialdehyde as a biomarker is that the biological significance of its production is partly understood (Thomas 1990). However, factors such as nutritional status and age can influence malondialdehyde formation (Wofford and Thomas 1988). Therefore, caution should be taken when evaluating malondialdehyde production in relation to chemical exposure.

Reproductive Hormones

Reproduction in teleosts is a complex process involving considerable physiological coordination under the control of reproductive hormones. Alterations in the circulating levels of the reproductive steroid hormones 17 β-estradiol, testosterone, and 11-ketotestosterone, and reproductive function have been demonstrated in several teleost species after exposure to chemical and physical stressors (Sangalang and Freeman 1974; Thomas 1989; Johnson et al. 1988). Chemicals may alter plasma concentrations of reproductive steroid hormones by several mechanisms. For example, chemicals may alter the secretion of gonadotropin, which regulates the synthesis of steroid hormones, or directly affect steroidogenesis itself. In addition, chemicals with estrogenic or antiestrogenic activity may alter reproductive endocrine function and plasma steroid levels by interfering with steroid feedback mechanisms at the brain or pituitary gland. The complexity of the interactions among various parts of the reproductive system is likely to limit use of reproductive steroid hormones as biomarkers (Thomas 1990).

Vitellogenin

Vitellogenin is a large molecular weight lipophosphoprotein synthesized by the liver in vertebrates regulated by 17 ß-estradiol. Among the invertebrates, the analogous compound is a lipoprotein synthesized by a variety of tissues. These lipoprotein and lipophosphoprotein molecules are released directly into the blood and are sequestered in the developing oocyte in response to gonadotropin and other hormones. In the vertebrate oocyte, vitellogenin is proteolytically cleaved to form yolk proteins. In teleosts, laboratory study results suggest that the impact of contaminants on plasma vitellogenin concentrations can be quite variable (Chen 1988; Johnson et al. 1988). Exposure to chemicals may decrease plasma concentrations of vitellogenin by inhibiting its production (Chen and Sonstegard 1984), or increase plasma vitellogenin by decreasing the uptake of vitellogenin by developing oocytes (Ruby et al. 1987). In a field study of starry flounder, vitellogenic females from contaminated and uncontaminated sites in San Francisco Bay showed no differences in plasma vitellogenin (Spies et al., unpublished data). Very little information is available on the effect of chemicals on invertebrate vitellogenin. Because of the limited information available a better understanding of the normal patterns of vitellogenin production in teleosts and invertebrates is required before plasma vitellogenin concentrations can be used as reliable biomarkers for evaluating chemical exposure and effects.

Polyamines

The polyamines spermidine and spermine, and their precursor putrescine, are ubiquitous components in many eukaryotic cells and appear to function in cellular proliferation and differentiation (Marton and Morris 1987). In mammals, a number of investigations have demonstrated accumulation of putrescine and spermidine in cells during growth and a large, rapid increase in the activity of ornithine decarboxylase following exposure to growth-promoting stimuli. On the other hand, depletion of polyamines during early developmental stages arrests embryogenesis when polyamine synthesis inhibitors are used. Recent reports have confirmed that the ability to induce ornithine decarboxylase activity is a common property of many tumor promoters. The reduction of enzymatic activity by polyamine biosynthesis inhibitors also inhibits the development of tumors induced by a range of chemical carcinogens. The importance of ornithine decarboxylase activity and polyamines in tumor development has led to many mammalian studies investigating the possibility that these may serve as useful carcinogenesis biomarkers (Pegg and McCann 1988). The available evidence indicates that polyamines may, in the future, serve as developmental toxicity or carcinogenesis biomarkers. However, much more information is required on the basic role of polyamines in cellular proliferation and differentiation before they can be considered as biomarkers in evaluating exposure and environmental effects of toxic compounds.

Higher plants exposed to suboptimal or stress-inducing environmental conditions respond by accumulating high putrescine concentrations (Galston 1989). However, the concentration of tri- and tetra-amines (spermidine and spermine) are unaffected by exposure to stress (see review by Smith 1985). This increase in diamine levels has been suggested to have an adaptive function in plants. In isolated protoplasts and excised intact leaves, exposure to exogenous polyamines reduced chlorophyll loss (Cohen et al. 1979) and inhibited increase in ribonuclease and protease activities (Kawr-Sawheney and Galston 1982). Putrescine, in particular, has been suggested to regulate the ion content of the symplasm (Young and Galston 1983).

The reverse correlation has also been reported, however. The application of exogenous diamines to barley (*Vicia faba*) resulted in injury (Richards and Coleman 1952). Increases in endogenous putrescine have been associated with reductions in cell division in oat protoplasts (Tiburcio et al. 1986). A result that may explain some of these discrepancies was reported by Ferrante et al. (1983). Ferrante et al. found that the protozoan *Plasmodium falciparum* suffered toxic effects when exposed to spermine and polyamine oxidase administered together, but no toxicity was observed when exposed to either compound alone. Since the action of polyamine oxidase on spermine can lead to the formation of aldehydes, hydrogen peroxide, and free radicals, the authors inferred that this mechanism gave rise to the toxic symptoms seen when spermine and polyamine oxidase were administered together.

DiTomaso et al. (1989) suggest that the discrepancies in the plant literature concerning the beneficial or harmful effects of polyamines arise from the respective subcellular localizations of putrescine and diamine oxidase, the enzyme that catalyzes putrescine breakdown into toxic, oxidizing, molecular species (Suresh and Adiga 1979). This breakdown process appears to occur within the cell wall and not within the boundaries of the symplastic cell proper (Flores and Filner 1985). Any experimental system in which the cell, with its plasmalemma intact, is removed from the cell wall environment is, therefore, unsuitable as a vehicle to study putrescine metabolism and its response to environmental stress. The use of protoplasts is appropriate under these conditions.

DiTomaso et al. (1989) present data consistent with the hypothesis that putrescine appears to elicit a "wounding response" in intact corn roots and that this wounding response involves the generation of free radicals and an increase in membrane leakage. Membrane potential and net ionic fluxes were adversely affected by application of exogenous putrescine. The addition of ascorbic acid, an antioxidant and a free radical scavenger, partially reversed the deleterious effects of putrescine.

Careful experimentation is needed to determine the subcellular location of the polyamines under particular conditions of environmental stress. Some protection may be afforded to organellar membranes, systems that are somewhat removed from the cell wall/plasmalemma environment. If free radicals arise as a consequence of the interaction of the diamine oxidases of the cell wall with diamines such as putrescine, however, it is highly probable that the cell as a whole will experience toxic symptoms. These unresolved issues and ambiguities make it uncertain whether polyamines in plants will be useful predictors of the physiological outcome of plant stress.

Stress Ethylene

Ethylene is evolved from the foliar tissues of higher plants in response to a wide range of stresses (Wang et al. 1990; Yang and Hoffman 1988). The major pollutants, virus infection, cadmium, chilling, and flooding have all been found to elicit what is sometimes called the "stress ethylene" response. This stimulation typically takes 10 to 30 min, and in many instances has been associated with a stimulation of the synthesis of 1-aminocyclopropane-1-carboxylic acid (ACC), a key intermediate in the ethylene biosynthetic pathway. Under extreme environmental stress, ethylene evolution is replaced by ethane evolution (Kimmerer and Koslowski 1982). Both ethane and ethylene evolution may occur independently of the ACC biosynthetic pathway, as a result of the oxidation of polyunsaturated fatty acids.

Mehlhorn and Wellburn (1987) reported data consistent with an interaction of stress ethylene with oxidizing air pollutants such as ozone in the substomatal cavity, resulting in the generation of toxic free radical species and, eventually, in increased visible injury. Cape et al. (1988) in a survey of forest decline across European countries, found the rate of ethylene emission was one of only three suitable predictors of the probability of forest decline of the 30 parameters tested.

Stress ethylene evolution is sometimes used as a biomarker for environmental and other stresses. However, cross-species comparisons are often not possible, because not all species tested showed a (good) correlation between susceptibility to stress and rates of ethylene emission. The caveats surrounding the use of stress ethylene as a biomarker were outlined previously. Species specificity is the most important attribute and potential pitfall of ethylene production. The interaction of ethylene with other factors may have great relevance to the degree of damage sustained. Much additional information is needed about these interactions.

SUMMARY

Ideally, endogenous biomarkers would indicate both exposure and environmental effects of toxic chemicals; however, such comprehensive biochemical and physiological indices are currently being developed and, at the present time, are unavailable for use in environmental monitoring programs. Continued work is required to validate the use of biochemical and physiological stress indices as useful components of monitoring programs.

Of the compounds discussed, only phytochelatins and porphyrins are currently in a useful state; however, glutathione, metallothioneins, stress ethylene, and polyamines are promising as biomarkers in environmental monitoring.

REFERENCES

Ahokas, J.T., H. Saarni, D.W. Nebert and L. Pelkonen. "The in vitro Metabolism and Covalent Binding of Benzo[a]pyrene to DNA Catalyzed by Trout Liver Microsomes," *Chem.-Biol Interact.* 25:103–112 (1979).

Alscher, R.G., and J.A. Amthor. "The Physiology of Free-radical Scavenging: Maintenance and Repair Processes," in *Air Pollution and Plant Metabolism,* S. Schulte-Hostede, Ed. (Amsterdam: Elsevier, 1988) pp. 91–126.

Alscher, R., J.L. Bower and W. Zipfel. "The Basis for Different Sensitivities of Photosynthesis to SO_2 in Two Cultivars of Pea," *J. Exp. Bot.* 38:99–108 (1987).

Alscher R.C., and A. Hausladen. "Ozone and Winter Injury: A Comparison of Field Physiology with Controlled Exposures," U.S. Forest Service Technical Report, (1991).

Anders, M.W., and L.R. Pohl. "Halogenated Alkanes," in *Bioactivation of Foreign Compounds,* M.W. Anders, Ed. (New York: Academic Press, Inc., 1985) pp. 283–315.

Aniya, Y., J.C. McLenithan and M.W. Anders. "Isozyme Selective Arylation of Cytosolic Glutathione S-transferase by [^{14}C]bromobenzene Metabolites," *Biochem. Pharmacol.* 37:251–257 (1988).

Asada, K. "Formation and Scavenging of Superoxide in Chloroplasts, with Relation to Injury by Sulfur Dioxide in Studies on the Effects of Air Pollutants on Plants and Mechanisms of Phytotoxicity," *Res. Rep. Natl. Inst. Environ. Stud.* 11:165–179 (1980).

Baer, K.N., and W.H. Benson. "Influence of Chemical and Environmental Stressors on Acute Cadmium Toxicity," *J. Toxicol. Environ. Hlth.* 22:35–44 (1987).

Bartolone, J.B., W.P. Beiraschmitt, R.B. Birge, S.G. Emeigh Hart, S. Wyand, S.D. Cohen and E.A. Khairallah. "Selective Acetaminophen Metabolite Binding to Hepatic and Extrahepatic Proteins: An In Vivo and In Vitro Analysis," *Toxicol. Appl. Pharmacol.* 99:240–249 (1989).

Bean, R.M., D.D. Dauble, B.L. Thomas, R.W. Hanf, Jr., and E.K. Chess. "Uptake and Biotransformation of Quinoline by Rainbow Trout," *Aquat. Toxicol.* 7:221–239 (1985).

Benson, W.H., K.N. Baer and C.F. Watson. "Metallothionein as a Biomarker of Environmental Metal Contamination: Species-Dependent Effects," in Biomarkers of Environmental Contamination, J.F. McCarthy and L.R. Shugart, Eds. (Chelsea, MI: Lewis), pp. 255–265 (1990).

Benson, W.H., and W.J. Birge. "Heavy Metal Tolerance and Metallothionein Induction in Fathead Minnows: Results from Field and Laboratory Investigations," *Environ. Toxicol. Chem.* 4:209–217 (1985).

Beyer, W.N., J.W. Spann, L. Sileo and J.C. Franson. "Lead Poisoning in Six Captive Avian Species," *Arch. Environ. Contam. Toxicol.* 17:121–130 (1986).

Birge, R.B., J.B. Bartolone, S.D. Goehn and E.A. Khairallah. "Comparison of the Proteins Covalently Bound to Acetaminophen and 2,6-Dimethyl Acetaminophen in Cultured Mouse Hepatocytes," *The Toxicologist* 9:49 (1989).

Brandt, I., J. Lund, A. Bergman, E. Klasson-Wehler, L. Poellinger and J.-A. Gustafsson. "Target Cells for the Polychlorinated Biphenyl Metabolite 4,4',-Bis(methylsulfonyl)-2,2',5,5',-Tetrachlorobiphenyl in Lung and Kidney," *Drug Metab. Dispos.* 13:490–496 (1985).

Brewster, M.A. "Biomarkers of Xenobiotic Exposures," *Ann. Clin. Labor. Sci.* 18:306–317 (1988).

Brooks, G.T. in "Biological and Environmental Aspects," *Chlorinated Insecticides. Vol. 2.* (Cleveland: CRC Press, 1974) p. 197.

Brown, D.A., R.W. Gossett, G.P. Hershelman, C.F. Ward, A.M. Wescott and J.N. Cross. "Municipal Wastewater Contamination in the Southern California [USA] Bight: Part I. Metal and Organic Contaminants in Sediments and Organisms," *Mar. Environ. Res.* 18:291–310 (1986).

Brown, D.A., R.W. Gossett and K.D. Jenkins. "Contaminants in White Croakers *Genyonemous lineatus* (Ayres, 1855) from the Southern California Bight: II. Chlorinated Hydrocarbon Detoxification/Toxification," in *Physiological Mechanisms of Marine Pollutant Toxicity,* W.B. Vernberg, A. Calabrese, F.P. Thurberg and F.J. Vernberg, Eds. (New York: Academic Press, 1982) pp. 197–214.

Brown, D.A., R.W. Gossett and S.R. McHugh. "Oxygenated Metabolites of DDT and PCBs in Marine Sediments and Organisms," in *Biological Processes and Wastes in the Ocean,* J.M Capuzzo, and D.R. Kester, Eds. (Malabar, FL: Robert E. Krieger Publishing Co., 1987) pp. 61–69.

Buben, J.A., N. Narasimhan and R.P. Hanzlik. "Effects of Chemical and Enzymic Probes on Microsomal Covalent Binding of Bromobenzene and Derivatives: Evidence for Quinones as Reactive Metabolites," *Xenobiotica* 18:501–510 (1988).

Burke, J.J., P.E. Gamble, J.L. Hatfield and J.E. Quisenberry. "Plant Morphological and Biochemical Responses to Field Water Deficits," *Plant Physiol.* 79:415–419 (1985).

Calleman, C.J., L. Ehrenberg, B. Jansson, S. Osterman-Golkar, D. Segerback, K. Svensson and C.A. Wachtmeister. "Monitoring and Risk Assessment by Means of Alkyl Groups in Hemoglobin in Persons Occupationally Exposed to Ethylene Oxide," *J. Environ. Pathol. Toxicol.* 2:427–442. (1978)

Cape, J.N., I.S. Patterson, A.R. Wellburn, J. Wolfenden, H. Mehlhorn, P.H. Freer-Smith and S. Fink. "Early Diagnosis of Forest Decline: A Report," NERC Institute of Terrestrial Ecology, Merlewood, Grange-over-Sands, U.K. (1988) p. 68.

Chen, T.T. "Investigation of Effects of Environmental Xenobiotics to Fish at Sublethal Levels by Molecular Biological Approaches," *Mar. Environ. Res.* 24:333–337 (1988).

Chen, T., and R. Stonstegard. "Development of a Rapid, Sensitive and Quantitative Test for the Assessment of the Effects of Xenobiotics on Reproduction in Fish," *Mar. Environ. Res.* 14:429–430 (1984).

Cohen A.S., R.B. Popovic and S. Zalik. " Effects of Polyamines on Chlorophyll and Protein Content, Photochemical Activity, and Chloroplast Ultrastructure of Barley Leaf Discs During Senescence," *Plant Physiol.* 64:717–720 (1979).

Collier, T.K., M.M. Krahn and D.C. Malins. "The Disposition of Naphthalene and its Metabolites in the Brain of Rainbow Trout (*Salmo gairdneri*)," *Environ. Res.* 23:35–41 (1980).

Colodey, A.G., L.E. Harding, P.G. Wells and W.R. Parker. "Effects of Pulp and Paper Mill Effluents on Estuarine and Marine Environments in Canada: A Brief Review," Presented at 17th Annual Aquatic Toxicology Workshop, Vancouver, B.C., November 5–7, 1990.

Conney, A.H. "Induction of Microsomal Enzymes by Foreign Chemicals and Carcinogenesis by Polycyclic Aromatic Hydrocarbons; G.H.A. Clowes Memorial Lecture," *Cancer Res.* 42:4875–4917 (1982).

Creaven, P.J., D.V. Parke and R.T. Williams. "A Fluorometric Study of the Hydroxylation of Biphenyl In Vitro by Liver Preparations of Various Species," *Biochem. J.* 96:879–885 (1965).

de Kok, L., and F.A. Oosterhuis. "Effects of Frost-hardening and Salinity on Glutathione and Sulfhydryl Levels and on Glutathione Reductase Activity in Spinach (*Spinacia oleracea* Cultivar Vroeg Reuzenblad) Leaves," *Physiol. Plant.* 58:47–51 (1983).

Dhindsa, R. "Glutathione Status and Protein Synthesis During Drought and Subsequent Rehydration in *Tortula ruralis*," *Plant Physiol.* 83:816–819 (1987).

Diamond, L., and H.F. Clark. "Comparative Studies on the Interaction of Benzo[a]pyrene with Cells Derived from Poikilothermic and Homeothermic Vertebrates. I. Metabolism of Benzo[a]pyrene," *J. Natl. Cancer Inst.* 45:1005–1011 (1970).

DiTomaso, J.M., J.E. Shaff and L.V. Kochian. "Putrescine-induced Wounding and its Effects on Membrane Integrity and Ion Transport Processes in Roots of Intact Corn Seedlings," *Plant Physiol.* 90:988–995 (1989).

Dunn, B.P., J.J. Black and A. Maccubbin. "[32]P-postlabeling Analysis of Aromatic DNA Adducts in Fish from Polluted Areas," *Cancer Res.* 47:6543–6548 (1987).

Ehrenberg, L., E. Moustacchi, S. Osterman-Golkar and G. Ekman. "Dosimetry of Genotoxic Agents and Dose-response Relationship of Their Effects," *Mutat. Res.* 123:121–182 (1983).

Elder, G.H., and A.J. Urquhart. "Porphyrin Metabolism as a Target of Exogenous Chemicals," in *Occupational and Environmental Chemical Hazards: Cellular and Biochemical Indices for Monitoring Toxicity*, V. Foa, E.A. Emmett, M. Maroni and A. Colombi, Eds. (New York: Wiley-Interscience, 1987) pp. 221–230.

Esterbauer, H., and D. Grill. "Seasonal Variation of Glutathione and Glutathione Reductase in Needles of *Picea abies*," *Plant Physiol.* 61:119–121 (1978).

Ferrante A., C.M Rzepczyk and A.C. Allison. "Polyamine Oxidase Mediates Intra-erythrocytic Death of *Plasmodium falciparum*," *Trans. R. Soc. Trop. Med. Hyg.* 77:789–791 (1983).

Flores, H.E., and P. Filner. "Polyamine Catabolism in Higher Plants: Characterization Pyrroline Dehydrogenase," *Plant Growth Reg.* 3:277–291 (1985).

Fox, G.A., S.W. Kennedy, R.J. Norstrom and D.C. Wigfield. "Porphyria in Herring Gulls: A Biochemical Response to Chemical Contamination of Great Lakes Food Chains," *Environ. Toxicol. Chem.* 7:831–840 (1988).

Galston, A.W. "Polyamines and Plant Responses to Stress," in *The Physiology of Polyamines, Vol. 1*, U. Bachrach, and Y.M. Heimer, Eds. (Boca Raton, FL: CRC Press, Inc., 1989) pp. 99–106.

Gingerich, W.H. "Tissue Distribution and Elimination of Rotenone in Rainbow Trout," *Aquat Toxicol.* 6:179–196 (1986).

Glickman, A.H., C.N. Statham, A. Wu and J.J. Lech. "Studies on the Uptake, Metabolism and Disposition of Pentachlorophenol and Pentachloroanisole in Rainbow Trout," *Toxicol. Appl. Pharmacol.* 41:649–658 (1977).

Goldstein, J.A., P. Hickman, H. Bergman and J.G. Vos. "Hepatic Porphyria Induced by 2,3,7,8-Tetrachlorodibenzo-p-Dioxin in the Mouse," *Res. Commun. Chem. Pathol. Pharmacol.* 6:919–928 (1973).

Gossett, R.W. "Measurement of Oxygenated Metabolites of DDTs and PCBs: A Caution," *Mar. Environ. Res.* 26:155–159 (1988).

Gossett, R.W., S.R. McHugh, P. Szalay, K.D. Rosenthal and D.A. Brown. "Xenobiotic Organics and Biological Effects in Scorpionfish Caged Near a Southern California Municipal Outfall," *Mar. Environ. Res.* 14:449–450 (1984).

Grill, E., W. Gekeler, E.-L. Winnacker and H.H. Zenk. "Homo-phytochelatins are heavy metal-binding peptides of homo-glutathione containing Fabales," *FEBS Letts*. 205:47–50 (1986).

Grill, E., E.L. Winnacker and M.H. Zenk. "Phytochelatins, a Class of Heavy-metal Binding Peptides from Plants are Functionally Analogous to Metallothioneins," *Proc. Nat. Acad. Sci. U.S.A.* 84:439–443 (1987).

Grill, E. "Phytochelatins in Plants," in *Metal Ion Homeostasis: Molecular Biology and Chemistry*, D.H. Hamer, and D.R. Winge, Eds. (New York: Alan R. Liss, 1989) pp. 283–300.

Guiney, P.D., M.J. Melancon, Jr., J.J. Lech and R.E. Peterson. "Effects of Egg and Sperm Maturation and Spawning on the Distribution and Elimination of a Polychlorinated Biphenyl in Rainbow Trout *(Salmo gairdneri)*," *Toxicol. Appl. Pharmacol.* 47:261–272 (1979).

Halliwell, B., and J.M.C. Gutteridge. *Free Radicals in Biology and Medicine*, (Oxford: Clarendon Press, 1985) p. 346.

Harris, C.C., B.F. Trump, R.C. Grafstrom and H. Autrup. "Differences in Metabolism of Chemical Carcinogens in Cultured Human Epithelial Tissues and Cells," *J. Cell Biochem.* 18:285–294 (1982).

Hausladen, A., N.R. Madamanchi, S. Fellows, R.G. Alscher and R.G. Amundson. "Seasonal Changes in Antioxidants in Red Spruce as Affected by Ozone," *New Phytologist* 115:447–458 (1990).

Heath, R.L. "The Biochemistry of Ozone Attack on the Plasma Membrane of Plant Cells," *Rec. Adv. Phytochem.* 21:29–54 (1987).

Henderson, P.T., R. Van Doorn, C.M. Leijdekkers and R.P. Bos. "Excretion of Thioethers in Urine after Exposure to Electrophilic Chemicals," *IARC Sci. Publ.* 59:173–187 (1984).

Jakoby, W.B., and J.H. Keen. "A Triple-threat in Detoxication: The Glutathione-S-transferases," *Trends Biochem. Sci.* 2:229–231 (1977).

Johnson, L.L., E. Casillas, T.K. Collier, B.B. McCain and U. Varanasi. "Contaminant Effects on Ovarian Development in English Sole *(Parophrys vetulus)* from Puget Sound, Washington," *Can. J. Fish. Aquat. Sci.* 45:2133–2146 (1988).

Juedes, M.J., W.H. Bulger and D. Kupfer. "Monooxygenase-mediated Activation of Chlorotrianisene (TACE) in Covalent Binding to Rat Hepatic Microsomal Proteins," *Drug Metab. Dispos.* 15:786–793 (1987).

Kan, B., I.M. London and D.H. Levin. "Role of Reversing Factor in the Inhibition of Protein Synthesis Initiation Caused by Oxidised Glutathione," *J. Biol. Chem.* 263:15652–15656 (1988).

Kawr-Sawheney, R., and A.W. Galston. "On the Physiological Significance of Polyamines in Higher Plants," in *Recent Developments in Plant Sciences*, S.P. Sen, Ed. (New Delhi: Today and Tomorrow's Printers and Publishers, 1982) pp. 129–144.

Kennedy, S.W., D.C. Wigfield and G.A. Fox. "Tissue Porphyrin Pattern Determination by High-speed High-performance Liquid Chromatography," *Anal. Biochem.* 157:1–7 (1986).

Kimmerer, T.W., and T.T. Koslowski. "Ethylene, Ethane, Acetaldehyde and Ethanol Production by Plants Under Stress," *Plant Physiol.* 69:840–847 (1982).

Kobayashi, K. "Metabolism of Pentachlorophenol in Fishes," in *Pentachlorophenol: Chemistry, Pharmacology, and Environmental Toxicology*, K.R. Rao, Ed. (New York: Plenum Press, 1978) pp. 89–95.

Kobayashı, K. "Metabolism of Pentachlorophenol in Fish," in *Pesticide and Xenobiotic Metabolism in Aquatic Organisms*, M.A.Q. Kahn, J.J. Lech and J.J. Menn, Eds. (Washington, D.C.: American Chemical Society, 1979) pp. 131–143.

Koss, G., S. Seubert, A. Seubert, W. Koransky, P. Krauss and H. Ippen. "Conversion Products of Hexachlorobenzene and Their Role in the Disturbance of Triporphyrin Pathways in Rats," *Int. J. Biochem.* 12:1003–1006 (1980).

Krahn, M.M., D.G. Burrows, W.D. MacLeod, Jr. and D.C. Malins. "Determination of Individual Metabolites of Aromatic Compounds in Hydrolyzed Bile of English Sole (*Parophrys retulus*) from Polluted Sites in Puget Sound, Washington," *Arch. Environ. Contam. Toxicol.* 16:511–522 (1987).

Krahn, M.M., L.J. Kittle, Jr., and W.D. Macleod, Jr. "Evidence for Exposure of Fish to Oil Spilled into the Columbia River," *Mar. Environ. Res.* 20:291–298 (1986).

Krahn, M.M., and D.C. Malins. "Gas Chromatographic-mass Spectrometric Determination of Aromatic Hydrocarbon Metabolites from Livers of Fish Exposed to Fuel Oil," *J. Chromatog.* 248:99–107 (1982).

Krahn, M.M., M.S. Myers, D.G. Burrows and D.C. Malins. "Determination of Metabolites of Xenobiotics in the Bile of Fish from Polluted Waterways," *Xenobiotica* 14(8):633–646 (1984).

Kruzynski, G.M. "Some Effects of Dehydroabietic Acid (DHA) on Hydromineral Balance and Other Physiological Parameters in Juvenile Sockeye Salmon (*Oncorhynchus nerka*). PhD Thesis, University of British Columbia, Vancouver, Canada (1979).

Lamoureux, G.L., and D.G. Rusness. "The Role of Glutathione and Glutathione-S-transferases in Pesticide Metabolism, Selectivity, and Mode of Action in Plants and Insects, in *Glutathione: Chemical, Biochemical, and Medical Aspects*, D. Dolphin, O. Avramovic and R. Poulson, Eds. (New York: Wiley-Interscience, 1989).

Lech, J.J. "Isolation and Identification of 3-trifluoromethyl-4-nitrophenyl glucuronide from bile of rainbow trout exposed to 3-Trifluoromethyl-4-nitrophenol," *Toxicol. Appl. Pharmacol.* 24:114–124 (1973).

Lee, R.F. "Metabolism and Accumulation of Xenobiotics Within Hepatopancreas Cells of the Blue Crab, *Callinectes sapidus*," *Mar. Environ. Res.* 28:93–97 (1989).

Lee, R.F., R. Sauerheber and G.H. Dobbs. "Uptake, Metabolism and Discharge of Polycyclic Aromatic Hydrocarbons by Marine Fish," *Mar. Biol.* 17:201–208 (1972).

Lund, J., I. Brandt, L. Poellinger, A. Bergman, E. Klasson-Wehler and J.-A. Gustafsson. "Target Cells for the Polychlorinated Biphenyl Metabolite 4,4'-bis(methylsulfonyl)-2,2',5,5'-tetrachlorobiphenyl. Characterization of High Affinity Binding in Rat and Mouse Lung Cytosol," *Mol. Pharmacol.* 27:314–323 (1985).

Malins, D.C., B.B. McCain, D.W. Brown, S.-L. Chan, M.S. Myers, J.T. Landahl, P.G. Prohaska, A.J. Friedman, L.D. Rhodes, D.G. Burrows, W.D. Gronlund and H.O. Hodgins. "Chemical Pollutants in Sediments and Diseases of Bottom-dwelling Fish in Puget Sound, Washington," *Environ. Sci. Technol.* 18:705–713 (1984).

Malins, D.C., M.S. Meyers and W.T. Roubal. "Organic Free Radicals Associated with Idiopathic Liver Lesions of English Sole (*Parophrys vetulus*) from Polluted Marine Environments," *Environ. Sci. Technol.* 17:679–685 (1983).

Marks, G.S. "Exposure to Toxic Agents: The Heme Biosynthetic Pathway and Hemoproteins as Indicator," *CRC Crit. Rev. Toxicol.* 15:151–179 (1985).

Marton, L.J., and D.R. Morris. "Molecular and Cellular Functions of the Polyamines," in

Inhibition of Polyamine Metabolism: Biological Significance and Basis for New Therapies, P.P. McCann, A.E. Pegg and A. Sjoerdsma, Eds. (New York: Academic Press, Inc., 1987) pp. 305–316.

Matringe, M., and R. Scalla. "Studies on the Mode of Action of Acifluorfen-methyl in Nonchlorophyllous Soybean Cells," *Plant Physiol.* 86:619–622 (1988).

Mehlhorn, H., and A.R. Wellburn. "Stress Ethylene Formation Determines Plant Sensitivity to Ozone," *Nature (London)* 327:417–418 (1987).

Melancon, M.J., Jr., and J.J. Lech. "Isolation and Identification of a Polar Metabolite of Tetrachlorobiphenyl from Bile of Rainbow Trout Exposed to ^{14}C-tetrachlorobiphenyl," *Bull. Environ. Contam. Toxicol.* 15:181–188 (1976).

Melancon, M.J., Jr., and J.J. Lech. "Uptake, Biotransformation, Disposition and Elimination of 2-Methylnaphthalene and Naphthalene in Several Fish Species," in *Aquatic Toxicology, Proceedings of the Second Annual Symposium on Aquatic Toxicology, ASTM STP 667,* L.L. Marking, and R.A. Kimerle, Eds. (Philadelphia: American Society for Testing and Materials, 1979) pp. 5–22.

Miller, A., III, M.C. Henderson and D.R. Buhler. "Cytochrome P-450 Mediated Covalent Binding of Hexachlorophene to Rat Tissue Proteins," *Mol. Pharmacol.* 14:323–336 (1978).

Miller, E.C., and J.A. Miller. "Mechanism of Chemical Carcinogenesis," *Cancer* 47:1055–1069 (1981).

Miller, E.C., and J.A. Miller. "Some Historical Perspectives on the Metabolism of Xenobiotic Chemicals to Reactive Electrophiles," in *Bioactivation of Foreign Compounds,* M.W. Anders, Ed. (Orlando, FL: Academic Press, 1985) pp. 3–28.

Murthy, M.S.S., C.J. Calleman, S. Osterman-Golkar, D. Segerback and K. Svensson. "Relationships Between Ethylation of Hemoglobin, Ethylation of DNA and Administered Amount of Ethyl Methanesulfonate in the Mouse," *Mutat. Res.* 127:1–8 (1984).

Mutoh, N., and Y. Hayashi. "Isolation of Mutants of *Schizosaccharomyces pombe* Unable to Synthesize Cadystin, Small Cadmium-binding Peptides," *Biochem. Biophys. Res. Commun.* 151:32–39 (1988).

Narasimhan, N., P.E. Weller, J.A. Buben, R.A. Wiley and R.P. Hanzlik. "Microsomal Metabolism and Covalent Binding of [^{3}H/^{14}C]-bromobenzene. Evidence for Quinones as Reactive Metabolites," *Xenobiotica* 18:491–499 (1988).

Nishanian, E.V., S.D. Cohen and E.H. Khairallah. "Glutathione S-transferases are not Major Acetaminophen Binding Proteins in Mouse Liver," *The Toxicologist* 9:49 (1989).

Notten, W.R.F., and P.T. Henderson. "Alterations in the Glucuronic Acid Pathway Caused by Various Drugs," *Int. J. Biochem.* 6:111–119 (1975).

Notten, W.R.F., and P.T. Henderson. "The Interaction of Chemical Compounds with the Functional State of the Liver. I. Alteration in the Metabolism of Xenobiotic Compounds and D-glucuronic Acid Pathway," *Int. Arch. Occup. Environ. Health* 38:197–207 (1977).

Oikari, A.O.J. "Metabolites of Xenobiotics in the Bile of Fish in Waterways Polluted by Pulpmill Effluents," *Bull. Environ. Contamin. Toxicol.* 36:429–436 (1986).

Oikari, A., and E. Anas. "Chlorinated Phenolics and Their Conjugates in the Bile of Trout (*Salmo gairdneri*) Exposed to Contaminated Waters," *Bull. Environ. Contamin. Toxicol.* 35:802–809 (1985).

Oikari, A., E. Anas, G. Kruzynski and B. Holmbom. "Free and Conjugated Resin Acids in the Bile of Rainbow Trout, *Salmo gairdneri,*" *Bull. Environ. Contam. Toxicol.* 33:233–240 (1984).

Oikari, A., and B. Holmbom. "Assessment of Water Contamination by Chlorophenolics and

Resin Acids with the Aid of Fish Bile Metabolites," in *Aquatic Toxicology and Environmental Fate, ASTM STP 921*, T.M. Poston, and R. Purdy, Eds. (Philadelphia: American Society for Testing and Materials, 1986) pp. 252–267.

Oikari, A., B. Holmbom, E. Anas, M. Millunpalo, G. Kruzynski and M. Castren. "Ecotoxicological Aspects of Pulp and Paper Mill Effluents Discharged to an Inland Water System: Distribution in Water, and Toxicant Residues and Physiological Effects in Caged Fish (*Salmo gairdneri*)," *Aquat. Toxicol.* 6:219–239 (1985).

Oikari, A., and T. Kunnamo-Ojala. "Tracing of Xenobiotic Contamination in Water with the Aid of Fish Bile Metabolites: A Field Study with Caged Rainbow Trout (*Salmo gairdneri*)," *Aquat. Toxicol.* 9:327–341 (1987).

Olsson, P.-E., C. Haux and L. Förlin. "Variations in Hepatic Metallothionein, Zinc and Copper Levels During an Annual Reproductive Cycle in Rainbow Trout, *Salmo gairdneri*," *Fish. Physiol. Biochem.* 3:39–47 (1987).

Overnell, J., R. McIntosh and T.C. Fletcher. "The Levels of Liver Metallothionein and Zinc in Plaice, *Pleuronectes platessa* L., During the Breeding Season, and the Effect of Oestradiol Injection," *J. Fish Biol.* 30:539–546 (1987).

Pain, D.J., and B.A. Rattner. "Mortality and Hematology Associated with the Ingestion of One Number Four Lead Shot in Black Ducks, *Anas rubripes*," *Bull. Environ. Contam. Toxicol.* 40:159–164 (1988).

Pegg, A.E., and P.P. McCann. "Polyamine Metabolism and Function in Mammalian Cells and Protozoans," *ISI Atlas Sci. Biochem.* 1:11–18 (1988).

Pereira, M.A., and L.W. Chang. "Binding of Chemical Carcinogens and Mutagens to Rat Hemoglobin," *Chem.-Biol. Interact.* 33:301–306 (1981).

Pereira, M.A., L.H.C. Lin and L.W. Chang. "Dose Dependency of 2-Acetylaminofluorene Binding to Liver DNA and Hemoglobin in Mice and Rats," *Toxicol. Appl. Pharmacol.* 60:472–478 (1981).

Perera, F.P. "Molecular Cancer Epidemiology: A New Tool in Cancer Prevention," *J. Natl. Cancer Inst.* 78:887–898 (1987).

Philip, D.H., and P. Sims. "Polycyclic Aromatic Hydrocarbon Metabolites: Their Reactions with Nucleic Acids," in *Chemical Carcinogens and DNA, Vol. 2*, P.L. Grover, Ed. (CRC Press, Boca Raton, FL, 1979) pp. 29–57.

Pompella, A., E. Maellaro, A.F. Casini, M. Ferrali, L. Ciccoli and M. Comporti. "Measurement of Lipid Peroxidation In Vivo: A Comparison of Different Procedures," *Lipids* 22:206–211 (1987).

Pritchard, J.B., A.M. Guarino and W.B. Kinter. "Distribution, Metabolism and Excretion of DDT and Mirex by a Marine Teleost, the Winter Flounder," *Environ. Health Perspect.* 4:45–54 (1973).

Pryor, W.A., Ed. *Free Radicals in Biology, Vol. V.* (New York: Academic Press, 1982) p. 283.

Rattner, B.A., W.J. Fleming and C.M. Bunck. "Comparative Toxicity of Lead Shot in Black Ducks (*Anas rubripes*) and Mallards (*Anas platyrhynchos*)," *J. Wild. Dis.* 25:175–183 (1989).

Reddy, M.V., R.C. Gupta, E. Randerath and K. Randerath. "^{32}P-postlabeling Test for Covalent DNA Binding of Chemicals In Vivo: Application to a Variety of Aromatic Carcinogens and Methylating Agents," *Carcinogenesis* 5:231–243 (1984).

Reddy, V R , W R Christenson and W N Piper. "Extraction and Isolation by High Performance Liquid Chromatography of Uroporphyrin and Coproporphyrin Isomers from Biological Tissues," *J. Pharmacol. Meth.* 17:51–58 (1987).

Rennenberg, H. "Aspects of Glutathione Function and Metabolism in Plants," in *Plant*

Molecular Biology, D. von Wettskein, and N.-H. Chua, Eds. (New York: Plenum Press, 1987) pp. 279–292.

Richards, F.J., and R.G. Coleman. "Occurrence of Putrescine in Potassium-deficient Barley," *Nature (London)* 170:460 (1952).

Rogers, I.H., I.K. Birtwell and G.M Kruzynski. "The Pacific Eulachon (*Thaleichthys pacificus*) as a Pollution Indicator Organism in the Fraser River Estuary, Vancouver British Columbia," *Sci. Total Environ.* 97/98: 713–727 (1990).

Roscoe, D.E., S.W. Nielson, A.A. Lamola and D. Zuckerman. "A Simple Quantitative Test for Erythrocytic Protoporphyrin in Lead-poisoned Ducks," *J. Wild. Dis.* 15:127–136 (1979).

Ruby, S.M., D.R. Idler and Y.P. So. "Changes in Plasma, Liver, and Ovary Vitellogenin in Landlocked Atlantic Salmon Following Exposure to Sublethal Cyanide," *Arch. Environ. Contam. Toxicol.* 16:507–510 (1987).

Sabourin, T.D., D.B. Gant and L.J. Weber. "The Influence of Metal and Nonmetal Stressors on Hepatic Metal-binding Protein Production in Buffalo Sculpin, *Enophrys bison*," in *Marine Pollution and Physiology: Recent Advances*, F.J. Vernberg, F.P. Thurberg, A. Calabrese and W. Vernberg, Eds. (Columbia, SC: University of South Carolina Press, 1985) pp. 247–266.

Sangalang, G.B., and H.C. Freeman. "Effects of Sublethal Cadmium on Maturation and Testosterone and 11-Ketotestosterone Production In Vivo in Brook Trout," *Biol. Reprod.* 11:429–435 (1974).

Schmidt, A., and K.J. Kunert. "Lipid Peroxidation in Higher Plants: The Role of Glutathione Reductase," *Plant Physiol.* 82:700–702 (1986).

Schmidt, G. and K.J. Kunert. *Antioxidative Systems: Defense Against Oxidative Damage in Plants*, UCLA Symp. Mol. Cell. Biol. New Ser. (New York: Alan R. Liss, 1987) pp. 401–414.

Segerback, D., C.J. Calleman, L. Ehrenberg, G. Loforth and S. Osterman-Golkar. "Evaluation of Genetic Risks of Alkylating Agents. IV. Quantitative Determination of Alkylated Amino Acids in Hemoglobin as a Measure of the Dose after Treatment of Mice with Methylmethanesulfonate," *Mutat. Res.* 49:71–82 (1978).

Shah, D.M., C.M. Hironaka, R.C. Wiegand, E.I. Harding, G.G. Krivi and D.C. Tiemeier. "Structural Analysis of a Maize Gene Coding for Flutathione-S-transferase Involved in Herbicide Detoxification," *Plant Mol. Biol.* 6:203–212 (1986).

Shimada, T. "Metabolic Activation of [^{14}C] Polychlorinated Biphenyl Mixtures by Rat Liver Microsomes," *Bull. Environ. Contam. Toxicol.* 16:25–32 (1976).

Shugart, L., J. McCarthy, B. Jimenez and J. Daniels. "Analysis of DNA Adduct Formation in the Bluegill Sunfish (*Lepomis macrochirus*) Between Benzo(a)pyrene and DNA of the Liver and Hemoglobin of the Erythrocyte," *Aquat. Toxicol.* 9:319–325 (1987).

Sikka, H.C., J.P. Rutkowski, C. Kandaswami, S. Kumar, R. Earley and R.C. Gupta. "Formation and Persistence of DNA Adducts in the Liver of Brown Bullheads Exposed to Benzo(a)pyrene," *Cancer Lett.* 49:81–87 (1990).

Smith, I.K., A.C. Kendall, A.J. Keys, J.C. Turner and P.J. Lea. "Increased Levels of Glutathione in a Catalase-deficient Mutant of Barley (*Hordeum vulgare* L.)," *Plant Sci. Lett.* 37:29–33 (1984).

Smith, I.K., A.C. Kendall, A.J. Keys, J.C. Turner and P.J. Lea. "The Regulation of the

Biosynthesis of Glutathione in Leaves of Barley (*Hordeum vulgare* L.)," *Plant Sci.* 41:11–17 (1985).

Smith, T.A. "Polyamines," *Ann. Rev. Plant Physiol.* 36:117–143 (1985).

Solbakken, J.E., K. Ingebrigtsen and K.H. Palmork. "Comparative Study on the Fate of the Polychlorinated Biphenyl 2,4,5,2',4',5'-hexachlorobiphenyl and the Polycyclic Aromatic Hydrocarbon Phenanthrene in Flounder (*Platichthys flesus*) Determined by Liquid Scintillation Counting and Autoradiography," *Mar. Biol.* 83:239–246 (1984).

Solbakken, J.E., and K.H. Palmork. "Metabolism of Phenanthrene in Various Marine Animals," *Comp. Biochem. Physiol.* 70C:21–26 (1981).

Statham, C.N., M.J. Melancon and J.J. Lech. "Bioconcentration of Xenobiotics in Trout Bile: A Proposed Monitoring Aid for Some Waterborne Chemicals," *Science* 193:680–681 (1976).

Statham, C.N., S.K. Pepple and J.J. Lech. "Biliary Excretion Products of 1-[1-^{14}C]naphthyl-N-methylcarbamate (Carbaryl) in Rainbow Trout (*Salmo gairdneri*)," *Drug Metab. Disp.* 3:400–406 (1975).

Steffens, J.C. "Heavy Metal Stress and the Phytochelatin Response," in *Stress Responses in Plants: Adaptation and Acclimation Mechanisms*, R.G. Alscher, and J.R. Cumming, (New York:Wiley-Liss, 1990) pp. 377–394.

Steffens, J.C., D.F. Hunt and B.G. Williams. "Accumulation of Non-protein Metal-binding Polypeptides (Gamma-glutamyl-cysteinyl)$_n$-glycine in Selected Cadmium-resistant Tomato Cells," *J. Biol. Chem.* 261: 13879–13882 (1986).

Stein, J.E., T. Hom, E. Casillas, A. Friedman and U. Varanasi. "Simultaneous Exposure of English Sole (*Parophrys vetulus*) to Sediment-associated Xenobiotics. II. Chronic Exposure to an Urban Estuarine Sediment with Added ^{3}H-benzo[a]pyrene and ^{14}C-polychlorinated Biphenyls," *Mar. Environ. Res.* 22:123–149 (1987).

Stickel, L.F. "Pesticide Residues in Birds and Mammals," in *Environmental Pollution by Pesticides*, Edwards, C.A., Ed. (London: Plenum Press, 1973) pp. 254–312.

Strik, J.J.T.W.A. "Porphyrins in Urine as an Indication of Exposure to Chlorinated Hydrocarbons," *Ann. N.Y. Acad. Sci.* 320:308–310 (1979).

Sunderman, F.W., Jr. "Biochemical Indices of Lipid Peroxidation in Occupational and Environmental Medicine," in *Occupational and Environmental Chemical Hazards: Cellular and Biochemical Indices for Monitoring Toxicity*, V. Foa, E.A. Emmett, M. Maroni and A. Colomi, Eds. (New York: Wiley-Interscience, 1987) pp. 151–158.

Suresh, M.R., and P.R. Adiga. "Diamine Oxidase of *Lathyrus sativus* Seedlings: Purification and Properties," *J. Biosci.* 1:109–124 (1979).

Tanaka, K., H. Saji and N. Kondo. "Immunological Properties of Spinach Glutathione Reductase and Inductive Biosynthesis of the Enzyme with Ozone," *Plant Cell Physiol.* 29:637–642 (1988).

Tannenbaum, S.R., P.L. Skipper, L.C. Green, M.W. Obiedzinski and F. Kadlubar. "Blood Protein Adducts as Monitors of Exposure to 4-aminobiphenyl," *Proc. Am. Assoc. Cancer Res.* 24:69 (1983).

Thakker, D.R., H. Yagi, W. Levin, A.W. Wood, A.H. Conney and D.M. Jerina. in *Bioactivation of Foreign Compounds*, Anders, M.W., Ed., (Orlando, FL: Academic Press, 1985) pp. 177–242.

Thomas, P. "Reproductive Endocrine Function in Female Atlantic Croaker Exposed to Pollutants," *Mar. Environ. Res.* 24:179–183 (1989).

Thomas, P. "Molecular and Biochemical Responses of Fish to Stresses and Their Potential Use in Environmental Monitoring," in *Biological Indicators of Stress in Fish*, S.M. Adams, Ed. (Bethesda, MD: American Fisheries Society, 1990) pp. 9–28.

Thorgeirsson, S., I.B. Glowinski and M.E. McManus. "Metabolism, Mutagenicity and Carcinogenicity of Aromatic Amines," *Rev. Biochem. Toxicol.* 5:349–386 (1983).

Tiburcio, A.F., M.A. Masdeu, F.M. Dumortier and A.W. Galston. "Polyamine Metabolism and Osmotic Stress: I. Relation to Protoplast Viability," *Plant Physiol.* 82:369–374 (1986).

van Doorn, C.M. Leijdekkers, R.P. Bos, R.M.E. Brouns and P.Th. Henderson. "Detection of Human Exposure to Electrophilic Compounds by Assay of Thioether Detoxification Products in Urine," *Ann. Occup. Hyg.* 24:77–92 (1981).

van Kreijl, C.F., D. deZwart and W. Slooff. "Bioconcentration, Detoxification, and Excretion of Mutagenic River Pollutants in Fish Bile," in *Aquatic Toxicology and Environmental Fate. ASTM STP 921*, T.M. Poston, and R. Purdy, Eds. (Philadelphia: American Society for Testing and Materials, 1986) pp. 268–276.

van Ommen, B., P.J. van Bladeren, J.H.M. Temmink, L. Brader and F. Muller. "The Microsomal Oxidation of Hexachlorobenzene to Pentachlorophenol; Effect of Inducing Agents," in *Hexachlorobenzene: Proceedings of an International Symposium*, C.R. Morris, and J.R.P. Cabral, Eds. (Lyons, France: International Agency for Research on Cancer, 1986) pp. 323–324.

Varanasi, U., and D.J. Gmur. "Metabolic Activation and Covalent Binding of Benzo[a]pyrene to Deoxyribonucleic Acid Catalyzed by Liver Enzymes of Marine Fish," *Biochem. Pharmacol.* 29:753–761 (1980).

Varanasi, U., and D.J. Gmur. "Hydrocarbons and Metabolites in English Sole (*Parophrys vetulus*) Exposed Simultaneously to (^{3}H)-benzo[a]pyrene and (^{3}H)-naphthalene in Oil-contaminated Sediment," *Aquat. Toxicol.* 1:49–68 (1981).

Varanasi, U., D.J. Gmur and P.A. Treseler. "Influence of Time and Mode of Exposure on Biotransformation of Naphthalene by Juvenile Starry Flounder (*Platichthys stellatus*) and Rock Sole (*Lepidopsetta bilineata*)," *Arch. Environ. Contam. Toxicol.* 8:673–692 (1979).

Varanasi, U., M. Nishimoto, W.L. Reichert and J.E. Stein. "Metabolism and Subsequent Covalent Binding of Benzo[a]pyrene to Macromolecules in Gonads and Liver of Ripe English Sole (*Parophrys vetulus*)," *Xenobiotica* 12:417–425 (1982).

Varanasi, U., W.L. Reichert, B.L. Eberhart and J.E. Stein. "Formation and Persistence of Benzo(a)pyrene-diol Epoxide-DNA Adducts in Livers of English Sole (*Porphrys vetulus*)," *Chem.-Biol. Interact.* 69:203–216 (1989a).

Varanasi, U., W.L. Reichert and J.E. Stein. "^{32}P-Postlabeling Analysis of DNA Adducts in Liver of Wild English Sole (*Parophrys vetulus*) and Winter Flounder (*Pseudopleuronectes americanus*)," *Cancer Res.* 49:1171–1177 (1989b).

Varanasi, U., J.E. Stein, M. Nishimoto and T. Hom. "Benzo[a]pyrene Metabolites in Liver, Muscle, Gonads and Bile of Adult English Sole (*Parophrys vetulus*)," in *Polynuclear Aromatic Hydrocarbons: Formation, Metabolism and Measurement, Seventh International Symposium*, M. Cooke, and A.J. Dennis, Eds. (Columbus, OH.: Battelle Press, 1983) pp. 1221–1234.

Wang, C.Y., S.Y. Wang and A.R. Wellburn. "Role of Ethylene Under Stress Conditions," in *Stress Responses in Plants: Adaptation and Acclimation Mechanisms*, R.G. Alscher, and J.R. Cumming, Eds. (New York: Alan R Liss, 1990) pp. 147–174.

Weller, P.E., N. Narasimhan, J.A. Buben and R.P. Hanzlik. "In Vitro Metabolism and Covalent Binding Among Ortho-Substituted Bromobenzenes of Varying Hepatotoxicity," *Drug Metab. Dispos.* 16:232–237 (1988).

Weston, A., R.M. Hodgson, A.J. Hewer, R. Kurode and P.L. Grover. "Comparative Studies of the Metabolic Activation of Chrysene in Rodent and Human Skin," *Chem.-Biol. Interact.* 54:223–242 (1985).

Wingate, V.P.M., M.A. Lawton and C.J. Lamb. "Glutathione Causes a Massive and Selective Induction of Plant Defense Genes," *Plant Physiol.* 87:206–210 (1988).

Wise, R.R., and A.W. Naylor. "Chilling-enhanced Photooxidation Evidence for the Role of Singlet Oxygen and Superoxide in the Breakdown of Pigments and Endogenous Antioxidants," *Plant Physiol.* 83:278–282 (1987).

Wofford, H.W., and P. Thomas. "Peroxidation of Mullet and Rat Liver Lipids In Vitro: Effects of Pyridine Nucleotides, Iron, Incubation Buffer, and Xenobiotics," *Comp. Biochem. Physiol. C Comp. Pharmacol. Toxicol.* 89C:201:206 (1988).

Yang, S.F., and N.E. Hoffman. "Ethylene Biosynthesis and its Regulation in Higher Plants," *Ann. Rev. Plant Physiol.* 35:155–189 (1984).

Young, N.D., and A.W. Galston. "Putrescine and Acid Stress," *Plant Physiol.* 71:767–771 (1983).

CHAPTER 3

DNA Alterations

Lee Shugart, John Bickham, Gene Jackim, Gerald McMahon, William Ridley, John Stein, and Scott Steinert

The exposure of an organism to genotoxic chemicals may induce a cascade of genetic events. Initially, structural alterations to DNA are formed. Next, the DNA damage is processed and subsequently expressed in mutant gene products. Finally, diseases result from the genetic damage. The detection and quantitation of various events in this sequence may be employed as biomarkers of exposure and effects in organisms exposed to genotoxic agents in the environment.

The methods to monitor exposure to genotoxic chemicals are based upon either: (1) direct measurement of DNA structural change; (2) the detection of their consequences by measuring DNA repair directly or indirectly; or (3) the production of mutations in the genome of the exposed organisms. Adducts are a type of structural change involving covalent attachments of a chemical or its metabolite to DNA. Several analytical procedures are currently used to monitor the formation of DNA adducts directly; the most prominent of these procedures is the ^{32}P-postlabeling technique. A more general approach involves the detection of DNA strand breaks that are produced either directly by the toxic chemical or by the processing of structural damage. The detection of this type of damage is facilitated by the alkaline unwinding assay. The current methods available for monitoring the consequences of DNA damage are based upon changes to chromosomes, such as aberrations, formation of micronuclei, detection of abnormal distribution of DNA within cells, and the activation of proto-oncogenes.

The activation of oncogenes and inactivation of tumor suppressor genes are areas of research that should receive greater emphasis in the future.

INTRODUCTION

Damage to DNA has been proposed as a useful variable for assessing the genotoxic properties of environmental pollutants (Kohn 1983; Committee on Biological Markers of the National Research Council 1987). The rationale underlying this statement is the observation that many of these pollutants are chemical carcinogens and mutagens with the capacity to cause various types of DNA damage (Wogan and Gorelick 1985). Recent studies have delineated many of the cellular mechanisms associated with DNA lesions produced by genotoxic chemicals. The cellular metabolism of genotoxic chemicals can be a relatively complex phenomenon, and the lack of complete detoxication can sometimes lead to the formation of highly reactive electrophilic metabolites (Phillips and Sims 1979; Harvey 1982; Phillips 1983; Beland and Kadlubar 1985). Nucleophilic centers in macromolecules such as lipids, proteins, DNA, and RNA undergo attack by these intermediates; this often results in cellular toxicity. Interaction with DNA is manifested primarily by structural alterations to the DNA molecule and can take the form of adducts, strand breakage, or chemically altered bases. The consequence of these structural perturbations can be innocuous due to repair of the damage or death of the cells containing the damaged DNA. However, those lesions not properly repaired may result in alterations that are fixed and eventually transmitted into daughter cells. Current theory suggests that the reaction of chemicals with DNA and the changes that result may cause deleterious pathological conditions, such as tumor formation (Harvey 1982).

Structural alterations caused by toxic chemicals are endpoints (biomarkers) of exposure. These lesions may potentiate irreversible changes to the DNA molecule and thereby result in the expression of other cellular responses such as chromosomal aberrations and oncogene activation. These responses can be considered biomarkers of deleterious effects on the continuum of cellular change from normal to transformed.

The concepts discussed can be depicted as a model (Figure 1) that reflects events and cellular processes related to DNA integrity (Shugart 1990a). The events and processes shown as Pathways 1 through 6 in Figure 1 are grouped into one of three categories: insult, repair, or synthesis.

DNA is present in cells as a functionally stable, double-stranded entity without discontinuity (strand breaks) or abnormal structural modifications. As such, it is considered to have high integrity. The rigid maintenance of this integrity is important for survival and is reflected in the low mutational rate observed in living organisms, estimated to be on the order of one mutation per average gene per 200,000 years

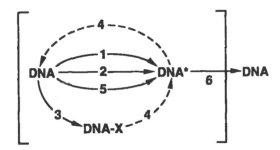

INSULT	REPAIR	SYNTHESIS
1. Normal wear and tear	4. Incision	5. Replication
2. UV and ionizing radiation	excision	6. Postreplication
(γ and X irradiation)	resynthesis	modification
3. Chemical	ligation	

DNA: Normal double-stranded DNA with no strand breaks.
DNA-X: Chemically modified DNA.
DNA*:   DNA with strand breaks.

Figure 1.   A schematic representation of the status of DNA in relation to insults that disrupt DNA integrity and cellular processes that maintain DNA integrity (Shugart 1990a).

(Alberts et al. 1989). DNA damage can occur as the result of the following insults:

- wear and tear of normal cellular events such as metabolism and random thermal collisions (Pathway 1)
- interaction with physical agents such as UV light and ionizing radiation (Pathway 2)
- interaction with chemical agents (Pathway 3)

These events give rise to structural alterations that, while usually repaired rapidly, produce a transient population of DNA with strand breaks and low integrity. Some chemicals work through free-radical mechanisms and cause strand breaks directly; others interfere with the fidelity of normal repair or DNA modifications (Pathway 6). The loss of bases from the DNA molecule (abasic sites) are frequent lesions that occur as a result of random thermal collisions or breakdown of chemically unstable adducts. Even the normal cellular process of replication (Pathway 5) produces DNA with transient strand breaks. Therefore, at any one time, a background level of DNA with low integrity (DNA with various types of structural alterations) may exist in the cell. Fortunately, most cells have DNA repair mechanisms (Sancar and Sancar 1988) that under normal circumstances efficiently eliminate DNA of low integrity (Pathway 4).

Table 1.   Concept of Biomarkers in Relation to Genotoxic Agents.

Event	Biomarker
Exposure	
(a) Presence of xenobiotic substance or interactive product	• Adduct formation
(b) Secondary concomitant reactions	• DNA strand breakage
	• Base composition
(c) Irreversible reactions	• Oncogene activation
	• Cytogenetic effect
Effect	
(a) Genotoxic diseases	• Tumorigenesis
	• Kurelec's syndrome

CURRENT METHODOLOGIES

This section considers methods currently used to detect exposure to genotoxic chemicals. These include adduct formation, secondary modifications, such as DNA strand breakage and changes in base composition, and irreversible events, such as oncogene activation and cytogenetic effects.

Background

Detection of DNA alterations as an in situ biomarker of environmental genotoxicity is a general approach that has had limited use in the past. Early attempts most often involved direct observation of neoplasms and chromosomal aberrations in various species including plants, wild terrestrial mammals, and aquatic vertebrates (Sandhu and Lower 1989). However, with the recent advent of knowledge gained from studies on mechanisms of chemical carcinogenesis at the cellular level, renewed and expanded efforts have been made. These efforts have been facilitated by new analytical techniques that detect and measure the interaction of toxic chemicals with DNA (Wogan and Gorelick 1985; Bartsch et al. 1988; Santella 1988).

Table 1 relates genotoxic agents to the concept of biomarkers. It shows the temporal occurrence of DNA alterations and the anticipated sequence of cellular events that ensues after exposure to toxic chemicals. The rationale for selecting a given DNA alteration or response for examination is also provided. We emphasize that the events listed are points on a continuum that are subject to change and redefinition as our knowledge in this area increases. An organism's inability (whether permanent or transient) to cope with DNA damage and to maintain DNA integrity provides the investigator the opportunity to test for genotoxicity of chemicals in the environment.

DNA Adducts

Soon after an organism is exposed to toxic chemicals, the presence of an exogenous substance or its interactive product may be detected in the form of a covalently bound DNA adduct. Detecting and quantifying DNA adducts are not simple tasks because the analytical techniques currently available are limited in their sensitivity or specificity. Difficulties arise because:

- the assay often has to accommodate the unique chemical and/or physical properties of the individual compound or its adduct form
- the percentage of the total chemical that becomes covalently attached to the DNA in the target tissue is very small, either because most of it is usually detoxicated and excreted, or the toxic effects of these chemicals cause the loss of the target cell and hence its complement of adducts
- not all adducts that form between the genotoxic agent and DNA are stable or are involved in the development of subsequent deleterious events in the organism
- the amount of DNA available for analysis from a given target tissue or organ is often quite limited

Currently, methods of varying sensitivity exist to measure DNA adducts and include ^{32}P-postlabeling, high performance liquid chromatography (HPLC)/ fluorescence spectrophotometry, and immunoassays using adduct-specific polyclonal or monoclonal antibodies.

^{32}P-Postlabeling

The ability to detect and quantitate DNA adducts in organisms exposed to complex mixtures of genotoxic compounds was recently made possible by the development of the ^{32}P-postlabeling technique (Randerath et al. 1981; reviewed by Watson 1987). Presently, this technique shows particular promise because it has a very low limit of detection (one adduct in 10^9 to 10^{10} nucleotides), does not require characterization of individual adducts before they are measured, and has been validated in field studies with fish (Dunn et al. 1987; Varanasi et al. 1989a). Another important advantage is that ^{32}P-postlabeling is a nonspecific procedure that can detect a variety of bulky aromatic adducts in animals exposed to a complex mixture of contaminants.

In the ^{32}P-postlabeling procedure (summarized by Gupta and Randerath 1988), DNA is isolated from cells or tissues by a procedure involving digestion of proteins and RNA with Proteinase K and RNases, respectively, and extractions with phenol and chloroform:isoamyl alcohol (24:1). This method provides DNA that is virtually free of RNA and protein, quickly and with good yield (1 to 2 μg/mg liver). After isolation, the DNA is enzymatically hydrolyzed to 3'-monophosphates of normal DNA nucleotides and adducts. The adducts are then enriched relative to the normal nucleotides either enzymatically, by removing the normal 3'-monophosphates using nuclease-P_1 (Reddy and Randerath 1986), or by extracting the adducts into n-butanol (Gupta 1985). Both techniques increase the sensitivity of the assay by allowing analysis of larger quantities of DNA (up to 60 μg rather than less than 0.2 μg). An advantage of adduct enrichment using n-butanol, however, is that it is less dependent on the chemical nature of the adducts than is adduct enrichment using nuclease-P_1 (Gupta and Early 1988). This finding is particularly relevant for organisms exposed to complex mixtures of contaminants. On the other hand, neither enrichment method is appropriate for use with adducts arising from low molecular weight compounds.

Following adduct enrichment, the adducted 3'-mononucleotides and any residual normal 3'-nucleotides are radiolabeled with ^{32}P by T4 polynucleotide kinase-catalyzed transfer of the terminal phosphate from [$\gamma^{32}P$] ATP to the 5' position of the 3'-mononucleotide leading to [5'-^{32}P]-3',5'-biphosphates. The remaining normal nucleotides and adducts are then separated by multi-dimensional, thin-layer chromatography (TLC) on polyethyleneimine cellulose TLC sheets. Finally, the adducts are detected by autoradiography and quantitated by scintillation counting.

The ^{32}P-postlabeling assay is slightly more laborious than some of the other procedures discussed here and involves the use of high specific activity ^{32}P, which necessitates special precautions to minimize radiation exposure. However, the high sensitivity of the ^{32}P-postlabeling assay for detecting a broad range of hydrophobic xenobiotic compounds adducted to DNA, as previously discussed, offsets some of these technical considerations.

Currently, the quantitation of adducts using the ^{32}P-postlabeling assay can be considered to be only semiquantitative and generally varies from one laboratory to another. However, the recent commercial availability of chemically synthesized DNA adducts (benzo[a]pyrene-diol-epoxide deoxyguanosine-3'-monophosphate and 5-methyl chrysene-diol-epoxide deoxyguanosine-3'-monophosphate) will allow incorporation of quality assurance procedures, both with adduct labeling and with TLC. Moreover, the recent development of improved radioactivity imaging technologies (e.g., the storage phosphor technology), which do not require the use of X-ray film, offer a greater range of linearity in response for detecting ^{32}P-derived radioactivity on the TLC sheets and allows for computer-assisted analysis and archiving of the distribution of the radioactivity on the individual chromatograms. These factors should improve the quantitation of adduct levels and patterns, as well as allow for retrospective analysis of the level of the radioactivity associated with individual spots or regions on the chromatograms. Currently, retrospective analysis of a DNA sample can only be done by reanalyzing another aliquot of DNA sample of interest by the full ^{32}P-postlabeling analysis. Finally, in TLC chromatograms of DNA from organisms exposed to complex mixtures of contaminants, a radioactive zone apparently representing multiple overlapping adducts is routinely observed, making it difficult to assign individual spots (adducts) to specific chemicals. Recent advances in labeling procedures, chromatographic conditions (Randerath et al. 1989; Gupta and Early 1988), and development of improved radioactivity imaging systems, suggest that the resolution and detection of multiple adducts can be enhanced, which should aid in characterizing individual adducts in organisms exposed to unknown mixtures of chemicals and, thus, may increase the chemical specificity of the ^{32}P-postlabeling assay.

Development of better methods for adduct detection and identification is continuing. One approach is to separate the ^{32}P-labeled adducts using HPLC instead of thin-layer chromatography to obtain better resolution and greater reproducibility (1988; Gorelick and Wogan 1989). Another method under investigation is to produce volatile derivatives of adducts for analysis and identification by gas chromatography/mass spectroscopy (Tomer et al. 1986). Liquid capillary separation technology is also being combined with fast atom bombardment/mass spectroscopy (FAB/MS) to directly identify adducted nucleosides (Weng et al. 1989). Capillary zone electrophoresis with various kinds of detectors is another very promising approach (Jackim and Norwood 1990). In theory, this latter technique should have the sensitivity of the ^{32}P-postlabeling technique but with greatly increased resolution.

Recent field studies (Dunn et al. 1987; Varanasi et al. 1989a; Stein et al. 1989) with benthic fish have validated the use of DNA adducts using the ^{32}P-postlabeling assay as a biomarker of exposure to genotoxic compounds. For example, comparing the levels of total hepatic DNA adducts in English sole from Puget Sound, Washington, to sediment levels of high molecular weight polycyclic aromatic hydrocarbons (PAHs) and bile levels of PAH-like metabolites revealed a general concordance among these variables, and suggests that the level of adducts reflects the degree of exposure (Varanasi et al. 1989a; Stein et al. 1989). Additionally, a recent study (Varanasi et al. 1989b) with English sole exposed to a single genotoxic PAH, benzo[a]pyrene, has demonstrated a linear dose response resulting in detection of hepatic DNA-benzo[a]pyrene adducts over a 50-fold dose range. In this study, no significant change in the level of DNA-benzo[a]pyrene adducts was observed for up to 60 d after a single exposure. These results, showing a linear dose response over a moderate dose range and the persistence of the adducts, indicate that the level of hepatic DNA adducts in wild fish may be representative of cumulative exposure to genotoxic compounds. This is an important feature of DNA adducts in fish because many xenobiotic compounds, such as PAHs, are readily metabolized and do not accumulate in tissues, and other biomarkers of exposure to PAHs, such as hepatic aryl hydrocarbon hydroxylase, rapidly return to basal levels after exposure to PAHs (Collier and Varanasi 1987).

The studies with English sole suggest that the postlabeling assay is a promising and sensitive method for assessing exposure of fish to genotoxic contaminants; however, the possibility exists that the efficacy of this assay for assessing contaminant exposure may be species dependent. Recent studies (Varanasi et al. 1989a; Stein et al. 1989) have evaluated the postlabeling assay with winter flounder, a fish species that appears to develop hepatic neoplasms from exposure to genotoxic environmental contaminants (Murchelano and Wolke 1985). The levels of total hepatic DNA adducts were generally related to the level of chemical contamination at the site of capture, as reflected by the sediment concentrations of high molecular weight PAH. These results were in general agreement with the results from studies with English sole, which show a relationship between exposure of anthropogenic chemicals and the presence of hepatic DNA adducts. The general similarity of results for winter flounder and English sole suggest that the use of hepatic DNA adducts as a biomarker of exposure to genotoxic compounds is not species dependent.

A recent study (Kurelec et al. 1989) found that five species of freshwater fish exhibited qualitatively similar adduct patterns irrespective of whether the fish were sampled from apparently unpolluted or polluted sites, leading the authors to suggest that DNA adducts observed were due to natural environmental factors. In this study, historical data on sediment levels of chemical contaminants and contamination of the fish were used to define polluted and unpolluted, rather than direct measurements on the fish used in the study. In contrast to the study by Kurelec et al. (1989), several recent studies (Dunn et al. 1990; Poginsky et al. 1990) and those cited above (Dunn et al. 1987; Stein et al. 1989; Varanasi et al. 1989a; Reichert et al. 1990) show that in several species of fish from both U.S. and European waters, substantially higher DNA adduct levels were detected in fish from contaminated sites than in fish from relatively uncontaminated sites. In fish clearly resident to relatively uncontaminated sites, the levels of DNA adducts detected by the ^{32}P-postlabeling method in fish from contaminated sites are of anthropogenic chemical origin, rather than being of natural-environmental or biologic origin. Overall these studies illustrate that careful selection of appropriate reference sites in conjunction with supporting analytical chemistry data for the organisms sampled is a critical factor in the current use of any biomarker.

HPLC/Fluorescence

An alternative method to the immunoassay or ^{32}P-postlabeling technique is detection of the binding of fluorescent xenobiotic chemicals to DNA (Rahn et al. 1982). This technique involves removal of the adduct from DNA (usually by acid hydrolysis) and separation by high performance liquid chromatography (HPLC) coupled with fluorescence analysis. Application of this technique is limited by the presence and release of an intact fluorescent moiety. Shugart et al. (1983) have shown that adduct formation at the femtomole level between the ubiquitous chemical carcinogen, benzo[a]pyrene, and DNA can be detected and quantitated by this method in both mice and fish exposed under laboratory conditions (Shugart and Kao 1985; Shugart et al. 1987). The technique has also been used in the field to detect benzo[a]pyrene adducts in the DNA of beluga whales from the St. Lawrence River in Canada (Martineau et al. 1988).

Immunological Techniques

The specificity and sensitivity of antibody-antigen reactions to detect DNA adducts have been exploited in studying DNA alterations. Both polyclonal and monoclonal antibodies have been produced against various types of chemicals adducted to DNA. Theoretically, the specificity of the antigen-antibody reactions should permit the detection of a single type of chemical adduct in DNA where many types exist.

Highly specific polyclonal and monoclonal antibodies have been developed to a number of modified DNAs with addition products varying in structure from ethyl and methyl groups to aromatic amines, polycyclic aromatic hydrocarbons, aflatoxin, and platinum-DNA complexes (Poirier 1984; Santella 1988). In general, antibody sensitivity is highest for the original antigen, but there can be significant cross-reactivity with structurally similar adducts (Everson et al. 1986).

Antibodies can be used in sensitive immunoassays to quantitate levels of specific adducts in large numbers of samples. For instance, competitive forms of radioimmunoassays (RIA), enzyme-linked immunosorbent assays (ELISA), or ultrasensitive enzymatic radioimmunoassays (USERIA) have all been used on samples from animals and humans for quantitation of adduct levels (Santella 1988). Assay sensitivity depends on the affinity constant of the antibody as well as the type of assay used. Specifically, competitive ELISA assays are able to detect attomole (10^{-18} mole) concentrations of adducts corresponding to about one adduct in 10^8 normal nucleotides (Poirier et al. 1982). Aside from this investigation, detection limits have generally fallen within the range of one adduct in 10^6 to 10^8 nucleotides (Sanders et al. 1986; Yang et al. 1987).

Adduct-specific antibodies have also been utilized for histochemical localization of DNA adducts. Using this approach, cell types possessing specific adducts can be identified. Studies of the formation of O^6-ethyl-deoxyguanosine adducts in rat brain identified these adducts at a concentration of one adduct in 10^6 nucleotides (Heyting et al. 1983). Studies of this type may yield valuable information toward identifying target cells as well as metabolic processes leading to adduct formation.

A major limitation of the antibody approach is the need to develop specific antibodies and establish the specificity of the antibody for each DNA adduct or class of adducts of interest. This is an important consideration when exposure is to a complex mixture of environmental chemicals. In addition, sufficient quantities of either the modified DNA or individual adduct must be available for immunization and antibody characterization. In certain cases broad specificity antibodies have been used to detect DNA adducts of a class of carcinogens such as benoz[a]pyrene, benz[a]anthracene, and chrysene (Santella et al. 1987). These limitations may explain why this technique has not been applied in the laboratory or the field on a large scale to detect DNA damage in aquatic or terrestrial species.

Other Methods

Several other analytical methods currently under development may eventually be used for measuring DNA adducts in organisms exposed to environmental contaminants. These include gas chromatography (GC), gas chromatography/mass spectroscopy (GC/MS), capillary zone electrophoresis-mass spectroscopy, and fluorescence line-narrowing techniques (Mohamed et al. 1984; Sanders et al. 1986). For the GC and GC/ MS analyses, adducts are first released from the DNA and then derivatized for increased volatility and thermal stability. Sensitivity of detection may be increased to the femtogram range through the use of halogenated derivatizing agents, such as pentafluorobenzoyl chloride. In the fluorescence line-narrowing technique, a laser is used for fluorescence excitation of samples at low temperature. Current sensitivities for fluorescent adducts are about one adduct in 10^6 nucleotides.

Secondary Modifications

Exposure to toxic chemicals may cause secondary concomitant types of DNA alterations. Besides direct adduct formation, damage may include strand breaks in the DNA polymer, changes in the DNA's minor base composition, or an increase in the level of unscheduled DNA synthesis (DNA repair).

Strand Breakage

A sensitive method for detecting strand breaks in DNA is the alkaline unwinding assay, which has been used with cells in culture to detect and quantify strand breaks induced by physical and chemical carcinogens (Kanter and Schwartz 1979; Ahnstrom and Erixon 1980; Kanter and Schwartz 1982; Daniel et al. 1985). The technique takes advantage of the characteristic that DNA strand separation under defined conditions of pH and temperature occurs at sites of single-strand breaks within the DNA molecule. The amount of double-stranded DNA remaining after a given period of alkaline unwinding is inversely proportional to the number of strand breaks present at the initiation of the alkaline exposure, provided that renaturation of the DNA is prevented (Daniel et al. 1985).

Shugart (1988a,b) has modified the alkaline unwinding method of Kanter and Schwartz (1982) for the analysis of strand breaks in DNA from environmental samples. The determination of the double-stranded DNA:total DNA ratio (F value), after a time-dependent partial alkaline unwinding of DNA, is facilitated by monitoring the DNA fluorometrically in the presence of Hoechst Dye 33258, a specific DNA-binding dye that fluoresces with double-stranded DNA at about twice the intensity as it does with single-stranded DNA (Kanter and Schwartz 1982; Shugart 1988a,b).

Rydberg (1975) has established the theoretical background for estimating strand breaks in DNA by alkaline unwinding, which is summarized by the equation:

$$\ln F = -(K / M)(t^b)$$

where K is a constant, t is time, M is the number average molecular weight of the DNA between two breaks, and b is a constant less than one that is influenced by the conditions for alkaline unwinding.

The relative number of strand breaks (N value) in DNA of organisms from sampled sites can be compared to those from reference sites as follows (Shugart 1988b):

$$N = (\ln F_s / \ln F_r) - 1$$

where F_s and F_r are the mean F values of DNA from the sampled sites and reference site, respectively. N values greater than zero indicate that DNA from the sampled sites has more strand breaks than DNA from the reference site; an N value of five, for example, indicates five times more strand breakage.

The method is sensitive and amenable to routine laboratory analysis. It has been used to examine the integrity of DNA from two aquatic organisms, the fathead minnow (*Pimephales promelas*) and the bluegill sunfish (*Lepomis macrochirus*), chronically exposed to benzo[a]pyrene in their water at a concentration of 1 μg/L. In both fish the assay detected an increase in strand breaks in the benzo[a]pyrene-exposed populations (Shugart 1988a,b).

Current laboratory research with this technique has focused on the effect that simultaneous exposure to nongenotoxic type stress conditions (such as temperature and hepatotoxic agents) and genotoxic chemicals will have on DNA integrity (Shugart 1990a).

The feasibility of utilizing this technique on environmental species as a general biomarker for pollution-related genotoxicity is being evaluated (Shugart 1990a,b; McCarthy et al. 1990). Also, analyses have been conducted on numerous environmental species, including oysters and mussels (Nacci and Jackim 1989), desert rodents (Shugart, personal communication), and turtles (Meyers et al. 1988).

Minor Nucleoside Content

In eukaryotic DNA, 5-methyl deoxycytidine is generally the only methylated deoxyribonucleoside present, and its level is enzymatically maintained (Razin and Riggs 1980; Ehrlich and Yang 1981; Holliday 1987). Chemical carcinogens have been shown to produce hypomethylation of DNA as a result of their effect on these enzymes (Boehim and Drahovsky 1983; Wilson and Jones 1983; Pfeifer et al. 1984). The base composition of DNA is easily determined by ion-exchange chromatography (Uziel et al. 1965; Shugart 1990c). As with the detection of strand breakage, this method measures a loss of DNA integrity but does not identify the chemical responsible.

Hypomethylation of DNA, as measured by the loss of 5-methyl deoxycytidine, was demonstrated in fish exposed to benzo[a]pyrene (Shugart 1990c). The onset and persistence of this phenomenon were found to be correlated with other types of DNA-damaging events, such as strand breaks and adduct formation. These observations suggest that hypomethylation may be a specific biological response to genotoxic agents.

Unscheduled DNA Synthesis

Unscheduled DNA synthesis measures non-semiconservative DNA replication resulting from DNA damage or alterations. It is a general indicator of genotoxic exposure. Its utility as a screening technique for potentially hazardous chemicals has been reviewed by the U.S. Environmental Protection Agency (Mitchell et al. 1983), which has recommended its use for genotoxic screening. The incorporation of [^{3}H]thymidine by cellular excision/repair mechanisms is measured either by autoradiographic grain counting or by liquid scintillation counting (LSC). Both methods have obvious advantages and disadvantages. Autoradiography allows the investigator to distinguish S-phase from nonS-phase cells and is useful in identifying tissue variations. However, processing, developing, and counting autoradiographs are very time-consuming tasks, while LSC is an inexpensive and rapid method. The LSC method, however, does not distinguish between cell types or cells in S-phase, although semiconservative replication can be arrested by the administration of hydroxyurea. This technique has been used in a variety of assays utilizing several cell lines, including cultured trout hepatocytes (Mitchell and Mirsalis 1984; Walton et al. 1987) and vivo in mice and rats (Mitchell and Mirsalis 1984). In these studies, significant increases in unscheduled DNA synthesis followed genotoxic exposure. The utility of this technique for environmental studies will depend on the animal system selected and on the sensitivity of different cell types to various genotoxic agents in laboratory studies.

Irreversible Events

As pointed out previously, the consequence of structural perturbations to the DNA molecule, such as adducts and secondary modifications, may result in subsequent lesions that become fixed. Affected cells often exhibit altered function indicative of a subclinical manifestation of genotoxic disease.

Cytogenetic Effects

Cytogenetic assays are well-accepted endpoints in laboratory mutagenicity assays and have been applied in field studies of environmental mutagens on several occasions (McBee and Bickham 1989). The cytogenetic assays considered in this review include standard (nondifferentially stained) chromosome analysis, sister-chromatid exchange, and the micronucleus assay. These three assays share the advantage of extensive validation and the disadvantage of being labor intensive. Progress in areas such as computer-based imaging systems and automatic metaphase finders should increase the ease at which these endpoints can be measured and should significantly reduce the cost of these assays.

Standard chromosome analysis is a relatively sensitive technique that is nonspecific in nature. Visible chromosomal aberrations are the result of mutagenic damage, and positive dose-response results have been documented in many laboratory studies. The procedure results in data that are easily quantifiable, but differences in scoring exist between laboratories. The use of historical controls and intercomparison of data among studies is risky, but the technique has high precision. Thus, some ambiguity of interpretation exists in scoring but not in the meaning of such data. Field studies using this procedure include McBee et al. (1987) and Thompson et al. (1988). Standard chromosome analysis is best used as a general indicator of exposure in environmental studies, and the procedure does not require further development.

Sister-chromatid exchange (SCE) is also a highly sensitive indicator of mutagenic damage that results in positive dose-response curves in laboratory studies. Although SCE is easily scored and is known to be caused by mutagens, the exact biological significance of this endpoint is ambiguous. This assay has been used by Tice et al. (1987) in field studies of rodents as well as by Pesch and Pesch (1980) with polychaetes.

The micronucleus assay has been shown to exhibit a dose-response relationship in laboratory studies. An advantage of this technique is that it reveals aneuploidy, which results from spindle anomalies as well as chromosomal breakage (clastogenicity). The technique is nonspecific and quantifiable in nature. It is a general indicator of exposure with relatively little ambiguity of interpretation, but its utility may be limited to particular species, life stages, or cell types (Schmid 1976). Compared to chromosome aberrations and SCE, this procedure is relatively insensitive. This is somewhat offset by the fact that many more cells can be assayed. The micronucleus assay has been applied in field studies with mammals by Tice et al. (1987) and with fish by Hose et al. (1987) and Metcalfe (1988).

Flow Cytometry

Flow cytometric measurement (FCM) is a technique that measures several cellular variables in suspended cells. Measurable variables include levels of DNA, RNA, protein, specific chemicals (using immunofluorescent probes), and numerous morphological attributes that affect time-of-flight and various light-scatter parameters (Shapiro 1988). Some flow cytometers can analyze as many as eight parameters from 10,000 cells per second. Cell sorting capabilities are available on many flow cytometers.

The application of flow cytometry to the study of environmental mutagenesis has been reviewed by Bickham (1990). The primary measurement of interest in such studies is DNA content, which can be measured with a high degree of precision and accuracy. Laboratory challenge experiments have shown that mutagenic chemicals and ionizing radiation result in a broader range of variation of nuclear or chromosomal DNA content in a positive dose-response relationship both in vivo (Bickham 1990) and in vitro (Otto et al. 1981). Thus, FCM has been extensively validated as a laboratory procedure for the evaluation of acute exposure to mutagenic chemicals. Field studies have demonstrated the efficiency of FCM in measuring the effects of chronic mutagen exposure to chemical pollutants (McBee and Bickham 1988) and low-level radioactivity (Bickham et al. 1988; Lamb et al. 1991).

Bickham (1990) concluded that FCM is a highly sensitive assay for detecting the effects of environmental mutagens. Advantages of FCM over other cytogenetic and cytometric techniques include low cost, speed, greater sensitivity due to the vast number of cells analyzed, and tremendous diversity of applications for which FCM is suitable. Virtually any tissue can be examined compared to chromosomal assays, which are limited to rapidly proliferating tissues such as bone marrow. This allows investigation on the effects of organ-specific mutagens. With the use of multiparameter analysis, specific cell types can be differentiated and analyzed. Moreover, FCM is easily adapted for use on species in which chromosomal analysis is difficult (Bickham et al. 1988; Lamb et al. 1991).

FCM also has identified a potential qualitative difference in the response of animals to chronic environmental exposure vs acute laboratory mutagen exposure. Aneuploid mosaicism was observed at low frequency in animals exposed to environmental mutagens in each of three studies (Bickham et al. 1988; McBee and Bickham 1988; Lamb et al. 1991). Such mosaicism was not observed in animals from control sites or in animals exposed to acute laboratory doses (Bickham 1990). This demonstrates the capability of FCM to identify both low frequency variant cells and multiple populations of cells that might have subtle differences in DNA content.

Mutations

Two areas worth considering within this topic are oncogene activation and mutation rates.

Oncogene activation. One important aspect of genotoxic exposure is the creation of mutations in DNA, which lead to alterations in gene functions. If such genetic aberrations correlate with abnormal cell morphology, these mutations may signify somatic events that occur during the formation of the abnormal cell. Potential sites in the DNA for such a genetic marker are limited. However, the study of proto-oncogenes and oncogenes as major factors in the control of normal cell proliferation and differentiation has had important consequences for studies in chemical carcinogenesis (Balmain and Brown 1988). One of the principal advances resulting from research on oncogenes is evidence that this group of genes provides, at the very least, potential targets for activation by carcinogens. A full spectrum of genetic changes have been noted in the DNA of cells treated with chemical carcinogens, including point mutation, gene translocation, and gene amplification, all of which have been implicated in the activation of particular proto-oncogenes. In animals exposed to chemical carcinogens, the most frequently detected oncogenes in tumors include the c-*ras* family of proto-oncogenes (c-Ha-*ras*, c-Ki-*ras*, and N-*ras*). This finding may be due, in part, to the highly selective nature of the assay system, which involves transfection of high molecular weight DNA into NIH3T3 mouse fibroblasts followed by formation of transformed foci or subcutaneous tumors in athymic mice. The mechanism of activation of this class of genes has been shown to be the result of single-base mutations in specific polypeptide-coding domains of the c-ras genes (Barbacid 1987). The emerging consensus on the function of normal c-ras proteins implicates them as membrane-associated polypeptides that modulate the transduction of external signals from the cell surface to the interior of the cell. Subtle mutations in the gene encoding the protein may lead to changes in properties relating to intrinsic guanosine 5¢-triphosphate (GTP) hydrolysis and resultant conformational changes that alter the specificity of reaction of the c-ras proteins with appropriate receptor or effector molecules.

Recent evidence suggests that the mutation of this class of genes may occur during "initiation" of carcinogenesis by direct interaction of the carcinogen with the target gene. More importantly, a wealth of information derived from DNA of malignant and premalignant lesions in animal models suggests a chemical specificity for the activating mutations (Balmain and Brown 1988). Chemicals with a high probability of reacting with guanine often show specificity in the 12th codon whereas those with adenine often show it in the 61st. Thus, the specificity is mainly with respect to the site of reaction with DNA, and this occurs within areas where mutations are detected. The creation of such nucleotide-specific changes is viewed as an interplay of the chemical-DNA adduct profile and the repair capacity of the cell to particular chemical-DNA adduct forms.

In general, the activation of particular c-ras oncogenes by single-base mutations creates a variant gene that is highly selective in the target cell, leading to the emergence of transformed cells containing these mutations. However, such somatic mutations can also serve to deregulate gene expression, leading to increased levels of oncogene proteins in the emergent tumor cell. More recently, somatic mutations have been shown to inactivate tumor suppressors such as the p53 protein or the retinoblastoma protein, Rb. In a more general sense, other such stable genetic events mediated by chemical-DNA interaction (including mutation) may operate in the creation of somatic variant genes that encode a variety of proteins involved in malignant progression.

Without the use of transformed cells derived by DNA-mediated transfer methods, the detection of such genetic alterations in a given cellular DNA preparation is predicated on the enrichment of particular cellular phenotypes, which predominate in the target cell population. To address this technical problem and to expedite the determination of mutations, PCR (polymerase chain reaction) technologies have been used to determine mutations in primary tissue samples containing preneoplastic and neoplastic cells. The PCR technique of Saiki et al. (1988a) utilizes synthetic DNA oligonucleotides (primers) specifically to amplify short genetic elements from a given DNA preparation using sequential cycles of primer annealing and polymerase extension. The analysis of mutations in PCR-amplified DNA has been accomplished for specific nucleotide changes by:

- restriction analysis (Saiki et al. 1985)
- oligonucleotide hybridization (Bos et al. 1987)
- direct DNA sequencing (McMahon et al. 1987)
- RNase mapping (Forrester et al. 1987)
- gel retardation (Cariello et al. 1988)

- plaque screening assay, (McMahon et al. 1990a)
- liquid hybrid selection (Kumar and Barbacid 1988)
- nonradioactive, restriction fragment length polymorphism (RFLP) (Kahn et al. 1990)

The diagnostic procedures listed differ in their individual potential to detect a given mutated gene, ranging from 1 mutant in 10 genes for direct DNA sequencing or dot-blot oligonucleotide hybridization procedures to 1 mutant in 100,000 genes or greater for gel retardation or liquid hybrid selection techniques. In addition, the relative ease and cost of these procedures vary substantially, from rapid dot-blot analysis with nonradioactive DNA probes (Saiki et al. 1988b) or RFLP (Kahn et al. 1990), to more laborious and time-consuming PCR DNA cloning and DNA sequence analysis methods.

Until recently, studies of oncogene activation in environmentally exposed animal populations were lacking. However, a study by McMahon et al. (1990b) of abnormal samples containing preneoplastic and neoplastic liver lesions in winter flounder from Boston Harbor, Massachusetts, indicated a high percentage of activated c-Ki-*ras* oncogene in individual liver samples measured by DNA transfection methods followed by selection of transformed NIH3T3 cells in athymic mice. In addition, Wirgin et al. (1989) have detected c-Ki-*ras* oncogenes in a high percentage of liver tumors derived from tomcod from the Hudson River. The detection of these genes is consistent with the high degree of conservation of c-Ki-*ras*, c-Ha-*ras*, and N-*ras* genes in fish DNA when compared to mammalian sources (McMahon et al., personal communication). In the Boston Harbor study, analysis of PCR-amplified DNA derived from transformants and primary liver DNA indicates a mutation in the 12th codon of the c-Ki-*ras* gene as determined by direct DNA sequence analysis and dot-blot techniques using oligonucleotide hybridization. The detection of oncogenes in the majority of abnormal fish livers in these two studies and the application of PCR-based techniques suggest that similar approaches may be sufficient to detect single-nucleotide modifications resulting from chemical exposure in wild fish.

Mutation rates. Various molecular genetic techniques have proven extremely valuable in identifying genetic markers for the diagnosis of human genetic diseases. Although these techniques have not yet been applied in studies of environmental mutagens, their potential application in population genetics represents an important new avenue of research.

It is well known that selective constraints vary tremendously among genetic loci. For example, histone genes are virtually immutable and are extremely similar or identical in base sequence in diverse taxa. Certain restriction sites, such as the Sac II restriction enzyme sites in the ribosomal genes of the mitochondrial DNA (mtDNA), are conserved in vertebrates from fish to humans (Moritz et al. 1987). On the other hand, DNA such as the D-loop region of mtDNA and the hypervariable satellite DNA of the nuclear genome are highly mutable and are useful in differentiating closely related individuals. Thus, techniques and appropriate genetic loci are available to measure changes in mutation rates in natural populations of sentinel animals exposed to environmental mutagens.

The detection of mutations and the measure of relative mutation rates can be undertaken through a population genetics approach. For example, direct DNA sequence data can be obtained from organisms using PCR amplification of a specific DNA segment. DNA isolated from animals collected from control and polluted sites could be subjected to PCR amplification and sequence data collected for 300 to 500 base pairs. A number of individuals would be sequenced to establish the range of variation within populations. If mutation rates were increased due to the presence of environmental mutagens, more sequence variation would be found among animals at the polluted sites than among animals from control sites. Mitochondrial DNA would be particularly sensitive to this effect because it is maternally inherited and not subject to recombination (Moritz et al. 1987). Moreover, this sensitivity may be due to higher levels of adduct formation in mtDNA compared to nuclear DNA (Backer and Weinstein 1982).

There are various molecular approaches that can be used to obtain the type of data just discussed (Appels and Honeycutt 1986). Relatively few labs are presently applying DNA sequencing in population genetic studies, but the number has increased since the advent of PCR. More labs are using restriction fragment length polymorphism (RFLP) analysis in which restriction enzymes are used to cut the DNA and resulting fragment patterns are analyzed. There are two general methodological approaches for RFLP analysis. Mitochondrial DNA can be isolated and purified based upon its physical properties; purified mtDNA can be cut with restriction enzymes, radioactively end-labeled, and electrophoretically analyzed to reveal fragment patterns. Alternatively, total DNA can be isolated, digested with restriction enzymes, and hybridized with radioactively labeled DNA probes, a method which also reveals alterations in fragment patterns.

All of the above techniques share a common disadvantage in being costly. The major advantage of sequence or RFLP data is that very high resolution is obtained and heritable mutations are demonstrated. Extensive research is needed to evaluate the validity of this approach, but two factors indicate a possibility of success. First, many labs are studying the levels of genetic variation within and among natural populations of organisms. Thus, background levels of variation are being established for many species that will be useful sentinels. Second, most environmental problems associated with mutagens result in chronic, low-level exposure of populations. The population genetics approach should be a sensitive procedure because it measures the accumulation of mutations that have occurred through generations of organisms exposed to the mutagens and therefore, would be appropriate for organisms with very short life spans and rapid reproduction rates.

POTENTIAL VALUE

Xenobiotic chemicals, in the forms found in the environment, often do not by themselves constitute a hazard to indigenous organisms. However, once exposure has occurred and substances are bioavailable, a sequence of biological responses may progress. Whether the well-being of the organism is eventually affected will depend upon many factors, some intrinsic (e.g., age, sex, health, and nutritional status of the organism) and others extrinsic (e.g., dose, duration, route of exposure to the contaminant, and the presence of other chemicals). These factors represent barriers to the assessment of exposure and subsequent risk from that exposure. However, biological markers (responses) can help circumvent these problems to a large extent by focusing on relevant molecular events that occur after exposure and metabolism.

Reliable and sensitive analytical methods are being used to clarify the relationship between exposure to xenobiotic compounds and their effect. This is particularly evident in the field of environmental genotoxicity where alterations to DNA serve as biomarkers.

It has to be assumed that most DNA adducts are damaging in one way or another. This assumption might not be entirely correct since some adducts might be readily repaired by excision without further consequences. Some adducts on nucleotides in inert regions of DNA also may not produce adverse effects. It should be noted that DNA adducts still provide evidence of specific exposure that has passed all of the toxicokinetic barriers. Thus, characterization of specific DNA adducts may ultimately lead to the identification of a group of genotoxic chemicals of significant environmental concern.

The levels of DNA alteration that induce cellular responses vary with chemical. Background levels of DNA adducts, strand breaks, mutations, and other DNA alterations occur as a result of natural phenomena, such as ionizing radiation and dietary components. These levels can vary among species and among tissues within a single species. The ability to measure contamination-induced DNA alterations is directly dependent on an accurate measurement of the background levels of such alterations. These DNA alterations may potentiate irreversible changes to the DNA molecule and result in expression of other cellular responses, such as chromosomal aberrations and oncogene activation. For some chemicals, the level of induced DNA alteration has been correlated with a genotoxic effect, such as tumorigenesis. The eventual prediction of a toxic cellular response from a single biomarker will be limited by the complex interactions that characterize biological systems. It is likely that the prediction of a toxic effect such as tumorigenesis will be enhanced through the use of a battery of biomarkers that span a range of cellular processes. Although not all DNA alterations will produce harmful effects, they have the potential for increasing the likelihood for long-term, deleterious effects, not only for the organism but also for future generations. Therefore, there is the obvious need to correlate DNA damage with genotoxic responses and endpoints other than cancer.

The various techniques described in this chapter for detecting DNA alterations are sufficiently different in approach and/or methodology that a separate discussion of the potential value for evaluating exposure and environmental effects of toxic chemicals is detailed for each.

DNA Adducts

The potential for using DNA adducts as a biomarker of exposure to complex mixtures of genotoxic chemicals has begun to be validated in field studies using the ^{32}P-postlabeling assay. Currently, the ^{32}P-postlabeling technique is semiquantitative, laborious, and moderate in cost. However, recent technical advances indicate that improvements will be forthcoming in each of these areas, thereby making this assay attractive for use as one of the tools in large environmental monitoring studies. Additionally, since ^{32}P-postlabeling assay has been shown to have the highest sensitivity for measuring bulky aromatic adducts (Gupta and Randerath 1988), this assay can provide information about whether adducts are present at levels sufficient for other techniques to be applicable. In principle, this technique can be applied to DNA from any organ or any species and requires only small amounts (100 mg) of tissue. In situations where high molecular weight PAHs are the predominant contaminants, a relatively easy fluorescence method is currently available for measuring a specific DNA-PAH adduct (i.e., DNA-BaPDE) when present at relatively high levels in the DNA. This method has been applied to environmental samples (Martineau et al. 1988). The detection of DNA adducts in environmental samples provides a means of investigating the qualitative and quantitative relationships between the formation of DNA adducts, subsequent DNA alterations, and resulting lesions in target tissues.

The sensitivity and specificity of polyclonal and monoclonal antibodies suggest that this approach may have great utility in DNA adduct analysis. The requirement for specific antibodies for each adduct has limited this technique to a relatively small group of chemicals (Santella 1988). Advances in the in vitro production of DNA adducts and the isolation of adducts from biological samples, coupled with existing technology for the generation of antibodies, could significantly affect the application of this technique for the detection of adducts. Alternatively, antibodies to specific adducts or classes of adducts might, for example, be utilized for the isolation of these adducts by immunoaffinity chromatography. In this approach, the antibodies could be utilized for a preparative isolation step to enhance other more sensitive identification techniques. It seems obvious from the above example that the immunological approach shows potential as a detection system as well as a tool for the enhancement of other techniques.

Because of the limitations outlined in the previous sections, other methods are in various stages of development. These include:

1.   improved separation of the ^{32}P-labeled adducts by HPLC (Dietrich et al. 1987; Gorelick and Wogan 1989)

2.   development of techniques to produce volatile derivatives of adducts for analysis by GC/MS (Mohamed et al. 1984)
3.   application of liquid capillary HPLC separation and detection of adducts by dynamic flow FAB/MS to improve adduct separation and characterization (Jackim, personal communication)
4.   investigations into the use of capillary zone electrophoresis to separate and identify adducted nucleotides (Jackim and Norwood 1990)
5.   application of fluorescence line-narrowing spectroscopy to characterize individual fluorescent DNA adducts (Jankowiak et al. 1988)

Secondary Modifications

Strand Breakage

Many toxic chemicals cause strand breaks in DNA either directly or indirectly. The alkaline unwinding assay can estimate the increase in the level of breaks above background resulting from exposure to these chemicals (Shugart 1990a,b). The technique can be applied to the analysis of many samples without the need for costly reagents or laboratory equipment. For field studies, laboratory analyses are performed on fresh or frozen tissues. Data is available within a few hours and is best interpreted in relation to data collected from other biomarkers. The method is ideally suited for routine, in situ monitoring of environmental species because of its ease and low cost. A positive result can be seen as a "red flag" since, in theory, exposure to any genotoxic chemicals will illicit such a response.

Minor Nucleoside Content

The feasibility of using hypomethylation of DNA as a biomarker for genotoxicity is being evaluated in environmental species (Shugart 1990c). Laboratory analyses are easy to perform and low in cost.

Unscheduled DNA Synthesis

This technique has potential as a general measure of genotoxic exposure, principally as an ancillary technique associated with other, more specific methodologies. As such, its potential has been realized in the area of toxic chemical screening, primarily due to the sensitivity of the cellular excision/repair process to a broad spectrum of genotoxic chemicals. In essence, rates of this process above appropriate control levels imply increased DNA damage and indicate genotoxic damage as a potential concern.

The utilization of this technique for environmental applications, however, will require considerable development. In most cases, it will be applied to nontraditional animal systems. Therefore, a majority of the initial effort will be devoted to the definition and identification of the relative strengths and pitfalls of that particular system. Unscheduled DNA synthesis, or for that matter, any other assay system, should not be considered a "stand alone" answer for environmental monitoring, nor appropriate under all conditions. This is where extensive laboratory studies become very important in defining a biomarker's limitations and potential.

Irreversible Events

Cytogenetic Effects

Chromosome analysis, measurement of sister-chromatid exchange, the micronucleus assay, and flow cytometry all have been validated as useful indicators of mutagenic damage in laboratory challenge tests. All four techniques have also been applied in field studies. In general, these techniques can be useful in both the initial screening for effects and the subsequent evaluation of damage caused by environmental mutagens.

For use as an initial screening procedure, flow cytometry has tremendous potential because of its low cost and high sensitivity. Hundreds of thousands of cells from scores of individuals can be analyzed quickly, in a matter of a few days if necessary. The development of antibodies to cells expressing oncogenes may allow one to look by this methodology for types of alterations which are linked to genotoxic chemicals. FCM also can be used to evaluate cells from any species and tissue type, so the degree of impact of an environmental insult can be extensively investigated.

The micronucleus assay also has potential as an initial screening technique, but because it is a relatively labor-intensive assay, its application is much more restricted. Additionally, a recent study (Metcalfe 1988) suggests that it may be species specific. The development of flow cytometric and image analysis systems to analyze micronuclei may greatly increase the utility of this biomarker.

Both chromosome analysis and sister-chromatid exchange have limited utility as screening techniques because they are labor-intensive procedures. Their greatest application would seem to be in the documentation of the kinds of mutagenic events being expressed in a population in which other screening techniques have indicated an impact. However, it might be appropriate to note that these techniques may have value in certain cases for relating the level of alterations found in environmental samples to the human population in that area.

Mutations

Research is needed to validate both oncogene activations and mutation rates as biomarkers of chemically exposed natural populations of animals.

Oncogene activations. A DNA-based assay to measure specific nucleotide changes in oncogenes and other appropriate genes of chemically exposed animals has two main attributes. First, the detection of mutant forms of normal genes that are exhibited only in abnormal cells are, by inference, deleterious. Second, different types of mutations can be correlated to specific chemical-DNA interactions because mutagenic events due to exogenous agents are a consequence of specific chemical-DNA adduct formation and endogenous repair. A current model (Bos 1988) suggests that chemicals mutate oncogenes in a subset of cells exposed to a given DNA-damaging chemical that later emerges after months or years of latency as a premalignant or malignant lesion from clonal expansion of this subset of cells. This latter consideration defines this biomarker in the context of previous chemical exposure, which leads to mutations associated with the cancer process. In this case, small amounts of frozen or fixed tissues (10,000 cells) derived from field specimens exhibiting preneoplastic or other abnormal morphology could be analyzed in the laboratory using polymerase chain reaction (PCR) DNA amplification. The PCR procedures require about a day to obtain a gene-specific DNA fragment population of interest. However, the time required for analysis of this DNA population varies greatly depending on the frequency of the mutated gene and the specificity of the diagnostic procedure employed. In the most favorable scenario, the processing of a tissue sample to detect a specific oncogene mutation can take one week. In the worst case, it could take a month.

In all cases, including PCR amplification, the technologies will require a delineation of specific nucleotide changes for a particular oncogene implicated in cells derived from a given species, target organ, histopathologic condition, and site of exposure. This important aspect will require extensive development and may require the cloning and sequencing of relevant genes, the optimization of the PCR procedure, and the development of expedient diagnostic procedures that would provide unequivocal detection of the mutations. In general, such technologies have been developed for other applications and would require appropriate modifications using the DNA region from the species of interest. In addition, this biomarker will require validation in field studies at several contaminated sites, as well as in laboratory studies using defined chemicals and complex mixtures of genotoxic chemicals.

The first steps to validate oncogene activation has already been made as shown by the *ras* family. Since this occurs in many chemically induced tumors, it will already serve as a good marker. Further studies will then be appropriate in environmental situations after mechanisms of activation of other oncogenes are better characterized. When sufficient knowledge of specific mutations and diagnostic technologies has been gained, relatively simple and inexpensive gene screening procedures may be employed. Further expansion of the technologies to address gene amplification or rearrangement (as in the case of the *myc* gene family) or inactivation of tumor suppressor genes (as in the case of the p53 protein) will require a substantial investiment of effort to delineate a relationship between environmentally induced preneoplasia and genetic change.

Mutation rates. The value of DNA sequencing and restriction fragment length polymorphism data in the study of environmental mutagens has yet to be investigated. The techniques are applicable to any organism and to a broad spectrum of gene loci, ranging from structural genes to nontranscribed DNA. Before the value of this technology as a biomarker can be assessed, research efforts will need to document increased mutation rates among populations to establish heritable mutations caused by environmental mutagens.

CONCLUSION

The detection of DNA structural changes and ensuing events has only recently been demonstrated and documented as a viable scientific tool for the in situ biological monitoring for genotoxicity of chemicals in the environment. The full potential of this concept and approach will be realized only if continued effort and emphasis is placed on (1) research concerned with those basic cellular mechanisms by which certain chemicals elicit a genotoxic responses, (2) development of new and sensitive analytical techniques, and (3) field-validation studies, particularly with respect to species specificity of individual tests and the selection of appropriate sentinel species (Shugart et al. 1989).

An important consideration for current and future work will be the archiving of samples. This practice will be extremely useful for several reasons. First, if data were lost or needed for rechecking, relevant samples would be available. Second, future research may suggest new or more appropriate indicators for biological monitoring, and the presence of archived samples would be invaluable for the documentation of a historical picture with the new techniques.

REFERENCES

Ahnstrom, G., and K. Erixon. "Measurement of Strand Breaks by Alkaline Denaturation and Hydroxyapatite Chromatography," in *DNA Repair, Vol. 1, Part A*, E. C. Friedbert, and P. C. Hanawalt, Eds. (New York: Marcel Dekker, Inc., 1980) pp. 403–419.

Alberts, B., D. Bray, J. Lewis, M. Raff, K. Roberts and J.D. Watson. *Molecular Biology of the Cell*, 2nd. ed. (New York:Garland Publishing Co., Inc., 1989) pp. 220–227.

Appels, R., and R.L. Honeycutt. "rDNA: Evolution Over a Billion Years," in *DNA Systematics*, S.K. Dutta, Ed. (Boca Raton, FL: CRC Press, 1986) pp. 81–135.

Backer, J.M., and I.B. Weinstein. "Interaction of Benzo[a]pyrene and its Dihydodiol-epoxide Derivative with Nuclear and Mitochondrial DNA in C3H10T1/2 Cells in culture," *Cancer Res.* 42:2764–2769 (1982).

Balmain, A., and K. Brown. "Oncogene Activation in Chemical Carcinogenesis," *Adv. Cancer Res.* 51:147–182 (1988).

Barbacid, M. "Ras Genes," *Ann. Rev. Biochem.* 56:779–827 (1987).

Bartsch, H., K. Hemminki and I.K. O'Neill, Eds. *Methods for Detecting DNA Damaging Agents in Humans: Application in Cancer Epidemiology and Prevention* (New York: Oxford University Press, 1988).

Beland, F.A., and F.F. Kadlubar. "Formation and Persistence of Arylamine DNA Adducts *In Vivo*," *Environ. Health Perspect.* 62:19–30 (1985).

Bickham, J.W. "Flow Cytometry as a Technique to Monitor the Effects of Environmental Genotoxins on Wildlife Populations," in *In situ Evaluation of Biological Hazards of Environmental Pollutants*, S. Sandhu, W.R. Lower, F.J. DeSerres, W.A. Suk, and R.R. Tice, Eds., Environmental Research Series Vol. 38 (New York: Plenum Press, 1990) pp. 97–108.

Bickham, J.W., B.G. Hanks, M.J. Smolen, T. Lamb and J.W. Gibbons. "Flow Cytometric Analysis of the Effects of Low Level Radiation Exposure on Natural Populations of Slider Turtles (*Pseudemys scripta*)," *Arch. Environ. Contam. Toxicol.* 17:837–841 (1988).

Boehim, T.L., and D. Drahovsky. "Alteration of Enzymatic Methylation of DNA Cystosines by Chemical Carcinogens: A Mechanism Involved in the Initiation of Carcinogenesis," *J. Nat. Can. Inst.* 71:429–433 (1983).

Bos, J. L. "The *ras* gene family and human carcinogenesis," *Mutation Res.* 195:255–271 (1988).

Bos, J. L., E.R. Fearon, J.R. Hamilton, M. Verlaan-de Vries, J.H. van Boom, A.J. van der Eb and B. Vogelstein. "Prevalence of *ras* Gene Mutations in Human Colorectal Tumors," *Nature* 327:293–297 (1987).

Cariello, N.F., J.K. Scott, A.G. Kat, W.G. Thilly and P. Keohavong. "Resolution of a Missense Mutant in Human Genomic DNA by Denaturing Gradient Gel Electrophoresis and Direct Sequencing Using In Vitro DNA Amplification," *Am. J. Human Genet.* 42:726–734 (1988).

Collier, T.K., and U. Varanasi. "Biochemical Indicators of Contaminant Exposure in Flatfish from Puget Sound, WA," in *Proceedings Oceans 87, Vol. 5.* (Washington, D.C.: IEEE, 1987) pp. 1544–1549.

Committee on Biological Markers of the National Research Council (CBMNRC). "Biological markers in Environmental Health Research," *Environ. Health Perspect.* 74:3–9 (1987).

Daniel, F.B., D.L. Haas and S.M. Pyle. "Quantitation of Chemically Induced DNA Strand Breaks in Human Cells via an Alkaline Unwinding Assay," *Anal. Biochem.* 144:390–402 (1985).

Dietrich, M.W., W.E. Hopkins II, K.J. Asbury and W.P. Ridley. "Liquid Chromatographic Characterization of the Deoxyribonuleoside 5'-Phosphate and Deoxyribonucloside-3',5'-Bisphosphate Obtained by ^{32}P-postlabeling of DNA," *Chromatographia* 24:545–551 (1987).

Dunn, B., J. Black and A. Maccubbin. "^{32}P-postlabeling Analysis of Aromatic DNA Adducts in Fish from Polluted Areas," *Cancer Res.* 47:6543–6548 (1987).

Dunn, B.P., J. Fitzsimmons, D. Stalling, A.E. Mccubbin, and J. J. Black. "Pollution-related Aromatic DNA Adducts in Liver from Populations of Wild Fish," *Proc. Am. Assoc. Cancer Res.* 31(Abstract 570):96 (1990).

Ehrlich, M., and R.Y.-H.Yang. "5-Methylcytosine in Eukaryotic DNA," *Science* 212:1350–1357 (1981).

Everson, R.B., E. Randerath, R.M. Santella, R.C. Cefalo, T.A. Avitts and K. Randerath. "Detection of Smoking-related Covalent DNA Adducts in Human Placenta," *Nature* (*London*) 231:54–57 (1986).

Forrester, K., C. Almoguera, K. Han, W.E. Grizzle and M. Perucho. "Detection of High Incidence of K-*ras* Oncogenes During Human Colon Tumorigenesis," *Nature* (*London*) 327:298–303 (1987).

Gorelick, N.J., and G.N. Wogan. "Fluoranthene-DNA Adducts: Identification and Quantification by an HPLC-^{32}P-postlabeling Method," *Carcinogenesis* 10:1567–1577 (1989).

Gupta, R.C. "Newly Detected DNA Damage at Physiological Temperature," *Proc. Am. Assoc. Cancer Res.* 30(Abstract 594):150 (1989).

Gupta, R.C., and K. Early. "^{32}P-postlabeling Assay: Comparative Recoveries of Structurally Diverse DNA Adducts in the Various Enhancement Procedures," *Nature (London)* 9:1687–1693 (1988).

Gupta, R.C., and K. Randerath. "Analysis of DNA Adducts by ^{32}P-labeling and Thin Layer Chromatography," in *DNA Repair, Vol. 3.,* E. Friedberg, and P.H. Hanawalt, Eds. (New York: Marcel Dekker, Inc., 1988) pp. 399–418.

Gupta, R.C., M.L. Sopori and C.G. Gairola. "Formation of Cigarette Smoke-induced DNA Adducts in the Rat Lung and Nasal Mucosa," *Cancer Res.* 49:1916–1970 (1989).

Harvey, R.C. "Polycyclic Hydrocarbons and Cancer," *Am. Scientist* 70:386–393 (1982).

Heyting, C., C.J. Van Der Laken, W. Van Raamsdouk and C. Pool. "Immunohistochemical Detection of O^6-ethyldeoxyguauosine in the Rat Brain after In Vivo Applications of N-ethyl-N-nitrosourea," *Cancer Res.* 43:2935–2941 (1983).

Holliday, R. "The Inheritance of Epigenetic Defects," *Science* 238:163–170 (1987).

Hose, J.E., J.N. Cross, S.C. Smith and D. Diehl. "Elevated Circulating Erythrocyte Micronuclei in Fishes from Contaminated Sites Off Southern California," *Mar. Environ. Res.* 22:167–176 (1987).

Jackim, E., and C. Norwood. "Separation and Detection of Benzo[a]pyrene Deoxyguanosyl-t-monophosphate Adduct by Capillary Zone Electrophoresis," *J. High Resolution Chrom.,* 13:195–196 (1990).

Jankowiak, R., R.S. Cooper, D. Zamzow, G.J. Small, G. Doskocil and A.M. Jeffrey. "Fluorescence Line Narrowing-nonphotochemical Hole Burning Spectrometry: Femtomole Detection and High Selectivity for Intact DNA-PAH Adducts," *Chem. Res. Toxicol.* 1:60–68 (1988).

Kahn, S.M., W. Jiang, and I.B. Weinstein. "Rapid Nonradiactive Detection of *ras* Oncogenes in Human Tumors," *Amplifications* 4:22–26 (1990).

Kanter, P.M., and H.S. Schwartz. "A Hydoxylapatite Batch Assay for Quantitation of Cellular DNA Damage," *Anal. Biochem.* 97:77–84 (1979).

Kanter, P.M. and H.S. Schwartz. "A Fluorescence Enhancement Assay for Cellular DNA Damage," *Mol. Pharmacol.* 22:145–151 (1982).

Kohn, H.W. "The Significance of DNA-damaging Assays in Toxicity and Carcinogenicity Assessment," *Ann. N.Y. Acad. Sci.* 407:106–118 (1983).

Kumar, R., and M. Barbacid. "Oncogene Detection at the Single Cell Level," *Oncogene.* 3:647–651 (1988).

Kurelec, B., A. Garg, S. Krca, M. Chacko, and R.C. Gupta. "Natural Environment Surpasses Polluted Environment in Inducing DNA Damage in Fish," *Carcinogenesis* 7:1337–1339 (1989).

Lamb, T., J.W. Bickham, J.W. Gibbons, M.J. Smolen and S. McDowell. "Genetic Damage in a Population of Slider Turtles (*Trachemyus scripta*) Inhabiting a Radioactive Reservoir," *Arch. Environ. Contam. Toxicol.,* 20:138–142 (1991).

Martineau, D., A. Legace, P. Beland, R. Higgins, D. Armstron, D. and L.R. Shugart. "Pathology of Stranded Beluga Whales (*Delphinapterus leucas*) from the St. Lawrence Estuary, Quebec, Canada," *J. Comp. Pathol.* 98:287–311 (1988).

McBee, K., and J.W. Bickham. "Petrochemical-related DNA Damage in Wild Rodents Detected by Flow Cytometry," *Bull. Environ. Contam. Toxicol.* 40:343–349 (1988).

McBee, K., and J.W. Bickham. "Mammals as Bioindicators of Environmental Toxicity," in *Current Mammalogy*, H.H. Genoways, Ed. (New York: Plenum Press, 1989) pp. 37–88.

McBee, K., J.W. Bickham, K.C. Donnely and K.W. Brown. "Chromosomal Aberrations in Native Small Mammals (*Peromyscus leucopus* and *Sigmodon hispidus*) at a Petrochemical Waste Disposal Site: I. Standard Karyology," *Arch. Environ. Contam. Toxicol.* 16:681–688 (1987).

McCarthy, J.F., B.D. Jimenez, L.R. Shugart and A. Oikari. "Biological Markers in Animal Sentinels: Laboratory Studies Improve Interpretation of Field Data," in *In situ Evaluation of Biological Hazards of Environmental Pollutants*, S. Sandhu, W.R. Lower, F.J. deSerres, W.A. Suk, and R.R. Tice, Eds., Environmental Research Series Vol. 38, (New York: Plenum Press, 1990) pp. 163–176.

McMahon, G., E. Davis and G.N. Wogan. "Characterization of c-Ki-*ras* Oncogene Alleles by Direct Sequencing of Enzymatically Amplified DNA from Carcinogen-induced Tumors," *Proc. Natl. Acad. Sci. U.S.A.* 84:4974–4978 (1987).

McMahon, G., E.F. Davis, L.J. Huber, Y. Kim and G.N. Wogan. "Characterization of c-K-*ras* and N-*ras* oncogenes in Aflatoxin B_1-induced Rat Liver Tumors," *Proc. Natl. Acad. Sci. U.S.A.*, 87:1104–1108 (1990a).

McMahon, G., L.J. Huber, M.J. Moore, J.J. Stegeman and G.N. Wogan. "Mutations in c-Ki-*ras* Oncogenes in Diseased Livers of Winter Flounder from Boston Harbor," *Proc. Natl. Acd. Sci. U.S.A.* 87:841–845 (1990b).

Metcalfe, C.D. "Induction of Micronuclei and Nuclear Abnormalities in the Erythrocytes of Mudminnows (*Umbra limi*) and Brown Bullheads (*Ictalurus nebulosus*)," *Bull. Environ. Contam. Toxicol.* 40:489–495 (1988).

Meyers, L.J., L.R. Shugart and B.T. Walton. "Freshwater Turtles as Indicators of Contaminated Aquatic Environments," Paper presented at the 9th Annual Meeting of the Society of Environmental Toxicology and Chemistry, Arlington, VA, November 15, 1988.

Mitchell, A.D., M.L. Casciano, D.E. Metz, R.H.C. Robinson, G.M. San, G.M. Williams and E.S. Von Halle. "Unscheduled DNA Synthesis Tests: A Report of the "Gene-Tox" Program," *Mut. Res.* 123:363–410 (1983).

Mitchell, A.D., and J.C. Mirsalis. "Unscheduled DNA Synthesis as an Indicator of Genotoxic Exposure," in *Single-Cell Mutation Monitoring Systems: Methodologies and Applications*, Chapter 8, A.A. Ansari, and F.J. DeSerres, Eds. (New York: Plenum Press, 1984) pp. 165–216.

Mohamed, G.B., A. Nazareth, M.J. Hayes, R.W. Giese and P. Vouros. "GC-MS Characteristics of Methylated Perfluoroaryl Derivatives of Cytosine and 5-Methylcytosine," *J. Chromatogr.* 314:211–217 (1984).

Moritz, C., T.E. Dowling and W.M. Brown. "Evolution of Animal Mitochondrial DNA: Relevance for Population Biology and Systematics," *Ann. Rev. Ecol. Syst.* 18:269–292 (1987).

Murchelano, R., and R. Wolke. "Epizootic Carcinoma in the Winter Flounder, *Pseudopleuronectes americanus*," *Science.* 228:587–589 (1985).

Nacci, D., and G. Jackim. "Using the DNA Alkaline Unwinding Assay to Detect DNA Damage in Laboratory and Environmentally Exposed Cells and Tissue," *Marine Environ. Res.* 28:333–337 (1989).

Otto, F.J., H. Oldiges, W. Gohde and V.K. Jain. "Flow Cytometric Measurement of Nuclear DNA Content Variations as a Potential in vivo Mutagenicity Test," *Cytometry* 2:189–191 (1981).

Pesch, G.G., and C.E. Pesch. "Neanthes Arenaceodentata (*Polychaeta annelida*): A Proposed Cytogenetic Model for Marine Genetic Toxicology," *Can. J. Fish. Aquat. Mar. Genet. Toxicol.* 37:1225–1228 (1980).

Pfeifer, G.P., D. Grungerger and D. Drahovsky. "Impaired Enzymatic Methylation of BPDE-modified DNA," *Carcinogenes.* 5:931–935 (1984).

Phillips, D. "Fifty Years of Benzo[a]pyrene," *Nature (London)* 303:468–472 (1983).

Phillips, D., and P. Sims. "PAH Metabolites: Their Reaction with Nucleic Acids," in *Chemical Carcinogens and DNA, Vol. 2,* P.L. Grover, Ed. (Boca Raton, FL: CRC Press, Inc., 1979) pp. 9–57.

Poginsky, B., B. Blomeke, A. Hewer, D.H. Phillips, L. Karbe, and H. Marquardt. "[32]P-postlabeling Analysis of Hepatic DNA of Benthic Fish from European Waters. *Proc. Am. Assoc. Cancer Res.* 31(Abstract 568):96 (1990).

Poirier, M.C. "The Use of Carcinogen-DNA Adduct Antisera for Quantitation and Localization of Genomic Damage in Animal Models and the Human Population," *Environ. Mutagenesis* 6:879–887 (1984).

Poirier, M.C., S.J. Lippard, L.A. Zwelling, H.M. Ushay, D. Kerrigan, C.C. Thill, R.M. Santella, D. Grunberger, S.H. Yuspa. "Antibodies Elicited Against cis-diammine Dichloroplatinum (II)-Modified DNA are Specific for cis-diammine Dichloroplatinum (II)-DNA Adducts Formed in vivo and in vitro," *Proc. Natl. Acad. Sci. U.S.A.* 79:6443–6447 (1982).

Rahn, R., S. Chang, J.M. Holland and L.R. Shugart. "A Fluorometric-HPLC Assay for Quantitating the Binding of Benzo[a]pyrene Metabolites to DNA," *Biochem. Biophys. Res. Commun.* 109:262–269 (1982).

Randerath, K., M. Reddy and R.C. Gupta. "[32]P-postlabeling Analysis for DNA Damage," *Proc. Natl. Acad. Sci. U.S.A.* 78:6126–6129 (1981).

Randerath, K., E. Randerath, T.F. Danna, K.L. van Golen and K.L. Putnam. "A New Sensitive ^{32}P-postlabeling Assay Based on the Specific Enzymatic Conversion of Bulky DNA Lesions to Radiolabeled Dinucleotides and Nucleoside 5′Monophosphates," *Carcinogenesis* 10:1231–1239 (1989).

Razin, A., and A.D. Riggs. "DNA Methylation and Gene Function," *Science.* 210:604–609 (1980).

Rydberg, B. "The Rate of Strand Separation in Alkali of DNA of Irradiated Mammalian Cells," *Radiat. Res.* 61:274–285 (1975).

Saiki, R.K., C. Chang, C.H. Lerenson, T.C. Warren, C.D. Boehm, H.H. Kazazian and H.A. Erlich. "Diagnosis of Sickle Cell Anemia and β-thylassemia With Enzymatically Amplified DNA and Nonradioactve Allele-specific Oligonucleotide Probes," *N. Engl. J. Med.* 319:537–541 (1988b).

Saiki, R.K., D.H. Gelfand, S. Stoffel, K.B. Mullis and H.A. Erlich. "Primer-directed Enzymatic Amplification of DNA With a Thermostable DNA Polymerase," *Science* 239:487–491 (1988a).

Saiki, R.K., J.S. Scharf, F. Faloong, K.B. Mullis, G.T. Horn, H.A. Erlich and N. Arnheim. "Enzymatic amplification of β-globin Genomic Sequences and Restriction Site Analysis for Diagnosis of Sickle Cell Anemia," *Science* 230:1350–1354 (1985).

Sancar, A., and G.B. Sancar. "DNA Repair Enzymes," *Ann. Rev. Biochem.* 57:29–67 (1988).

Sanders, M.J., R.S. Cooper, R. Jankowiak, G.J. Small, V. Heising and A.M. Jeffrey. "Identification of Polycyclic Aromatic Hydrocarbon Metabolites and Adducts in Mixtures Using Fluorescence Line-narrowing Spectrometry," *Anal. Chem.* 58:816–820 (1986).

Sandhu, S.S., and W.R. Lower. "In situ Assessment of Genotoxic Hazards of Environmental Pollution," *Toxicol. Indust. Health* 5:73–83(1989).

Santella, R.M. "Application of New Techniques for the Detection of Carcinogen Adducts to Human Population Monitoring," *Mutat. Res.* 205:271–282 (1988).

Santella, R.M., F. Gasparo and L. Hsieh. "Quantitation of Carcinogen-DNA Adducts with Monoclonal Antibodies," *Prog. Exp. Tumor Res.* 31:63–75 (1987).

Schmid, W. "The Micronucleus Test for Cytogenetic Analysis," in *Chemical Mutagens: Principles and Methods for Their Detection, Vol. 6,* A. Hollaender, Ed. (New York: Plenum Press, 1976) pp. 31–53.

Shapiro, H.M. *Practical Flow Cytometry,* 2nd Edition (New York: Alan R. Liss, Inc., 1988) p. 353.

Shugart, L.R. "An Alkaline Unwinding Assay for the Detection of DNA Damage in Aquatic Organisms," *Marine Environ. Res.* 24:321–325 (1988a).

Shugart, L.R. "Quantitation of Chemically Induced Damage to DNA of Aquatic Organisms by Alkaline Unwinding Assay. *Aquatic Toxicol.* 13:43–52 (1988b).

Shugart, L.R. "Biological Monitoring: Testing for Genotoxicity," in *Biological Markers of Environmental Contaminants,* J.F. McCarthy, and L.R. Shugart, Eds. (Boca Raton, FL: Lewis Publishers, Inc., 1990a) pp. 205–216.

Shugart, L.R. "DNA Damage as an Indicator of Pollutant-induced Genotoxicity," in *13th Symposium on Aquatic Toxicology Risk Assessment,* W.G. Landis, and W.H. van der Schalie, Eds. (Philadelphia: ASTM Publishers, 1990b) pp. 348–355.

Shugart, L.R. "5-Methyl Deoxycytidine Content of DNA from Bluegill Sunfish (*Lepomis macrochirus*) Exposed to Benzo[a]pyrene," *Environ. Toxicol. Chem.* 9:205–208 (1990c).

Shugart, L.R., S.M. Adams, B.D. Jimenez, S.S. Talmage and J.F. McCarthy. "Biological Markers to Study Exposure in Animals and Bioavailability of Environmental Contaminants," in *ACS Symposium Series No. 382, Biological Monitoring for Pesticide Exposure: Measurement, Estimation, and Risk Reduction,* R.G.M. Wang, C.A. Franklin, R.C. Honeycutt and J.C. Reinert, Eds. (Washington, D.C.: American Chemical Society, 1989) pp. 86–97.

Shugart, L.R., J.M. Holland and R. Rahn. "Dosimetry of PAH Carcinogenesis: Covalent Binding of Benzo[a]pyrene to Mouse Epidermal DNA," *Carcinogenesis* 4:195–199 (1983).

Shugart, L.R., and J. Kao. "Examination of Adduct Formation In Vivo in the Mouse Between Benzo[a]pyrene and DNA of Skin and Hemoglobin of Red Blood Cells," *Environ. Health Perspec.* 62:223–226 (1985).

Shugart, L.R., J.F. McCarthy, B.D. Jimenez and J. Daniel. "Analysis of Adduct Formation in the Bluegill Sunfish (*Lepomis macrochirus*) Between Benzo[a]pyrene and DNA of the Liver and Hemoglobin of the Erythrocyte," *Aquatic Toxicol.* 9:319–325 (1987).

Stein, J.E., W.L. Reichert, M. Nishimote and U. Varanasi. "^{32}P-postlabeling of DNA: A Sensitive Method for Assessing Environmentally Induced Genotoxicity," in *Procedings Oceans 89, Vol. 7.* (Washington, D.C.: IEEE, 1989) pp. 385–390.

Thompson, R.A., G.D. Schroeder and T.H. Connor. "Chromosomal Aberrations in the Cotton Rat *Sigmodon hispidus* to Hazardous Waste," *Environ. Mol. Mutagen.* 11:359–367 (1988).

Tice, R.R., B.G. Ormiston, R. Boucher, C.A. Luke and D.E. Paquette. "Environmental Biomonitoring with Feral Rodent Species," in *Short-term Bioassays in the Analysis of Complex Environmental Mixtures, Vol. V*, S.S. Sandhu, D.M. Demanine, M.J. Mass, M.M. Moore and J.L. Mumford, Eds. (New York: Plenum Press, 1987).

Tomer, K.B., M.L. Gross and M.L. Deinzer. "Fast Atom Bombardment and Tandem Mass Spectrometry of Covalently Modified Nucleosides and Nucleotides: Adducts of Pyrrolizidine Alkaloid Metabolites," *Anal. Chem.* 58:2527–2534 (1986).

Uziel, M., C.K. Koh and W.E. Cohn. "Rapid Ion-exchange Chromatographic Microanalysis of Ultraviolet-absorbing Materials and its Application to Nucleosides," *Anal. Biochem.* 25:77–98 (1965).

Varanasi, U., W.L. Reichert, B.-T. Eberhart and J. Stein. "Formation and Persistence of Benzo[a]pyrene-diolepoxide-DNA Adducts in Liver of English Sole (*Parophrys vetulus*)," *Chem.-Biol. Interact.* 69:203–216 (1989b).

Varanasi, U., W.L. Reichert, amd J. Stein. "^{32}P-postlabeling Analysis of DNA Adducts in Liver of Wild English Sole (*Parophrys vetulus*) and Winter Flounder (*Pseudopleuronectes americanus*)," *Cancer Res.* 49:1171–1177 (1989a).

Walton, D.G., A.B. Acton and H.F. Stich. "DNA Repair Synthesis in Cultured Fish and Human Cells Exposed to Fish S9-activated Aromatic Hydrocarbons," *Comp. Biochem. Physiol.* 86C:399–404 (1987).

Watson, W.P. "Post-radiolabeling for Detecting DNA Damage," *Mutagenesis.* 2:319–331 (1987).

Weng, Q.-M., W.M Hammargren, D. Slowikowski, K.H. Schram, K.Z. Borysko, L.L. Worting and L.B. Towsand. "Low Nanogram Detection of Nucleotides Using Fast Atom Bombardment-Mass Spectrometry," *Anal. Biochem.* 178:102–106 (1989).

Wilson, V.L., and P.A. Jones. "Inhibition of DNA Methylation by Chemical Carcinogens *In Vitro, Cell* 32:229–246 (1983).

Wirgin, I.I., D. Currie, C. Gorunwald and S.Y. Garte. "Molecular Mechanisms of Carcinogenesis in a Natural Population of Hudson River Fish," *Proc. AACR Mtg.* 30:194 (1989).

Wogan, G.N., and N.J. Gorelick. "Chemical and Biochemical Dosimetry to Exposure to Genotoxic Chemicals," *Environ. Health Perspec.* 62:5–18 (1985).

Yang, X.Y., V. Deleo and R.M. Santella. "Immunological Detection and Visualization of 8-Methyl Oxypsoralen-DNA Photoadducts," *Cancer Res.* 47:2451–2455 (1987).

CHAPTER 4

HISTOPATHOLOGIC BIOMARKERS

David E. Hinton, Paul C. Baumann, George R. Gardner, William E. Hawkins, Jerry D. Hendricks, Robert A. Murchelano, and Mark S. Okihiro

ABSTRACT

Histopathologic alterations in fish tissues are biomarkers of effect of exposure to environmental stressors. This category of biomarkers has the advantage of allowing one to examine specific target organs and cells as they are affected under in vivo conditions. Furthermore, for field assessment, histopathology is the most rapid method of detecting adverse acute and chronic effects of exposure in the various tissues and organs comprising an individual finfish or shellfish.

Although very numerous, environmental pollutants act through a finite number of ultimately toxic mechanisms to produce a finite number of histopathologic lesions. While only broad generalizations regarding the specific etiologic agent(s) responsible for such lesions can be made, the ability to determine the magnitude of toxic impairment strengthens efforts to predict eventual impact on the survival of the affected individual and, in some instances, the population. Confounding issues for histopathologic biomarkers include distinguishing changes caused by anthropogenic toxicants from those due to infectious disease, normal physiologic variation, or natural toxins.

In this chapter, we emphasize the use of biomarkers of toxicant-induced histopathologic change and attempt to distinguish these lesions from those caused by other etiologies. We limited present biomarkers (those ready for immediate use) to reliable, well-documented lesions, based on experience to date, in liver, ovary, musculoskeletal system, and skin. Other, potential or future biomarkers are also identified. These lesions may have strong field relevance but lack laboratory toxicologic support data, or they may have been seen in the laboratory but not in the field. We also describe an additional complement of research tools, available to the experimental pathologist, which may provide even more meaningful biomarkers. These powerful, integrative, laboratory-proven approaches correlate structural with biochemical and physiological alteration and yield quantitative endpoints amenable to statistical analysis. Many, however, lack field verification. The chapter also details the proper application of histopathologic biomarkers, identifies research needs to increase sensitivity and use of these approaches, and presents approaches for collection of biota, which increase precision of analysis and interpretation of results.

INTRODUCTION

Histopathologic biomarkers are lesions that signal effects resulting from prior or ongoing exposure to one or more toxic agent(s). Because myriad environmental toxicants act through a finite number of final pathways, an individual lesion type is rarely pathognomonic of exposure to a single toxicant. However, when field investigations detect a lesion, in higher than anticipated prevalence at a highly localized site, those toxicants consistently shown to be associated with that lesion in laboratory exposures emerge as suspects. Correlated analytical chemistry is then recommended and driven by the host biomarker response. We define biomarker as any contaminant-induced physiological or biochemical change in an organism that leads to the formation of a lesion in cells, tissues, or organs.

By selecting the appropriate target organ, a variety of morphologic indicators of toxicity become biomarkers of effect.

How do histopathologic biomarkers relate to other biomarkers? Some aspects of integration of the various biomarkers are shown in Figure 1. Histopathologic biomarkers are higher level responses and often signify prior metabolism, and macromolecular binding. Most chemicals that are potentially genotoxic require metabolic activation to an ultimate form that binds covalently, forming adducts to DNA. If the adduct is not repaired and persists, subsequent changes lead to a multistep process, that could result in acute toxicity (cell death) or perhaps abnormal growth and tumor formation. In the latter case, the histopathologic biomarker is a higher level response following chemical and cellular interaction.

Similarly, exposure to a xenobiotic might induce the formation of a specific enzyme. Subsequent exposure could lead to increased metabolism by the induced enzyme, resulting in levels of toxic intermediates that exceed cellular protective mechanisms. In this way, induction and metabolism could lead to cellular toxicity and

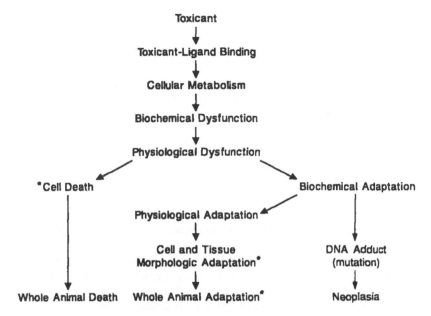

Figure 1.   Integration of histopathologic biomarkers with physiologic and biochemical approaches.

death, subsequently detected as tissue necrosis, a histopathologic biomarker. Direct inhibition of the plasma membrane enzyme ATPase leads to increased sodium levels in cytosol and resultant cell swelling. The histopathologic biomarker, hepatocytomegaly, may represent a higher level response signifying prior enzyme inhibition.

Finally, to withstand disease, host defense mechanisms (immune system) must be operative. Xenobiotic exposure may reduce immune competence, leading to infectious disease, possibly neoplasia or death. Certain histopathologic biomarkers may reflect prior contaminant-induced reduction in host defenses.

From the ongoing discussion, it should be obvious that histopathologic biomarkers are higher level responses, reflecting prior alteration in physiological and/or biochemical function. The integration of these functions is illustrated in Figure 1.

We consider it important to distinguish between present and future biomarkers. Almost without exception, current biomarkers, recommended by us, are lesions shown in laboratory studies to be due to chemical toxicity and verified by detection in tissues of bony fishes and selected shellfishes inhabiting polluted sites. Other (i.e., future) biomarkers have been reported in histopathologic investigations of laboratory toxicant exposures. Field investigations, however, have yet to demonstrate the relevancy of these potential biomarkers for them to be accepted as useful. One lesion, atypical vacuolation in the liver parenchyma of winter flounder (*Pseudopleuronectes americanus*) and other flatfish species, detected in field studies, has been recommended as a present biomarker, but is without verification of its toxicopathic basis in laboratory exposure studies.

Biomarker responses may involve all levels of biological organization. For example, changes may be seen in the distribution of molecules, such as glycoproteins on cell surfaces; organelle number, volume, morphology or distribution; cell number, volume, morphology, or distribution; and organ volume or relative weight. Morphologic expression of prior exposure, therefore, is an integrator of biochemical and physiological change.

Basic Considerations in Histopathologic Analysis

How does the histopathologist distinguish between toxicant-induced cell death and changes related to inadequate or delayed fixation? Death of cells without death of the organism (as in some forms of acute toxicity) is followed by a series of cellular reactions. Cells targeted by the toxicant and lethally injured reveal necrotic changes in nuclear morphology and cytoplasmic staining. Surviving cells may respond by phagocytosis of dead or dying cells. In addition, inflammatory cells may respond to chemotactic substances released by dying cells. In this instance, toxicant targets would be recognized as necrotic cells within an organ showing otherwise normal histology or inflammatory alterations. When tissues are properly fixed immediately after the animal is euthanized, toxicant-induced antemortem necrosis can be differentiated from postmortem changes (Trump et al. 1980) seen in the overall organ.

The extent of toxicant-induced necrosis within a target organ may be used to quantitatively estimate toxic impairment, but certain considerations are essential. Since some organs are collections of functional units containing remarkably heterogeneous (Bolender 1979) cell types in highly localized zones (e.g., proximal tubule of the renal nephron), the toxicant target may be a single cell type within highly localized region(s) of the organ. Other regions may be unaffected. In this instance, the lesion is specific and potentially life threatening, but does not comprise a major portion of the total organ volume.

When the concentration of a toxicant is sufficient to result only in cellular injury but not death, sublethal (adaptive) changes may be observed in affected cells. For a review of these types of alterations, see Trump et al. (1980). Quantitative changes within specific organelles of affected cells are often the structural basis of adaptive cellular injury (Trump et al. 1980; Dixon 1982). Sublethal changes include hydropic degeneration (cellular swelling), accumulation of cytoplasmic inclusions, and changes in cell and nuclear volume. Accumulation of triglyceride in the cytoplasm of affected cells is a common indicator of acute, subacute, and chronic toxicities (Trump et al. 1980; Dixon 1982). This lipid accumulates within vacuoles often occupying large portions of cytoplasm. Glycogen depletion, another example of sublethal injury, often affects hepatocyte staining since this complex carbohydrate does not stain with conventional methods and normally occupies large portions of the cytoplasm.

Inherent Advantages of the Histopathologic Approach

An understanding of in situ cellular, tissue and organ system organization and their spatial relationships is required to appreciate biological effects associated with toxicity in localized portions of an organ and subsequent derangements in fluids, tissues, or cells at other locations. While homogenates of entire organs may be used to chemically detect and estimate levels of marker enzymes and be reflective of change in specific organelles, detection of the cell(s) responsible for such change is not possible due to the inclusion and mixing of various cell types within the homogenate (Bolender 1978). Contrast this with the histologic section which retains in situ relationships and may, in the case of liver, reveal alteration within biliary epithelial cells without obvious change in hepatocytes and other cell types. These results may steer the physiologist and biochemist to appropriate tests (e.g., biliary retention times or identification of metabolites or serum analyses of substances normally cleared in the bile). When such integration between disciplines is achieved in toxicologic investigations, relevant biomarker effects and mechanistic considerations are likely to follow (Hinton and Laurén 1990a; 1990).

No other category of biomarker enables the researcher to examine so many potential sites of injury so rapidly. This feature was particularly well illustrated by Wester and Canton (1986) who studied the medaka (Oryzias latipes) after long-term exposure to B-hexachlorocyclohexane, an isomer of the insecticide, lindane. This estrogen mimetic, caused alterations in gonads, liver, kidney, pituitary, thyroid, spleen, and heart. Affected portions of organs were those known to be targets of estrogen or involved in the metabolism of estrogen-stimulated products.

Histopathologic analysis is particularly relevant to field investigation. This approach rapidly detects in vivo toxicity thereby helping to prioritize sites for more detailed analysis. With fish not easily captured or reared, histopathology represents the only viable alternative to other types of analysis. Histopathologic analysis yields data on a number of organ systems and permits localization of lesions within specific cell types. When methods such as liquid nitrogen preservation for biochemical assays are not available, this may be particularly important. The other emerging importance of histopathologic methods relates to their ability to allow assessment of changes in animals that are simply too small to dissect for biochemical studies. In addition to eggs and larvae, the adults of many small fish species that have value as sentinel organisms may be studied in this way. Using conventional methods of histopathology, it is routinely possible not only to evaluate target organ alteration but to determine the sex and reproductive status of affected animals.

Sources of Error in Using Histopathologic Biomarkers

Assuming that the microscopic anatomy of various organs and tissues of fishes to be evaluated is known, alterations from the expected pattern form the basis for detection of chemically-induced injury. Detection of these alterations depends on proper fixation, processing and staining of tissues. Proper interpretation of histopathologic lesions is also highly dependent on the experience of the investigator. Studies range among a variety of approaches at varying levels of resolution and among lesions in an assortment of tissues. When specific criteria for classification of lesions are applied uniformly, variations in diagnosis are reduced. The subjective nature of morphologic studies has made correlation with other, more quantitative physiological and biochemical approaches difficult (Weibel 1979; 1980; Bolender 1981; Reide and Reith 1980; Rohr et al. 1976; Loud and Anversa 1984). Use of microcomputers and appropriate software, however, now allows acquisition of quantitative morphologic data (i.e., morphometry and stereology) within a reasonable timeframe. Amenable to statistical analysis, this data can be integrated with other quantitative approaches.

Seasonal and hormonal changes. Inter- and intraspecific anatomical variations that fall in the range of normality need to be considered to prevent errors in histopathologic analysis. To ensure accuracy of diagnoses, normal seasonal variations, gender, and hormonal differences are important. However, variations of this nature are reflected in only one or two levels of structural organization and do not negate the acquisition of meaningful data at other levels. For example, glycogen and lipid levels in hepatocyte cytoplasm are likely to change seasonally or in relation to the reproductive cycle. Despite this, the basic architectural pattern of the liver is not altered and detection of many important biomarkers is not compromised.

Infectious disease and parasitism. Whenever histologic lesions are detected in wild populations of fish, other potential etiologies must first be considered before implicating a chemical one. These include infectious disease and parasitism. Viruses, bacteria, fungi, protozoa, and metazoan parasites all can cause degenerative and necrotizing lesions in various organs. Fortunately, the causes of these lesions can be determined histologically by visualization of the offending organism or the resultant inflammatory response. Some viral infections can, however, result in severe parenchymal and epithelial necrosis with minimal or no inflammation. Careful light microscopical examination for inclusions (nuclear and cytoplasmic) and electron microscopical (EM) examination for viral particles will help to eliminate the latter as potential etiologic agents for necrotizing lesions.

Viruses are also possible inducers of neoplastic lesions. Oncogenic viruses cause epithelial tumors in masu salmon (*Onchorhynchus masou*) (Kimura et al. 1981a; Kimura et al. 1981b; Yoshimizu et al. 1987) and are highly suspected as the etiologic agent of epizootic lymphoma in northern pike (*Esox lucius*)(Mulcahy and O'Leary 1970; Papas et al. 1976, 1977). Examination of tumors with EM for viral particles is the first step in establishment of a viral etiology, but with two major caveats: (1) oncogenic retroviruses are often nonproductive once infected cells have undergone neoplastic transformation, and (2) identification of viral particles does not necessarily imply causation, as is the case with papillomas of eels (*Anguilla anguilla*). Although viruses were associated with the lesion, they have not been shown to be oncogenic (McAllister et al. 1977; Ahne and Thomsen 1985). Assays for reverse transcriptase and transmission studies, in addition to EM, will help to rule out viruses as etiologic agents of neoplasms in wild fish.

In addition to virally induced neoplasms, there are two pseudoneoplastic conditions, lymphocystis and X-cell pseudotumors, which can be confused with chemically induced epizootics in wild fish. Lymphocystis occurs in both marine and freshwater fish and is characterized grossly by focal to multifocal, white, multinodular skin lesions and histologically by massive, subepidermal hypertrophy of fibroblasts. Hypertrophied fibroblasts are distinctive with peripheral hyalin capsules, marked karyomegaly, and occasional basophilic cytoplasmic inclusions. The causative agent is a transmissible iridovirus and lymphocystis can be diagnosed with routine LM and EM screening.

X-cell pseudotumors are common and have been reported in at least 23 species of marine fish from three orders; Pleuronectiformes, Perciformes, and Gadiformes (Harshbarger 1984). Lesions occur primarily in the skin, but have also been found in pseudobranch of Gadiform species. Cutaneous lesions are characterized grossly by raised, multinodular, verrucous masses which on section are limited to the epidermis and composed of thick pleated folds. Histologically, there is massive hyperplasia of the epidermal squamous epithelium with infiltration by dense sheets of amorphous, round to polygonal cells with coarsely granular acidophilic cytoplasm and large nuclei with prominent nucleoli (X-cells). There is often severe inflammation in the adjacent dermis, subcutis, and skeletal muscle. X-cell pseudotumors were long believed to represent cutaneous papillomas, but have since been shown to probably represent parasitic infestation by an amebic protistan possibly belonging to the order Hartmannelidae. X-cell pseudotumors can be differentiated from true neoplasms by identification of characteristic X-cells (Dawe 1981) and the accompanying inflammatory reaction.

Ectoparasites (ciliated protozoa and metazoa) are another potential source of diagnostic uncertainty because they can induce lesions in the fins, skin, and gills that mimic those induced by either waterborne or sediment-deposited chemicals. Cutaneous hyperplasia, erosion, and ulceration, are caused by a variety of ectoparasites. Similarly, ectoparasites in the gill have been associated with mucous and epithelial cell hyperplasia, lamellar capillary aneurysms, and lamellar clubbing and fusion. Examinations of the skin and gill with wet mount preparations, at the time of collection, will help eliminate ectoparasites as the cause of noninflammatory cutaneous and gill lesions. Certain lesions, such as respiratory epithelial hyperplasia in gill filaments and lamellae, can be the result of past ectoparasitic infestations (e.g., copepods), so that the absence of infectious or parasitic agents even as demonstrated in a wet mount concurrently taken with the tissue for sectioning does not fully eliminate them as etiologic agents of the condition.

All residual lesions from both infectious disease and parasitism should be regarded as additional sources of uncertainty. Macrophage aggregates, commonly found in the kidney, spleen, liver, and heart of fish exposed to chemical toxicants, are also abundant in fish with healing bacterial, fungal, or protozoan infections. Microsporidian and myxosporidian infections (Hoffman et al. 1962) can cause musculoskeletal lesions that may mimic deformities induced by a number of different heavy metals and organophosphate pesticides.

A final source of uncertainty is disease (infectious or neoplastic) secondary to the debilitation and immunosuppression caused by a toxicant. Fish culturists are very aware that a variety of environmental stressors affect the immune system and result in disease outbreaks and mortality (Anderson 1990). In reviewing immunosuppression and disease, Anderson (1990) cites various challenge tests (metals and corticosteroid drugs) that demonstrated suppression of disease resistance in exposed fish. Interestingly, Wolf (1988), after citing Pierce et al. (1978), who surveyed occurrence of tumors in wild fish populations and linked tumor prevalence with the presence of environmental contaminants, suggested that the tumor incidence may have in part reflected immunosuppression and uncontrolled proliferation of cells. Although increased prevalence of certain tumors, hepatocellular and cholangiocytic in liver and specific skin neoplasms signal exposure to xenobiotic initiators and possibly promoters of carcinogenesis, other factors such as toxicant induced reduction in neoplastic cell recognition and killing may be extant.

It is not our intent to review biomarkers of immunotoxicity in fish. This is the subject of an accompanying chapter (Weeks et al., Chapterr 5 in this book). However, some histopathologic alterations due to disease states may have resulted from a primary chemical toxicity. Despite the importance of linking toxicant exposure to immunosuppression and to epizootics of infectious disease in feral fishes, attempts to do so have not been successful. Anderson (1990) regards the large number of factors involving host, infectious agents, and environment as explanation for this failure. To make this linkage, Anderson (1990) recommends a multitiered approach including observations on fish populations in the field, studies of caged fish in the field, in vivo laboratory exposures, and in vitro laboratory assays. Field investigations with redbreast sunfish (*Lepomis auritus*) in a Tennessee stream receiving a single point source industrial discharge revealed greatest alteration in a battery of bioindicators at the discharge site and a downstream gradient at three subsequent sites. Hepatic volume occupied by parasites was greatest and functional hepatic parenchymal index lowest at the discharge site suggesting depression in host resistance (Adams et al. 1989). Despite apparent associations between contaminants and immunocompetency, histopathologic manifestations, restricted to infectious disease, cannot be reliably used as a present biomarker of toxicant-induced, immunocompetence reduction in fish.

PRESENT HISTOPATHOLOGIC BIOMARKERS

Although the liver is not the only site for present and future biomarkers as previously defined, it is certainly the primary locus based on current experience (Meyers and Hendricks 1985; Murchelano and Wolke 1985; Wolke et al. 1985; Myers et al 1987; Harshbarger and Clark 1990; Kranz and Dethlefsen 1990; Vogelbein et al. 1990). There are several reasons for this. First, the liver of teleosts is the major site of the cytochrome P450-mediated, mixed-function oxidase system (Stegeman et al. 1979). This system inactivates some xenobiotics, while activating others to their toxic forms. Secondly, nutrients derived from gastrointestinal absorption are stored in hepatocytes and released for further catabolism by other tissues (Walton and Cowey 1982; Moon et al. 1985). Third, bile synthesized by hepatocytes (Schmidt and Weber 1973; Boyer et al. 1976) aids in the digestion of fatty acids and carries conjugated metabolites of toxicants (Gingerich 1982) into the intestine for excretion or enterohepatic recirculation. Fourth, the yolk protein, vitellogenin, destined for incorporation into the ovum, is synthesized entirely within the liver (Vaillant et al. 1988). Receptors in the liver must bind the hormone, estradiol, for initiation of the signal to begin synthesis of this essential reproductive component. Given the liver's role in various key functions and its metabolic capacity, the hepatotoxic effect of various toxicants is not surprising. It has been the subject of recent reviews including mammalian (Arias et al. 1988) and teleost (Hinton and Lauren 1990a) species.

Hepatic Biomarkers

Hepatocellular Necrosis and Sequelae

Coagulative necrosis, associated with sudden cessation of blood flow to an organ and damage by toxic agents, represents a meaningful biomarker of exposure. With coagulative necrosis, shapes of cells and their tissue arrangement are maintained, facilitating recognition of the organ and tissue. Necrotic changes occur after cell death and represent the sum of degradative processes. These changes, useful in determining which cells died and underwent necrosis prior to death of the animal or fixation of its organs, are biomarkers of exposure. Coagulative hepatic necrosis must be distinguished from necrosis due to postmortem change. Here the timely administration of fixative is needed. In cases where this was not achieved, postmortem change would be reflected in all organs of the individual while coagulative hepatic necrosis would be anticipated to be focal or multifocal and within target organ(s). In addition, the process of coagulative necrosis may release chemotactic factors resulting in a localized inflammatory response. Inflammation that is spatially related to focal or multifocal necrosis would signify that the necrotic process preceded death of the host.

A substantial amount of information has been accumulated associating coagulative hepatocellular necrosis with exposure to anthropogenic environmental toxicants in both mammals and fish (Wyllie et al. 1980; Meyers and Hendricks 1985; Pitot 1988). As information from correlated biochemical and morphologic studies increases, refinement of this important biomarker may provide additional ways to identify general classes of causative agents. While this biomarker signifies prior exposure to toxic chemicals, we cannot rule out the possibility that natural, plant-derived toxins (Hendricks et al. 1981) may be etiologic agents. Currently, hepatocellular coagulative necrosis is a useful biomarker of anthropogenic toxicant exposure.

Hyperplasia of Regeneration

Following necrosis, surviving cells (presumptive stem cells) undergo hyperplasia, thereby regenerating needed hepatocytes to replace those lost. Regenerating cells are small and basophilic, forming small islands of irregular shape. Regenerative islands may be quantified either by using colchicine and establishing the ratio of metaphase to prophase nuclei, by using tritiated thymidine followed by autoradiography (Zuchelkowski et al. 1981; 1986), or by immunohistochemistry using monoclonal antibodies directed against the synthetic thymidine analogue, bromodeoxyuridine (Droy et al. 1988; Miller et al. 1986). In the absence of evidence of prior infection, this biomarker has high toxicological significance.

Hyperplasia is indicative of extensive prior necrosis from either toxicant exposure or infectious disease processes. In the absence of evidence for parasitic infestation (e.g., fibrotic tracks or cystic spaces) or prior infection by microorganisms, regenerative hyperplasia is a good biomarker of exposure to toxicants. The presence of specific types of inflammatory cells may also be useful in differentiating infectious from toxicant-derived etiologies. These characteristics plus the concurrent presence of remaining necrotic hepatocytes facilitate differentiation of regenerative hyperplasia from basophilic focal and nodular change associated with neoplasia (discussion to follow).

A good example of the utility of regenerative foci as a histopathologic biomarker was a recent epizootic of mortalities in Atlantic salmon (*Salmo salar*) maintained in sea pen culture in Puget Sound, Washington (Kent et al. 1988). Salmon in a sea pen at one site showed normal growth and development, while mortality occurred in a similar stock fed identical rations but maintained in a different location. Gross necropsy, assays for infectious disease, and histopathology indicated that hepatic necrosis was the underlying cause of death. Survivors from the affected pen, held for weeks under standard laboratory conditions, developed regenerative islands of small basophilic hepatocytes with increased mitotic activity. The histopathologic finding of hepatic necrosis followed by compensatory regenerative repair in surviving fish illustrates the usefulness of hepatocellular regeneration as a prolonged indicator of prior necrosis.

Bile Ductular/Ductal Hyperplasia

Recent investigations (Hampton et al. 1985; 1988; 1989) in trout have shown that this liver is normally enriched in biliary epithelial cells. However, profiles of bile ductules (cuboidal epithelium) and ducts (columnar epithelium with mucus) are infrequent except near the porta hepatis. With ductular/ductal hyperplasia, profiles of these biliary passageways are numerous and contiguous, with abundant branching and coiling. Cytologic features of hyperplastic epithelial cells are normal. This lesion is of a chronic duration and has been a consistent finding in wild fish from chemically contaminated sites (Murchelano and Wolke 1985; Hayes et al. 1990). In western Lake

Ontario, proliferative biliary diseases (cholangiohepatitis and cholangiofibrosis) of white suckers (*Catostomus commersoni*) were associated with bile duct neoplasms in polluted harbors (Hayes et al. 1990). The proliferative biliary disease was less severe in livers of fish from reference sites. These authors suggested that proliferative bile duct epithelial changes could predispose fish to initiation and promotion of bile duct neoplasia (Hayes et al. 1990).

Hepatocytomegaly

Hepatocellular hypertrophy is a type of hepatocytomegaly characterized by organelle hyperplasia within the cytoplasm of hepatocytes, with enlarged cellular diameter but without nuclear changes. This condition may lead to a net gain in the dry mass of a tissue or organ. One way in which hepatocytes undergo hypertrophy is through proliferation of endoplasmic reticulum. EM studies have shown this after exposure of mullet (*Mugil cephalus*) to the polynuclear aromatic hydrocarbon, 3-methylcholanthrene (Schoor and Couch, 1979), and in channel catfish (*Ictalurus punctatus*) after subacute exposure to Aroclor 1254 (Klaunig et al., 1979). Histologic preparations of liver with this condition show swollen hepatocytes with eosinophilic, hyalinized cytoplasm resembling ground glass. Nuclei typically are unaltered.

Megalocytosis is a second type of hepatocytomegaly and is characterized by marked cellular and nuclear enlargement. Enlarged nuclei often contain false and real inclusions, and multinucleated megalocytes may be seen. A condition involving megalocytosis, termed megalocytic hepatosis, is the most frequently encountered idiopathic lesion in the liver of English sole (*Parophrys vetulus*) from contaminant-laden sites within Puget Sound, Washington (Myers et al. 1990). These authors have interpreted megalocytosis as manifestation of chronic toxicity of these sediment contaminants. Megalocytosis was seen in fish from chemically contaminated sites in the Kanawha River of West Virginia (Hinton and Lauren, unpublished observations), and in sea pen cultures of Atlantic Salmon in Puget Sound (Kent et al., 1988). Megalocytosis has been produced in the laboratory in trout (*Oncorhynchus mykiss*) exposed to pyrrolizidine (senecio) alkaloids (Hendricks et al., 1981) and medaka (*Oryzias latipes*) exposed to diethylnitrosamine (Hinton et al., 1988). Megalocytes are probably sublethally injured hepatocytes and are able to survive for months (Groff, J., personal communication; Kent et al., 1988). With light microscopy, megalocytes are from three to five times larger than hepatocytes and their enlarged nuclei frequently show eosinophilic inclusions.

A third type of hepatocytomegaly arising from marked swelling of perinuclear endoplasmic reticulum cisternae is seen in vacuolated cells of liver (Bodammer and Murchelano 1990). Histologically, affected cells possess a clear cytoplasm, small compact nuclei and are markedly vacuolated. In the initial description, a high prevalence of winter flounder (*Pseudopleuronectes americanus*) from Boston Harbor, greater than 25 cm body length, were said to contain groups of vacuolated cells in acinar and tubular patterns which organizationally and cytochemically seemed more like ductal epithelial cells than hepatic parenchymal cells (Murchelano and Wolke 1985). Moore et al. (1989) examined younger specimens from the same harbor and described aspects of the pathogenesis of the lesion. They concluded that the earliest lesion, abnormal vacuolation, was in biliary preductular (Hampton et al. 1988) epithelial cells. In older fish (i.e., longer than 30 cm), they reported vacuolated hepatocytes as well as biliary epithelial cells. Bodammer and Murchelano (1990) described ultrastructural features in two affected female flounder between 38 and 46 cm in length. They reported vacuolation was restricted to hepatocytes. Due to the apparent involvement of both cell types, we refer to this condition as hepatocellular vacuolation. Hepatocellular vacuolation of winter flounder (Murchelano and Wolke 1985; Gardner et al. 1989b; Moore et al. 1989; Bodammer and Murchelano 1990) and windowpane flounder (*Scophthalmus aquosus*) (Murchelano and Wolke 1985) may be regarded as a variant of hydropic degeneration. Particularly in winter flounder, the condition is encountered in high prevalence in Boston Harbor, Massachusetts, and nearby estuaries where it is highly correlated with cholangiocytic neoplasms, less well with hepatocellular neoplasms (Harshbarger and Clark, 1990) and may be seen in livers free of neoplasia. The lesion has also recently been detected in rock sole (*Lepidopsetta bilineata*) and starry flounder (*Platicthys stellatus*) from contaminated sites in Puget Sound, Washington (Stehr et al., 1990). In the case of this lesion, the magnitude and unique nature of the cellular alterations, along with the relatively high prevalence of fish affected at contaminated sites point to this as a specific biomarker. Even in the absence of laboratory studies demonstrating induction of similar changes by exposure to toxicants, we recommend its use.

Foci of Cellular Alteration — Staining or Tinctorial Change

An early stage in the stepwise histogenesis of hepatic neoplasia is the formation of foci of cellular alteration. This term includes those foci detected by conventional hematoxylin and eosin preparations (foci of tinctorial or staining alteration) and foci detected by enzyme histochemical procedures (foci of enzyme alteration). Only limited use has been made of enzyme histochemistry in fish carcinogenesis studies (Nakazawa et al. 1985; Hinton et al. 1988; Lauren et al. 1990) and extension from laboratory to field investigations has not been made. For this reason, we discuss foci of staining alteration here and present these as present biomarkers. Enzyme altered foci are discussed under "future biomarkers".

A review of serial progression studies in laboratory exposures of fish to chemical carcinogens generally reveals loss of hepatocyte glycogen and necrosis initially (Scarpelli et al. 1963; Stanton 1965; Egami et al. 1981; Couch and Courtney 1987; Hinton et al. 1988; Lauren et al. 1990; Hinton et al. 1991). At some time after this initial toxicity, foci of tinctorially-altered hepatocytes appear. When viewed after conventional paraffin processing as above, cells of these foci may be basophilic (Egami et al. 1981), basophilic or eosinophilic (Stanton 1965), and basophilic or eosinophilic or clear (Couch and Courtney 1987; Hinton et al. 1991). In addition, a fatty vacuolated focus is described below. Foci are usually spherical to oval and show identical architecture with that of surrounding cells.

Cells of basophilic foci appear to differ from other hepatocytes only in the intensity of their staining which is apparently due to glycogen depletion (Scarpelli et al. 1963; Lauren et al. 1990) and reduced cellular volume with real or apparent increase in granular endoplasmic reticulum and mitochondria. The nucleic acids of ribosomes have affinity for the basic dye, hematoxylin, and tumor cells of trout liver contain increased levels of cytoplasmic RNA (Scarpelli et al. 1963). Taken together, these cytologic alterations are sufficient to explain enhanced cytoplasmic staining with hematoxylin.

Cells of eosinophilic foci frequently show variation in size as well as their altered tinctorial properties. Hendricks et al. (1984) reviewed focal alterations in rainbow trout (*Oncorhynchus mykiss*) after exposure to one of various carcinogens. Eosinophilic foci were invariably small but component hepatocytes were hypertrophic and contained enlarged and abnormally shaped nuclei. These hepatocytes were apparently reduced in glycogen content. Other, larger, eosinophilic foci contained enlarged hepatocytes with homogeneous, hyalinized cytoplasm. Peripheral regions of these foci were sites of lymphocytic infiltration. In figures 10 and 12 of the Hendricks et al. (1984) report, basophilic portions of hepatocyte cytoplasm and the hepatocyte nuclei appear to have been rearranged toward the periphery of the cell. Given this appearance, smooth endoplasmic reticulum proliferation may account for the enhanced eosinophilia and the peripheral basophilic zone. In the sheepshead minnow (*Cyprinodon variegatus*) exposed to diethylnitrosamine, Couch and Courtney (1987) found cells of eosinophilic foci were more pleomorphic than those of basophilic foci.

Clear cell foci are apparently collections of cells that are enriched in glycogen. Hendricks et al. (1984) showed glycogen storage "nodules", cells of which resembled glycogen enriched cells. Hinton et al. (1991) studied serial sections through clear cell foci produced in medaka by exposure to diethylnitrosamine. Cells of such foci were strongly positive by the periodic acid Schiff's reagent preparation for glycogen.

The other type of focal alteration encountered in conventional preparations is that containing vacuolated cells. The round margins of the vacuoles and their positive reactions by fat stains (Hinton et al., unpublished observations) distinguish these from the vacuolated cells or "vacuolar foci" of Murchelano and Wolke (1985). Focal fatty vacuolation of hepatocytes is a response associated with exposure of fish to a variety of different carcinogenic agents (Hendricks et al. 1984), and, with the other focal alterations, apparently precedes other changes in the development of neoplasia.

Focal fatty vacuolation should be distinguished from diffuse fatty change. Generalized (diffuse) fatty change is seen after a variety of hepatotoxic insults but this condition is also seen in vitellogenic females, in fish which store abundant lipids (van Bohemen et al. 1981), and is influenced by nutritional state (Segner and Möller, 1984; Segner and Juario, 1986; Segner and Braunbeck, 1988). Before diffuse fatty change can be reliably used as a biomarker of chemical exposure, the appearance of the liver in the same species of fish, at the same time of year, under normal conditions must be taken into consideration. Therefore, until additional research on the mechanism of fatty liver in representative teleosts and additional field verification is achieved, diffuse fatty change alone is not recommended for inclusion as a present biomarker. It should also be noted that focal fatty change is seen in aging control medaka although its prevalence is greater in diethylnitrosamine treated medaka (Hinton et al. 1991).

Field investigations confirming the relevance of focal alterations have linked these lesions to contaminated sites usually where hepatic neoplasms have also been found, and in satellite lesions in tumor bearing liver. Kranz and Dethlefsen (1990) reported a multiyear investigation of liver alterations in the bottom dwelling dab (*Limanda limanda*) of the southern North Sea. Histologically examining only those livers which showed gross alterations, they noted lesions of cellular alteration within a group of larger lesions termed "neoplastic changes". They did not refer to the lesions as foci but termed them nodules of varying size (<1.5 mm, 1.5 to 10 mm, and >10 mm). Smallest nodules were concentric areas of basophilic, eosinophilic, or vacuolated hepatocytes. These small nodules were usually multiple structures within the same liver and some phenotypes of individual nodules were mixed. Nodules and overt neoplasms were found in dab from the more contaminated sites. Both feral winter and windowpane flounder from contaminated sites in Boston Harbor contained basophilic and eosinophilic foci (Murchelano and Wolke 1985). Hayes et al. (1990) reported on liver lesions in white suckers (*Catostomus commersoni*) from contaminated harbors and reference sites of Lake Ontario. Focal cellular alterations were restricted to basophilic and clear cell phenotypes. They cited rodent carcinogenesis studies suggesting that the exclusion of the eosinophilic cytoplasmic differentiation typical of the "resistant" nodule phenotype may signify different initiating and/or promoting xenobiotics in those versus other sites. Baumann et al. (1990) reported on lesions of feral brown bullhead (*Ictalurus nebulosus*) from two Lake Erie tributaries. Focal hepatocellular alterations were well differentiated lesions usually less than one mm in diameter. Staining differently than surrounding tissue, these lesions were usually basophilic, occasionally acidophilic (eosinophilic), and occasionally clear staining. Myers et al. (1987; 1990) have performed histologic multiyear studies of English sole collected from contaminated and reference sites within Puget Sound, Washington. Basophilic, eosinophilic, and clear cell foci have been reported. Histologically all are reduced in cytoplasmic iron, and rarely contain other liver structures, such as blood vessels, macrophage aggregates and pancreatic acini. Vogelbein et al. (1990) reported on an epizootic of liver neoplasia in the mummichog (*Fundulus heteroclitus*) from a site the sediment of which is heavily contaminated with polycyclic aromatic hydrocarbons. Focal cellular alterations included basophilic, eosinophilic, and clear cells. Apparently a fourth (fatty vacuolated cell phenotype) was included in the clear cell focus.

Hendricks et al. (1984) reviewed results in the rainbow trout model and concluded that, for trout, basophilic foci were the more important foci of the early altered cells. They regarded basophilic foci as microcarcinomas. Hinton et al. (1988) studied medaka foci with conventional and enzyme histochemical procedures. Diethylnitrosamine exposure was associated with formation of basophilic foci whose enzyme and tinctorial properties were identical to cells of eventual hepatocellular neoplasms. Recently, Hinton et al. (1991) showed that the basophilic phenotype was expressed in the majority of the neoplasms. Eosinophilic and clear cell phenotypes were apparently restricted to focal lesions in medaka.

Foci of cellular alteration including basophilic, eosinophilic, clear, and fat vacuolated are associated with exposure of various fish to chemical carcinogens in the laboratory. Their environmental relevance has been confirmed in field investigations. Foci of cellular alteration are recommended as present biomarkers.

Hepatic Adenoma

This lesion, in rainbow trout, is thought to represent an enlargement of the basophilic focus previously described (Hendricks et al. 1984; Nunez et al. 1991). Apparently clear cell and eosinophilic variants also occur in English sole (Myers et al. 1987) and in brown bullheads (Baumann et al. 1990). Hepatocytes appear essentially normal retaining their normal architecture (Myers et al. 1987; Nunez et al. 1991). Proliferation is evident, through compression of surrounding hepatocytes in larger adenomas, although mitoses are rare. This lesion may be microscopic in size but often forms a bulge at the surface.

Baumann et al. (1990) present a different classification. From hepatocellular focal alteration, they describe a larger, more clearly defined subpopulation of hepatocytes, the "hepatocellular nodule". This intermediate stage bridges hepatocellular focal alterations to frank hepatocellular carcinoma. They (Baumann et al. 1990) prefer not to use the term "adenoma" and discuss reasons for this. They do accept "hepatoma" as a substitute for hepatocellular nodule. Necropsies performed at 6 or 9 months after initiation of hepatocarcinogenesis in rainbow trout (Nunez et al. 1991) and compared to necropsies after 12 or 18 months indicate that the adenoma is a transition lesion. Baumann et al. (1990) agree, stating that the bridging lesion may become large enough to bulge at the capsular surface and that it can progress further to become a full blown hepatocellular carcinoma. Agreement exists that early, intermediate, and endpoint stages of the process of hepatocarcinogenesis may be seen in the same organ.

This chapter is not intended to serve as the definitive source for nomenclature on fish liver neoplasia. Rather, we present salient features of lesions which we feel may be used as biomarkers of exposure and effect. Since we are recommending the use of focal cellular alterations (the proximate lesions), the bridging lesions (adenoma or hepatocellular nodule or hepatoma), and the endpoint (hepatocellular carcinoma) as present biomarkers, the reader should not be discouraged. All the above lesions are recommended as present biomarkers.

Hepatocellular Carcinoma

Depending on the age of the tumor at necropsy, this lesion can vary from microscopic to very large, often occupying a major portion of the organ. Cells of trout hepatocellular carcinoma (HCA) are predominantly basophilic (Nunez et al. 1991). In the brown bullhead, tumor cells are usually basophilic but may be eosinophilic or rarely, clear cell (Baumann et al. 1990). The margin is less well defined and invasion into otherwise normal parenchyma is common. In spite of invasiveness, rapid proliferation of this lesion usually causes severe compression of surrounding hepatocytes. The masses of cells are usually solid or trabecular in pattern, resembling engorged tubules. Mitotic figures are numerous and can be bizarre. Tumor cells may be quite pleomorphic with some assuming spindle shapes while others are small, polyhedral in shape and arranged as tight collections with distinct intercellular space (Hendricks et al. 1984; Myers et al. 1987; Couch and Courtney 1987; Hinton et al. 1988; Vogelbein et al. 1990). Infrequently, hepatocellular carcinomas metastasize (Baumann et al. 1990).

Cholangioma

These bile duct tumors are characterized by retention of their ductular architecture, presence of distinct margins resulting in a nodular, well-defined mass. Focal ductal elements may be present as well. Ducts may be cystic with papillary projections, but the lining epithelium is cuboidal to low columnar, simple (i.e., single row), and well differentiated (Hendricks et al, 1984; Myers et al., 1987; Baumann et al. 1990).

Cholangiocarcinoma

This lesion is larger, the component cells are pleomorphic, mitotic figures are common and invasion of surrounding tissues is common. At the margin columns of invading cancer interdigitate with nontumorous liver tissue (Baumann et al. 1990). Biliary epithelial cells of tumor can form sheets, undifferentiated into ductules (Hendricks et al. 1984; Myers et al. 1987).

Mixed Hepato-cholangiocellular Carcinoma

In the rainbow trout, mixed hepato-cholangiocellular carcinomas are seen at an equal or even greater frequency than hepatocellular carcinoma (Nunez et al. 1988; Nunez et al. 1991). Review of other reports on hepatic tumors in laboratory and feral fishes reveals similar findings. The close association of hepatocytes and biliary epithelial cells in the tubular teleost liver may account for this duality of components (Hampton et al. 1988). Nunez et al. (1991) discuss the possible origin of this neoplasm and present evidence to suggest that both cell types are indeed neoplastic and not just caught up in the expansion of one or the other neoplastic cell type.

Both laboratory experience and field investigations support the use of cholangiocelluar neoplasms as present biomarkers. A partial listing of species that developed cholangiocytic tumors after exposure to various established carcinogens includes trout (Hendricks et al. 1984; Nunez et al. 1991), estuarine sheepshead minnow

(Couch and Courtney 1987), rivulus (*Rivulus marmoratus*, Koenig and Chasar 1984), danio (Stanton 1965), and guppy (*Lebistes reticulatus*, Simon and Lapis 1984). Dawe et al. (1964) reported cholangiocellular neoplasms in white suckers of a fresh water lake in Maryland. Subsequently, Hayes et al. (1990) have found similar lesions in the same species in Lake Ontario. Winter flounder on the east coast (Murchelano and Wolke 1985; Harshbarger and Clark 1990) and English sole on the west coast (Myers et al. 1987) collected from sites where sediments are contaminated have also developed these lesions.

Field verification. Many carefully studied near shore-marine environments and freshwater lakes are polluted with chemicals as indicated primarily by measurements of certain organics and metals in sediments. For finfish especially, there are correlations between the occurrence and/or prevalence of cancerous diseases of the liver and the degree of chemical contamination in the environment they inhabit. For each of the biomarkers (previously discussed) laboratory toxicopathic link and environmental relevance have been demonstrated. After reviewing epizootiology of neoplasms of bony fish of North America, Harshbarger and Clark (1990) noted that hepatocellular neoplasms were strongly correlated with exposure to chemical contaminants. Carcinogenic risk for the host fish is related to life history and perhaps to species sensitivity.

Ovarian Biomarker

Oocyte atresia

Tolerance to stress is likely to be lower in the reproductive tract than in any other organ system in fish (Gerking 1980). Population decline stemming from reproductive impairment is potentially the most serious biological impact of a toxicant-compromised environment. Despite this potential, only one real reproductive biomarker has been identified: oocyte atresia.

Oocyte or follicular atresia is definitely a normal occurrence in the ovaries of all fish species, but can become pathologic following exposure to xenobiotic compounds (e.g., Johnson et al. 1988; Cross and Hose 1988; 1989; Kurugagran and Joy 1988; Lesniak and Ruby 1982; McCormick et al. 1989; Stott et al. 1981; Stott et al. 1983). Oocyte atresia is characterized by degeneration and necrosis of developing ova, and subsequent infiltration by macrophages. Implementation of oocyte atresia as a biomarker was demonstrated by McCormick et al. (1989). Fathead minnows (*Pimephales promelas*) were exposed to different levels of lowered pH in river water in a controlled field experiment. Morphometric analysis of ovaries from those fish demonstrated reproductive impairment when the ratio of atretic oocyte follicles to total ovarian volume exceeded 20%. This linkage of a histopathologic biomarker (oocyte atresia) to a potential population impact (lowered reproductive success), following a toxic event (lowered environmental pH), should encourage other investigators to test these observations with other species and toxicants.

Musculoskeletal System Biomarker

Vertebral Anomalies

Vertebral deformities (scoliosis, kyphosis, and lordosis) have frequently been reported in wild fish populations (Bengtsson 1974; 1975; Valentine 1975; Van de Kamp 1977; Sloof 1982; Baumann and Hamilton 1984) and are the most promising biomarkers in the musculoskeletal system. Although commonly referred to as "broken back" syndrome, not all spinal deformities are directly attributable to fracture of vertebral bodies.

Two mechanisms have been proposed for development of spinal deformities. The first centers on alteration of structurally critical biological processes involved with the collagenous matrix or mineral content of vertebrae. Causative agents implicated include hereditary defects, defective embryonic development induced by high water temperature or low dissolved oxygen, radiation, vitamin C or B_{12} deficiency, heavy metals, and xenobiotics (Bengtsson 1974; 1975; Mayer et al. 1977; Murai and Andrews 1978).

Vitamin C is an especially critical factor in bone development because of its key role in the production of hydroxyproline, an essential component of the collagenous matrix of bones. Vitamin C deficiency may be caused by a number of factors including oxidant dependent depletion, dietary lack, increased utilization, or decreased absorption. Cadmium depletes vitamin C by increasing utilization or decreasing absorption in mullet (*Mugil cephalus*) (Thomas et al. 1982). In addition, di-2-ethyl hexyl phthalate (DEHP) and several organochlorine compounds (toxaphene, kepone, mirex, Aroclor 1254, and 2,4-DMA) have been shown to alter bone structure by interfering with vitamin C metabolism (Mayer et al. 1977; Mehrle and Mayer 1977). Some fish species lack the ability to synthesize vitamin C and may therefore be more sensitive to dietary deficiency (Yamamoto et al. 1978).

In addition to contributing to decreased vitamin C levels, cadmium may also inhibit the uptake of calcium across the gill (Verbost et al. 1987; Reid and McDonald 1988), which in turn may lead to decreased plasma levels and reduced bioavailability for bone deposition (Roch and Maly 1979). Cadmium (Muramoto 1981), lead (Varanasi and Gmur 1978), and zinc (Saiki and Mori 1955; Sauer and Watabe 1984) may also weaken the structural integrity of bone by displacing calcium from normal binding sites.

The second mechanism leading to vertebral deformity is muscular tetany. Muscular tetany has been induced in the laboratory by electrical current (Spencer 1967), acute temperature change (Brungs 1971), parasitic infestation (*Myxosoma cerebralis*;Hoffman et al. 1962), heavy metals (zinc, cadmium, lead)(Bengtsson 1974;

Bengtsson 1975; Holcombe et al. 1976), organochlorine pesticides (toxaphene, chlordecone) (Merhle and Mayer 1975; Couch et al. 1977; Stehlik and Merriner 1983), trifluralin (Couch et al. 1979b), and organophosphate pesticides (parathion and malathion) (Weis and Weis 1989). Organophosphates cause spinal fractures by inhibiting cholinesterase, which leads to a buildup of acetylcholine at nerve endings, and muscular tetany. Organochlorines appear to cause both muscular tetany and vitamin C depletion.

Spinal deformities are a gross finding, only rarely requiring histopathologic confirmation, and are recommended for inclusion as a biomarker of effect. The pleuripotential status of etiology may limit its use. However, nutritional deficiency may be expected to cause defects in other bones and possibly cartilage. Gill deformities with vertebral defects may argue for a more systemic, metabolic etiology. The spinal curvature accompanying acute neuromuscular spasms after exposure of embryos and larvae to some pesticides (Weis and Weis 1989), are apparently diminished when fish are placed in clean water. Therefore, with careful attention to the age of affected individuals, historical observations of gross deformities in the species, monitoring after placement in clean water, attention to reference populations controlled for age and sex, biomechanical tests, and X-ray analysis, the majority of non-toxic etiologies can be ruled out.

Skin Biomarker

Neoplastic lesions

While epidermal and dermal neoplasms occur in a wide variety of freshwater and marine species, some of these lesions may have a viral etiology. Two others, lymphocystis disease and X-cell lesions in flatfish, cod, goby, and many other genera and species are actually pseudoneoplasms (Harshbarger 1984) and must be differentiated from the toxicant-induced cutaneous neoplasms which we are proposing as biomarkers.

The pseudoneoplasms were discussed above and will receive only brief attention here. Lymphocystis is a virally-induced cutaneous disease characterized histologically by massive hypertrophy of squamous epithelial cells (Weissenberg 1965). X-cell pseudotumors in pleuronectid flatfish from the Pacific coast were initially diagnosed as true neoplasms (epidermal papillomas) (Wellings et al. 1963; 1964; Brooks et al. 1969), but have subsequently been shown to have a Protistan parasitic origin (Dawe 1981). Furthermore, no associations between these lesions and exposure to xenobiotics have ever been demonstrated.

True cutaneous neoplasms, suspected of having a toxic etiology, include epidermal papillomas, squamous cell carcinomas, and chromatophoromas. Although initial reports of lip papillomas in white suckers (*Catostomus commersoni*) from the Great Lakes described the presence of virus (Sonstegard 1977), later studies failed to confirm these findings (Smith et al. 1989). The tumors are now believed to be associated with contaminated habitats. Papillomas occurred in 39% of white suckers from heavily polluted Hamilton Harbor in Lake Ontario and were found in only low prevalences at three reference sites (6% in eastern Lake Ontario, less than 1% in Lake Superior and Lake Huron) (V. Cairns, personal communication; Baumann and Whittle 1988). Hayes et al. (1990) studied epidermal neoplasms in white suckers from industrially polluted areas in Lake Ontario and compared them to lesions in fish from less polluted sites in the Great Lakes. When compared to adjacent skin and skin of nontumorous fish, tumors contained increased levels of glutathione peroxidase and glutathione reductase. Their observations suggest sediment contaminants may be promotional factors for the development of these lesions. High prevalences of epidermal papillomas have also been reported in white suckers from western Lake Erie and the Niagara River (Black 1983a).

Elevated prevalences of lip neoplasms (described as epidermoid carcinomas) were also reported in brown bullhead (*Ictalurus nebulosus*) from polluted areas of the Schuylkill and Delaware rivers, Pennsylvania (Lucke and Schlumberger 1941; Schlumberger and Lucke 1948). In addition, both lip and skin neoplasms have also been found in brown bullhead from the Black River, Ohio (Baumann et al. 1987). Approximately 60% of these lip neoplasms were diagnosed as epidermal papillomas and about 40% as papillary or squamous carcinomas. Although these neoplasms occurred in less than 1% of two-year-old fish from the Black River (N=263), four-year-old fish (N=50) had a 32% prevalence of lip neoplasms and an 18% prevalence of skin neoplasms. Brown bullhead three years or older (N=78) from Buckeye Lake, a reference location, had no skin tumors and only a 1.5% frequency of lip tumors (Baumann et al. 1987). Sediment in the Black River is highly contaminated with polynuclear aromatic hydrocarbons (PAH) (Fabacher et al. 1988). When extracted from this sediment, these PAH compounds induced skin papillomas when painted on the skin of mice (Black et al. 1985). Black (1983b) also induced skin papillomas by painting PAH-rich extract of Buffalo River, New York, sediment on the heads of brown bullhead catfish.

Skin neoplasms were also found in neotenous tiger salamanders (*Ambystoma tigrinum*) from a sewage treatment pond in Texas at a frequency of 15 to 50% over a nine-year period (Rose 1977; 1981). Black bullhead (*Ictalurus melas*) from a sewage treatment pond in Alabama were first reported by Grizzle et al. (1984) as having a high prevalence (73%) of oral papillomas. Caged black bullhead placed in the pond also developed neoplasms, and their aryl hydrocarbon hydroxylase activity was elevated compared to that of bullhead from a nearby pollution-free pond (Tan et al. 1981). While a precise etiology was not postulated, the prevalence of neoplasms declined significantly after chlorination had been reduced by approximately one-third (Grizzle et al. 1984).

Dermal pigment cell neoplasms, described as chromatophoromas or neurilemmomas, occurred in freshwater drum (*Aplodinotus grunniens*) from western Lake Erie and the Niagara River (Black 1983a). Drum from five polluted sites had a significantly

(P<0.05) higher tumor frequency (8.8%, N=305) than did drum collected from two reference areas (2.2%, N=891). However, much variation in tumor frequency occurred among the five polluted sampling sites, somewhat weakening the evidence for a chemical etiology.

Two marine species of drum, nibe (*Nibea mitsukurii*) and koichi (*Nibea albiflora*), found off the Pacific coast of Japan, were also found to have elevated prevalences of chromatophoromas (Kimura et al. 1984). Tumor prevalences as high as 75% were associated with marine environments adjacent to pulp and paper factories, and extracts of sediment and wastewater were found to be mutagenic with both the Rec and Ames assays. Kimura et al. (1984) were able to chemically induce chromatophoromas in laboratory-reared nibe using subcutaneous injections of the PAH, 7,12-dimethyl-benz[a]anthracene, oral administration of N-methyl-N'-nitro-N-nitrosoguanidine (MNNG), and bath immersion with MNNG or nifurpirinol. Kinae et al. (1990) induced melanophore hyperplasia in 70 to 100% of marine catfish (*Plotosus anguillaris*) and chromatophoroma in one *N. mitsukurii* using effluent from a kraft pulp mill which has been epidemiologically linked with identical lesions in wild fish of the same species. This work is especially important because it is one of the few examples where the hypothesis that chromatophoromas in feral fish are caused by exposure to chemical carcinogens has, at least partially, been tested under controlled laboratory conditions.

Chromatophoromas have also been found in two species of Hawaiian butterflyfish (*Chaetodon multicinctus*) and (*C. miliacis*) at prevalences of 50% and 5%, respectively (Okihiro 1988). At least circumstantial evidence supported the possibility of a chemical etiology. Pacific rockfish were sampled from Cordell Bank, 37 km off the coast of central California from 1985 to 1990. Hyperplastic and neoplastic cutaneous lesions, involving dermal chromatophores, were observed in five species: yellowtail (*Sebastes flavidus*), bocaccio (*S. paucispinis*), olive (*S. serranoides*), widow (*S. entomelas*), and chilipepper rockfish (*S. goodei*) (Okihiro et al. 1991). Electron microscopy did not reveal the presence of virus and etiology has not been determined. Yamashita et al. (1990) presented the following evidence suggesting that pigment cell disorders in croaker (*Nibea mitsukurii*) and sea catfish (*Plotosus anguillaris*) may be useful as biomarkers to monitor coastal water pollution for carcinogens: (a) known carcinogens induced the lesions in both species, (b) effluents from a kraft pulp mill located near the contaminated site contained mutagens subsequently identified (four chloroacetones and three alpha dicarbonyl compounds, and (c) effluents induced skin pigment cell hyperplasia in 70 to 100% of laboratory exposed fish.

Papillomas, squamous carcinomas, and chromatophoromas have all been induced by carcinogens in the laboratory and have been found in elevated frequencies on benthic feral fishes associated with contaminated sediments. Thus, these lesions represent useable biomarkers, as long as care is taken to distinguish them from skin lesions having a viral or Protistan parasitic etiology. Field verifications for chromatophoromas and papillomas appears stronger than that for squamous carcinomas. Additional field studies are needed.

Shellfish Biomarkers

Toxicopathic Lesions

Historical information. Knowledge of noninfectious diseases and the effects of toxic substances on shellfish (bivalve molluscs) remains limited and fragmented (Mix 1986a; 1986b). Historically, investigations of the pathologic, physiologic, and biochemical alterations induced in shellfish by chemical toxicants have been limited to laboratory investigations. Field studies have only recently attempted to correlate organ, tissue, and cellular alterations with body burdens of chemical contaminants (Couch 1985; Gardner et al. 1986; 1987; 1991a; 1991b; Gardner and Pruell 1987; Mix et al. 1979).

Laboratory investigations. Past toxicological investigations with shellfish focused on water column exposures to heavy metals, crude oil, refined petroleum products, and some laboratory-proven carcinogens. Recent investigations have also assessed the effect of some organic compounds (i.e., N-nitroso) on shellfish in water column tests. Cellular responses of shellfish to toxic chemicals generally include atypical hyperplasia, atrophy, vacuolation, necrosis, and inflammation in organs/tissues of the excretory, digestive, respiratory, cardiovascular, and reproductive systems (Couch et al. 1979a; Couch 1984; Akberali and Trueman 1985; Gardner and Pruell 1987; Yevich 1989; Gardner et al. 1991a).

The heavy metals cadmium (Cd), silver (Ag), arsenic (As), copper (Cu), and lead (Pb) induce many of the lesions previously described in at least eight bivalve species evaluated in acute and chronic water column tests. Other pathological changes caused by metals include melanization of penaeid shrimp gills, black gill disease, resulting from Cd exposure (Couch 1977; 1978) and copper exposure (Lightner, Personal Communication). Silver deposition (argyrosis) has been reported in basement membranes of digestive diverticula of blue mussels, *Mytilus edulis*, (Calabrese et al. 1984) and Atlantic slippersnail, *Crepidula fornicata*, (Nelson et al. 1983) after chronic exposures. Accumulation of silver was also observed in ocean quahogs, *Arctica islandica*, surveyed at ocean dump sites highly contaminated with Ag and other heavy metals. Chromium produces changes in antennal glands, hepatopancreas, gill, and midgut of grass shrimp (Doughtie and Rao 1984), and midgut and hepatopancreas of penaeid shrimp (Lightner 1983; 1987). Chromium has also been shown to cause gross alterations of "shell disease" (Doughtie et al. 1983). Copper exposure has been associated with inflammation, necrosis, and other alterations in a variety of crustaceans and molluscs (Betzer and Yevich 1975; LaRoche et al. 1973; Calabrese 1984; Yevich and Yevich 1985; Sunila 1986; 1987; 1988; Fujiya 1960). In addition, mineralized concretions that form in the kidney tubules of some bivalve molluscs (*Mercenaria* spp.) after exposure to heavy metals have been shown to correlate with gradients of pollution (Doyle et al. 1978).

Lesions in the epithelium of digestive glands in softshell clams, *Mya arenaria*, exposed to 250 ppb of #2 fuel oil for one year include atrophy and metaplastic change from columnar to squamous pavement epithelium with pronounced basophilia (Gardner et al. 1991b). Degenerative changes were also observed in gonadal follicles and germinal cells. Biological responses of bivalve molluscs to petroleum hydrocarbons in field-controlled (Neff et al. 1987) and actual oil spills, like the Amoco Cadiz (Berthou et al. 1987), are similar to those observed in controlled laboratory studies. In addition, petroleum promotes pathologic alterations in the cardiovascular system of some molluscs including oysters (*Crassostrea virginica*), scallops (*Argopecten irradiens*), hard clams (*Mercenaria mercenaria*), and softshell clams (*Mya arenaria*) (Gardner et al. 1975; 1989b; 1991b). To gain a wider understanding of the toxicological processes that lead to or influence invertebrate toxicopathic diseases, investigations have begun to focus on effects of complex chemical mixtures including sewage sludge, industrial effluent, and contaminated harbor sediments (Rogerson et al. 1985; Gardner et al. 1987). Chemical contaminants characterized in sediment of some estuaries near industrial centers along the New England coastline (i.e., Black Rock Harbor (BRH), Bridgeport, Connecticut) include substances known to be genotoxic, carcinogenic, cocarcinogenic, and tumor promoting (Gardner et al. 1987; 1989b; 1991b). BRH sediment contains PCB, aliphatic hydrocarbons and cycloalkanes, two-ring and polycyclic aromatic hydrocarbons, and metals (As, Cd, Cu, Pb, Cr, Mn, and Ni). The Ames test, Chinese hamster V-79/sister chromatid exchange, and V-79 metabolic cooperation assays confirmed that potentially genotoxic and tumor-promoting chemicals were present in contaminated BRH sediments (Gardner et al. 1987; 1991a).

Field investigations. Bivalve molluscs can serve as an indicator organism of pollutant exposure because they are found as natural populations in both relatively clean and polluted waters around the world (i.e., "Mussel Watch"). The ease with which these organisms can be manipulated in laboratory and field (i.e., caged in situ) makes them a valuable tool in ecosystem risk evaluations. Oysters, blue mussels, *Mytilus edulis*, and hardshell and softshell clams are being used in laboratory and/or field monitoring programs in some regions of the U.S. because they are sensitive indicators of chemical contamination from point and nonpoint sources. Bivalve histopathology is being used to assess the health status of ecosystems because this approach provides information on frequency and distribution of infectious and toxicopathic diseases. Examples of this histopathologic approach to field assessment are provided by surveys on the U.S. coasts of the Gulf of Mexico and New England. Evaluation of oysters in three Gulf of Mexico estuaries provided information on frequencies of known and new diseases and the general relationship between those diseases and human activity (Couch 1985). Softshell clams were similarly surveyed in Quincy Bay, Boston Harbor (Gardner and Pruell 1987). The types of nonspecific, nonproliferative lesions that serve as general indices of chemical exposures in the laboratory have been observed in the field (Couch 1984; 1985; Gardner et al. 1987; 1991a; Yevich 1989). In addition to the nonspecific lesions observed in Quincy Bay softshell clams (i.e., atypical cell hyperplasia, partial to advanced loss of cilia, and inflammation in gills), asynchronous development of gonads was seen. Although reproductive organs appeared normal histologically, ova development was restricted to early stages while spermatid development was advanced. Perhaps these histopathologic observations offer some explanation for the observed significant decline in the softshell clam fishery in Quincy Bay.

At least one histologic lesion is pathognomonic for a specific metal. Silver accumulation in basement membrane and connective tissue of the mollusc, *Crepidula fornicata*, and the blue mussel, *Mytilus edulis*, was seen following chronic (18 months) laboratory exposure to 1 ppb Ag. Furthermore, the condition resembled clinical argyria of mammals (Nelson et al. 1983; Calabrese et al. 1984). In the field, argyria was verified in hardshell clams collected near an offshore dumpsite (Philadelphia, Pennsylvania) where bottom sediments are highly contaminated with heavy metals.

We recommend three categories of toxicopathic histopathologic biomarkers with bivalve molluscs. The conditions for acceptance of these nonproliferative disorders are the same as those used with finfish (i.e., laboratory exposure and field verifications). These include (1) alterations in digestive, excretory and respiratory epithelia, (2) alterations of reproductive tract, and (3) alterations in cardiovascular system.

While we recommend the above toxicopathic lesions as present biomarkers, it should be noted that this decision is controversial. Additional linkage of body burdens of specific contaminants to histopathologic alterations are needed. We encourage additional studies to photographically document necrotic and other lesions. Appropriate statistics to support possible associations between certain lesion types and relative pollutant exposure are research criteria not uniformly met.

FUTURE BIOMARKERS

Liver

Foci of Cellular Alteration — Enzymes

Histochemical changes comprise the first of a spectrum of lesions associated with the progressive development of hepatic neoplasia in rodents (Popp and Goldsworthy 1989; Pitot et al. 1989). With histochemical procedures to localize selected enzymes altered phenotypes of putatively "carcinogen-initiated" cells can be demonstrated. In one model of rat liver carcinogenesis, the volume of enzyme-altered foci increased in a dose-dependent manner with application of compounds that promote liver tumors (Pitot et al. 1989).

Little application has been made of enzyme histochemistry in laboratory studies of hepatocarcinogenesis in fish (Hinton et al. 1988; Hinton et al. 1991; Hendricks et al. 1984; Nakazawa et al. 1985), and we have found no reports in field investigations. Because conventional cryostat sections reacted for enzyme histochemistry were compared to paraffin-embedded adjacent slices stained by conventional H&E methods, only larger lesions were compared in two studies (Hendricks et al. 1984; Nakazawa et al. 1985). Hinton et al. (1988, 1991) developed techniques which permit enzyme histochemistry and conventional staining on adjacent 4 μm sections of liver and permit comparison of serially-sectioned individual lesions including foci. Enzyme altered foci appeared first after exposure of medaka to diethylnitrosamine (DEN) (Hinton et al. 1989; 1991) and a greater number of foci of cellular alteration were detected by enzyme versus conventional H&E techniques. Furthermore, serial sampling of DEN-exposed medaka suggested a progression of lesions including foci followed by areas (lesions larger than one lobule in diameter), nodules, and tumors. Phenotype of resultant hepatocellular neoplasms (basophilic, enriched in gamma glutamyl transpeptidase and quinone oxido-reductase) was similar to that of foci and areas suggesting linkage of the earlier lesions to eventual tumor (Hinton et al. 1991).

From recent reviews in hepatic carcinogenesis (Farber and Sarma 1987), major steps in tumor formation involve initiation, promotion, and progression. Enzyme altered foci are indicators of initiation, and for a given concentration of initiator, their growth is related in a dose-dependent manner to concentration of promoting agent (Pitot et al. 1989). Since one known group of potential carcinogens, polycyclic aromatic hydrocarbons (PAH), are ubiquitous in sediments, they are likely candidates as initiators of carcinogenesis in livers of certain bottom-dwelling feral fish (Varanasi 1989). However, whether a tumor actually results from this exposure may be due to the presence or absence of xenobiotic substances with promotional capacity. The plasticizer and mammalian promoter, bis-(2-ethylhexyl) phthalate, is ubiquitous in the aquatic environment and found in high levels in the liver of the starry flounder (Spies et al. 1988). It follows that the incorporation of enzyme altered foci as a biomarker may allow detection of feral fish exposed to both initiating and promoting chemicals (Hinton 1989), and, by focusing on more proximate events in tumor progression, enable meaningful analysis of younger feral fish at contaminated sites.

Spongiosis Hepatis

Following hepatocyte necrosis in teleosts, spaces are sometimes formed which initially are confined to the hepatic tubule, but may expand to encompass large regions within the liver. This lesion, seen after laboratory exposure of medaka (Hinton et al. 1984; 1988) and sheepshead minnow *Cyprinodon variegatus*, (Couch and Courtney 1987) to carcinogens, closely resembled (via light microscopy, electron microscopy, and histochemistry) the condition first described as spongiosis hepatis by Bannasch et al. (1981). Interestingly, similar lesions were reported in English sole from contaminated sites within Puget Sound, Washington (Myers et al. 1987). Subsequently, the condition has been seen in older, control medaka of our colony (Marty, personal communication) and that of Hawkins et al. (personal communication). In addition, this condition and/ or a highly similar peribiliary edema with necrotizing hepatocyte change, has been seen in various bottom dwelling flatfishes of Puget Sound, Washington from reference, not highly contaminated sites (Myers, personal communication).

Rarely, spongiosis hepatis may progress into a possible pericytoma (Couch and Courtney 1987). Additional descriptions and reports on this and similar lesions are encouraged and more research on the fate and etiology of spongiosis hepatis in fish are needed. The material below is provided to review structural features of this interesting, potentially future biomarker.

Extensions of stellate, perisinusoidal, fat-storing, and putatively vitamin A containing cells of Ito (Ito et al. 1962) represent a stromal framework or scaffolding that possibly provide support for the hepatic parenchymal cells, and form an extensive network of cellular extensions not appreciated with conventional light microscopy of normal fish. However, when freeze fracture is followed by SEM (Fujita et al. 1986), these cells can be readily seen. Ito cells possess potential for fibroblastic transformation and are active in hepatic fibrosis (Yamamoto et al. 1986). Following chronic toxic injury to hepatocytes, spaces appear in the liver parenchyma which contain a homogeneous, lightly eosinophilic material. The spaces are lined and frequently partitioned by thin cellular extensions. EM analysis in *Oryzias latipes* (Hinton et al. 1984a) and *Cyprinodon variegatus* (Couch and Courtney, 1987) has shown numerous intermediate filaments and junctional complexes in these extensions. These ultrastructural features are consistent with Ito cells.

Spleen

Macrophage Aggregates

Macrophage aggregates (MA), including those in the spleen, have been proposed and used as indicators of contaminant exposure and more often as a generalized nonspecific response to several stressful stimuli (e.g., starvation, heat stress) in a number of studies including Blazer et al. (1987), Herraez and Zapata (1986), and Wolke et al. (1985). Tissue breakdown and age (Brown and George 1985) appear to be major factors contributing to the formation of MA. Increases in MA area/density/ frequency in diseased fish collected from degraded environments support its use as a biomarker. However, MA can be the result of a number of factors not related to toxicant exposure, and these need to be considered and controlled for.

Recently, splenic area occupied by MA in winter flounder collected from eight New England coastal and urban embayments of varied environmental degradation was correlated with chemical contamination of surface sediments (Benyi and Gardner 1989; Gardner et al. 1989). Based on evaluations of Georges Bank winter flounder, MA area was not age dependent. Because levels of polychlorinated biphenyls, polycyclic aromatic hydrocarbons and trace metals measured in surface sediment at these sites correlated and covaried in the same way, benzo(a)pyrene (BaP) was used as an index of exposure against which MA area was compared (Gardner et al. 1989b). The frequency of MA and their total area was found to increase with increasing levels of BaP (Benyi et al. 1989; Gardner et al. 1989). For example, mean MA percent area in spleen of flounder from offshore locations (i.e., Martha's Vineyard, Massachusetts; Georges Bank) was 0.4 and corresponding sediment BaP levels were less than 0.001

μg/g. The same areal parameter in flounder from polluted urban estuaries was greater than 12% and BaP levels in surface sediment approached 9 μg/g. Interestingly, the area of spleen MA was greater in fish with hepatic neoplasms. Flounder from Quincy Bay (Boston Harbor, Massachusetts) without hepatic noeplasms had an average value of 8% while those with neoplasms approached 50%. Linkage of increased splenic MA area to contaminated sediment was demonstrated in the same species using laboratory exposures to contaminated sediment (Gardner et al. 1987).

Although the above demonstrate that some of the factors considered responsible for MA areal increases may be controlled for in field investigations thereby increasing sensitivity of this biomarker, MA are best regarded as sensitive but non-specific indicators of stress (Blazer et al. 1987). We encourage additional laboratory exposure studies to quantify splenic MA. It should be noted that inferences about numerical density or size of foci from tests based on two-dimensional observations can be misleading. If hypothesis tests based on estimates of density and size of MAs are to form the basis for a bioassay, then the power of the statistical tests used to identify treatment effects should be investigated (Morris 1989). Volume percentages of the spleen occupied by MAs are recommended as an endpoint since they can be calculated using methods that do not depend on transections of MAs being circles or the individual MAs being spheres (Pitot et al. 1989).

Gastrointestinal Tract

Neoplasms

In a single communication (Black, J., personal communication) and a limited number of reports (Gardner and Pruell 1987; Gardner et al. 1989b), fish from heavily polluted freshwater and marine environments may develop noninvasive papillary adenomas or polyps in the mucosa of the glandular stomach and to a lesser degree, in the esophagus, pyloric stomach, and small intestine. We could find no publications on laboratory exposures resulting in gastrointestinal tract neoplasms. We encourage investigators to include this organ system in necropsy and subsequent histopathologic assessment.

Skin

Non-neoplastic Lesions

One of the largest organs in the teleost body, skin is in direct contact with the external environment. One, if not the primary function, of the skin is to protect the internal environment from toxic compounds. Not surprisingly, chemical contaminants cause responses in skin which may be used as an indicator of injury. As we shall show, certain skin lesions are classified as "pollution-associated diseases" (Sindermann 1990) but the co-association of opportunistic microorganisms makes it difficult to determine the specific etiologic agent and whether xenobiotics were the principal or simply a contributory etiologic factor. Nevertheless, certain of these lesions signal a contaminated environmental site and call for more intensive study in such locations. For this as well as other reasons, these lesions have strong bioindicator potential.

Ulcerative lesions represent the breakdown of the skin as a protective organ. Such lesions may be somatic or may occur on the fins as "fin rot". The etiology for such lesions is probably multifactorial, involving immune system deficiencies, fungal and bacterial pathogens, and toxic contaminants (Wellings et al. 1976). Thus, ulcers are a generalized response, and are not useful in defining a causative agent. They have been positively associated with contaminated marine environments in red hake (*Urophysis chuss*) from the New York Bight (Murchelano and Ziskowski, 1979), cod (*Gadus morhua*) and dab (*Limanda limanda*) from the North Sea (Dethlefsen, 1980), and in cod from Danish coastal waters (Jenson and Larsen, 1978). Elevated frequencies of such lesions have also been noted in brown bullhead (*Ictalurus nebulosus*) from the Cuyahoga and Black rivers, Ohio (Baumann, personal communication). Therefore, they have been well-documented as indicators of contaminated habitat in field investigations. Sindermann (1990) reviewed various laboratory exposures to verify etiology of fin erosion. Oil, Aroclor 1254, and lead produced these lesions in laboratory exposures of various fish. However, additional verification of possible etiology should be obtained before these lesions are regarded as other than future biomarkers.

Perhaps the most characteristic response of skin is the production and secretion of mucus. Physical trauma (Mittal and Munshi, 1974; Pickering et al., 1982), exposure to toxic metals (Varanasi et al., 1975; Lock and van Overbeeke, 1981; Miller and Mackay, 1982), or to acid pH (Anthony et al., 1971; Daye and Garside, 1976) lead to enhanced mucus production. Morphometric studies (Zuchelkowski et al., 1981; 1986; Schwerdtfeger, 1979a; 1979b) have shown differences in both the amount of mucus produced and mucous cell morphology between the sexes. Under acute conditions of exposure to acid pH (sulfuric acid addition to aquarium water, final pH 4.0) the volume of brown bullhead (*Ictalurus nebulosus*) skin occupied by mucous granules increases (Zuchelkowski et al., 1986). Immature females respond by increased size (hypertrophy) of mucous cells while immature males increase both the number of mucous cells (hyperplasia) and their size (hypertrophy). The shape (spherical) and staining characteristics (magenta) of mucous granules (alcian blue-PAS reaction) lend themselves to rapid quantification and constitute a realistic indicator of certain environmental stressors.

Since males and females differ in mucous cell responses, field studies should be designed to evaluate the sexes separately. Mucous cell alterations are best used as a trigger for further research or as part of a suite of indicators rather than as a biomarker.

Kidney

General Considerations

The kidneys are organs of obvious critical, though varied, function in the highly diverse fish species. One would assume that the renal tissues would be at major toxicologic risk since they receive large volumes of blood flow from both the renal portal venous system and the renal arteries. In addition, urine produced collectively or individually through glomerular filtration, tubular reabsorption, or tubular secretion, serves as a major route of excretion for metabolites of various xenobiotics to which fish have been exposed (for review see Pritchard and Renfro 1982).

In spite of these renal characteristics, however, we have no histopathologic markers in the kidneys that would provide a reliable indicator of the effects of toxicants in the environment. Recent laboratory experiments by Reimschuessel et al. (1990) are of interest. They exposed goldfish (*Carrasius auratus*) to hexachlorobutadiene, a potent nephrotoxin. New nephrons developed several weeks following termination of exposure. Stereologic quantification determined the volume percent of the kidney occupied by developing nephrons (greater in treated than in controls). These experiments are a solid laboratory foundation and should encourage field investigations of nephrotoxicity and repair.

Fewer histopathological studies in kidneys probably contribute to the lack of biomarkers in this important organ. However, it may also reflect either less exposure to toxic metabolites or greater resistance to their effects. As our interest and understanding in renal toxicity increases, a battery of tissue alterations described in a variety of experimental or field studies may provide future biomarkers. These histopathological changes include tubular necrosis and regeneration, glomerular changes (hyaline degeneration, mesangial lysis and fibrosis, visceral or parietal epithelial necrosis), eosinophilic proteinaceous droplets in proximal tubules, and neoplasia.

Neoplasms in kidneys of wild fish populations are extremely rare. Only a single renal cell adenocarcinoma has been described in wild fish (Gardner 1975) and only one in laboratory fish (Hawkins et al. 1989). In contrast, embryonal kidney neoplasms (nephroblastomas) are readily induced in the laboratory in rainbow trout and other species by a variety of chemicals such as N-methyl-N'-nitro-N-nitrosoguanidine (MNNG), methylazoxymethanol acetate (MAM) and dimethylbenzanthracene (DMBA) (Hawkins et al. 1989). Their occurrence in the wild, however, is too rare to be of value as a biomarker.

Researchers are encouraged to examine renal lesions with greater frequency and diligence because one or more of these may prove valuable as a biomarker of chemical exposure.

Reproductive System

Recent studies suggest high potential for other reproductive biomarkers, provided additional research is done. In anticipation of enhanced use of these reproductive biomarkers, the major histopathologic lesions reported in reproductive tissues are described, their principal cellular components and characteristics discussed, and relevant studies reviewed.

As with histopathologic biomarkers in other tissues, response to a wide variety of toxicants and conditions produces relatively few histopathologic lesions (Glaiser 1986). In the male, these include testicular atrophy, which may be either related to age or induced by toxicants; sperm reduction, which also may result from exposure to toxicants, especially those that inhibit mitosis; and inflammation due to bacteria, protozoa, or metazoa. Key lesions in the female include ovarian atresia, failure of ovulation often associated with fibrosis and adhesions, and inflammation following bacterial, protozoan, and metazoan infections.

Identification of reproductive system biomarkers first requires an understanding of the principal cellular components and their normal range of variation. The wide variety and complicated nature of breeding strategies in fish include the typical male/female pattern, self-fertilizing hermaphroditism, and gynogenesis, in which the male sperm stimulates but does not contribute to ovulation (Ferguson 1989). Exploitation of this rich diversity in reproductive strategy may yield important dividends, especially if the proper baseline studies are conducted. Because the synchrony of gametogenesis in seasonal breeders reduces the difficulty of distinguishing toxicant effects from normal gametocyte turnover, seasonal rather than continuous breeders offer advantages. Mottet and Landolt (1987) proposed basic studies in reproductive biology, toxicology, and physiology with seasonal breeders. Furthermore, many fish species exhibit sex reversal, which may be either natural or induced by hormonal or environmental factors.

The testis of the male includes the following cellular components that are of histopathologic concern: Sertoli cells, constituting the blood-testis barrier; Leydig cells, which are interstitial cells involved in male hormone production; and developing germ cells including, in order of differentiation, spermatogonia, spermatocytes, spermatids, and spermatozoa. In some species, secretory cells involved in production of spermatic fluid are part of the testis.

The ovary includes the germinal epithelium, which produces oogonia that become oocytes, and the follicular epithelium, which supports oogenesis. For a review of fish reproductive pathology, see Ferguson (1989).

Several toxicants have been shown to damage or impair testicular epithelium or sperm. Mercury damages sperm and decreases their motility probably by interfering with flagellar function (McIntyre 1973; Mottet and Landolt 1987). Cadmium induced testicular damage in goldfish following intraperitoneal (IP) injection (Tafanelli and Summerfelt 1975), and in brook trout following long-term exposure to sublethal doses as low as 0.1 ppm (Sangalang and O'Halloran 1972). Sangalang et al. (1981) observed testicular abnormalities in cod fed the PCB, Aroclor 1254. The lesions included disorganization of lobules and spermatogenic elements, inhibition of spermatogenesis, fibrosis in lobule walls, and fatty necrosis in testes of sexually mature specimens and those undergoing rapid spermatogenic proliferation, but not in specimens in which the testes were sexually mature or regressed. The fact that this study demonstrated specific "biomarker" effects in the male reproductive system of a fish following exposure to a contaminant is of considerable environmental importance and indicates the potential utility of lesions in this system to indicate toxicant exposure. A different type of histological effect, induction of intersexuality as indicated by the appearance of ovarian follicles in testis, occurred in medaka exposed to the _-isomer of lindane, _-hexachlorocyclohexane (Wester and Canton 1986), and in redear sunfish after pond exposure to the herbicide Hydrothol 191 (Eller 1969).

Histopathology of female reproductive tissues has resulted from exposure to several heavy metals, including cadmium in herring (Westernhagen et al. 1974), selenium in bluegills (Sorenson et al 1984), and arsenic in bluegills (Gilderhaus 1966).

The relationship between the occurrence of neoplasms of reproductive organs and toxic exposure is not clear. Two field surveys identified gonadal neoplasms in Great Lakes fish and suggested that their occurrence was related to pollution. These tumors involved germ cells and stroma in goldfish-carp hybrids (Leatherland and Sonstegard 1978) and tumors of gonadal-supporting elements in yellow perch (Budd et al. 1975). The occurrence of dysgerminoma and seminoma in medaka might be related to exposure to some toxicants, such as trichloroethylene. Our conclusion is that too little data exists to consider reproductive neoplasm in fish as reliable biomarkers of chemical exposure.

Swimbladder

Neoplasms

The swimbladder, a derivative of gut endoderm, exists in either a physostomic (with a pneumatic duct opening to the esophagus) or physoclistic (with no pneumatic opening) condition in various fish species. Its major function is to regulate buoyancy and, thus, depth of the fish in the water column. The volume of air in the organ is controlled through the action of the rete mirabile and gas gland. Few lesions of swimbladders have been reported in fish subjected to toxicants in either laboratory or field situations. Laboratory exposure of rainbow trout to MNNG, MAMA, or DMBA has produced benign papillary adenomas of the epithelial mucosa. Since this lesion has not been reported from epizootiologic studies, it cannot be designated as a biomarker at this time. However, in experimental studies with rainbow trout exposed as embryos, papillary adenomas of swim-bladder nearly always occur simultaneously with gastric adenomas (Hendricks, personal communication). Thus, if the latter are observed, the swimbladder should be carefully examined. One of the reasons swimbladder tumors may not have been reported is that they are easy to overlook, especially if the swimbladder is punctured and deflated. Thus, extreme care needs to be taken to keep the bladder inflated for examination of tumors, as seen through the transparent bladder wall.

Central Nervous System

An obvious paradox exists between the importance of the central nervous system (CNS; brain and spinal cord) and the attention it has been given as a site of histopathologic alterations. In anticipation that the CNS of fish will soon become recognized as a highly sensitive and predictive indicator of environmental stress, we provide here an overview of CNS organization related to its toxicologic pathology. We describe the principal features of the fish CNS, and review the few studies and recent data that indicate the CNS might be a good target organ for demonstrating histopathologic biomarkers of contaminant effects.

The CNS is an anatomically and functionally complex organ that is fundamentally important in maintaining homeostasis (Glaiser 1986). A blood-brain barrier prevents entry of certain chemicals into the CNS and glial cells further restrict entry of macromolecules. Lipid soluble compounds such as solvents, however, readily gain entry into the CNS. The nervous system is poor in toxicant-metabolizing systems yet is sensitive to many toxicants because of its high metabolic activity. The major components of the CNS that may be of toxicological interest are neurons, supporting cells (oligodendrocytes, Schwann cells, and astrocytes), and cells of the vascular support system. Any number of these can be the target of a toxicant. Degeneration is the principal toxic response observed in the nervous system. Although behavioral responses are some of the earliest indications of toxicity, those responses may or may not be reflected in structural changes.

The fish brain has essentially the same components as other vertebrates (Ferguson 1989). Cerebral hemispheres (telencephalon) of fish are reduced mainly to olfactory lobes and bulbs. The optic lobes representing the mesencephalon are the largest part of the brain. A feature that may eventually be of biomarker interest is the fact that ependymal cells of at least some teleost species are pluripotent, giving rise to supporting cells and, in some cases, neurons, even in post embryonic specimens . In contrast to higher organisms, the fish CNS is capable of at least some regeneration (Bernstein 1970).

Little is known about CNS histopathology, either induced by toxic or infectious agents, probably because of the complexity of the system and lack of studies (Ferguson 1989). Given the importance of the CNS to an organism's survival, identification of CNS biomarkers, whether histopathologic, biochemical, or molecular, should be a priority. The little data that exists suggests that CNS lesions in fish are both identifiable and analyzable. Gliosis is a common response to injury of fish CNS (Ferguson 1989), and vacuolization of optic lobes has been observed in salmonids after exposure to the pesticide carbaryl (Walsh and Ribelin 1975) and in optic and olfactory cortex of medaka (Hinton and Hawkins, personal communication). The herbicide 2,4-D caused vascular congestion in the brain of bluegills (*Lepomis macrochirus*) (Cope, 1970). Walsh and Ribelin (1975) also observed vascular congestion and hyperemia in brains of coho salmon (*Oncorhynchus kisutch*) and lake trout (*Salvelinus namaycush*) exposed to a variety of pesticides, but only 2,4-D exposure was it sufficient to have diagnostic value.

Although a rather large number of neoplasms, arising from various elements in the CNS and affecting a wide variety of species, has been reported (Masahito et al., in press), a toxicopathic etiology and thus their use as biomarkers is as yet undetermined. Smith (1984) observed massive hyperplastic lesions originating from the primitive meninx (fish equivalent of the dura mater and arachnoid) of fathead minnows following longterm ammonia exposures.

Neurosensory System

The neurosensory system in fish includes the olfactory organ, lateral line, and eyes. The consensus of a 1974 workshop stated that "histology of sensitive tissues, such as the olfactory and lateral line organs of fish can provide a simple, fast, semi-quantitative bioassay procedure in laboratory experiments and receiving waters" (Marine Technological Society 1974). Unfortunately, these systems have received little attention as sites of histopathologic biomarkers. Because the olfactory and lateral line organs act as an interface between the environment and the central nervous system, they should be excellent monitors of exposure to neurotoxicants. Compromise of organs so important in feeding, homing, and prey avoidance could have severe impact at both the individual and population level.

Olfactory organ and lateral line

The olfactory organ lies in the bases of pits on the head region of fish and consists of folds of tissue lined with neuroepithelial cells. The apices of the neuroepithelial cells have abundant microvilli and cilia, which are bathed in ambient water. The bases of the neuroepithelial cells are connected to the brain via the olfactory nerve.

The lateral line system consists of a series of neuromasts located in a canal between the epaxial and hypaxial musculature. Sensory cilia of the neuroepithelial cells respond to vibratory stimuli and relay information to the brain. Similar to that of the olfactory organ, the apical surfaces of the neuroepithelial cells are bathed in ambient water.

The few studies that have examined olfactory organ and lateral line effects in fish indicate that changes in these organs might not only provide a window to nervous system effects generally, but may also help to explain some of the behavioral abnormalities that often accompany toxic exposures. Gardner (1975) examined the effects of several chemicals on neurosensory structures in the mummichog (*Fundulus heteroclitus*) and found that copper, mercury, and silver caused degeneration of the anterior lateral line and the olfactory organ, while cadmium and zinc caused no changes in these structures. Methoxychlor caused only lateral line lesions in mummichogs. In the Atlantic silverside (*Menidia menidia*), crude oil exposure resulted in hyperplasia of the olfactory sustentacular epithelium. Behavioral changes in silversides were related to structural damage to the lateral line, cephalic sinuses, olfactory organ, and inner ear. Crude oil and its water soluble fractions caused hyperplasia of non-neural cells of the olfactory organs and necrosis of both neurosensory and sustentacular epithelium of the inland silverside (*Menidia beryllina*) (Solangi and Overstreet 1982). In the hogchoker (*Trinectes maculatus*), those substances caused severe necrosis of both neurosensory and sustentacular cells of the olfactory mucosa. DiMichele and Taylor (1978) exposed mummichogs to naphthalene and observed damage in neurosensory cells of the taste buds, olfactory organ, and lateral line. Clearly, the sensitivity of these organs to the few compounds tested, their importance to the well-being of the organism, and their unique anatomical organization, all warrant further investigations into their utility as biomarkers of environmental exposures.

Eye

A great deal of diversity exists in the morphological organization of fish eyes, but retinal form and function are comparatively conservative (Wilcock and Dukes 1989). Pathological responses of corneal and lens tissues to chemical and other agents are highly nonspecific. Cataract is a common ocular lesion that can be caused by numerous factors, thus limiting its value as a biomarker of a specific exposure; its use as an indicator of general environmental deterioration, however, is not affected.

The sensory portion of the retina of fish is an extension of the brain and is unique among vertebrates in that it continues to add neurosensory cells throughout the life of the organism. The site where new retinal cells proliferate was believed to be the site of origin for neoplastic lesions in medaka exposed to methylazoxymethanol acetate (Hawkins et al. 1986). Hose et al. (1984) found microphthalmia, patent optic fissure, depressed sensory retinal mitotic rates, retinal folding, and poor retinal differentiation in rainbow trout exposed as juveniles to BaP. Whereas ocular tumors probably will not be good biomarkers because of their generally rare occurrence and high compound-specificity, studies of other histopathological effects in retina such as those seen by Hose et al. (1984) are certainly worth pursuing.

Gill

In constant contact with water, the gill is a sensitive primary target for a variety of insults including low pH (McDonald 1983), transition metals (Laurén and McDonald 1985), heavy metals (Verbost et al. 1987), detergents (Abel and Skidmore 1975), and polycationic agents (Greenwald and Kirschner 1976). The gills are so sensitive, in fact, that laboratory exposure to any of the agents just listed, at concentrations comparable to those found in the wild, causes sufficient ionoregulatory disruption to result in death of the fish often without the formation of histologic lesions. Only at acutely lethal concentrations, with low pH and some metals, were mucous coagulation, necrosis of lamellar epithelial cells, and epithelial lifting observed (Packer and Dunson 1972; Skidmore and Tovell 1972; Daoust et al. 1984). Chevalier et al. (1985) have reported epithelial lifting and chloride cells degeneration in trout from an acidified lake.

At sublethal concentrations, a variety of toxicants have been shown to induce chloride cell hyperplasia (Laurent et al. 1985; Perry and Wood 1985; Avella et al. 1987). The proliferation of chloride cells is apparently a compensatory response to ion loss, and chloride cell hyperplasia may be a good biomarker of adaptation to ionoregulatory stress.

In contrast, hyperplasia of undifferentiated epithelial cells (resulting in clubbing and fusion of lamellae) is a much less specific lesion, one associated with a wide variety of unrelated insults inducing infection by microorganisms, ectoparasitism, phenols, heat, NH_3, and metals. Mucous cell hyperplasia has been reported in trout exposed to low pH (Daye and Garside 1976), but has not been seen in wild fish from acidified lakes (Chevalier et al. 1985). Detection of hyperplastic lesions in the absence of inflammation, infection, and parasitism has increased significantly and can be quantified using morphometrical analysis (Hughes et al. 1979).

A potentially useful biomarker in the gill is the presence of cytochrome P450E (P451AI) in pillar cells. This enzyme is inducible in both scup (*Stenotomus chrysops*) and rainbow trout (*Onchorhynchus mykiss*) and is detectable with the use of monoclonal antibodies (Miller et al. 1989). P450E induction may serve as a biomarker of exposure to a variety of organic (i.e., PAH) contaminants in the environment.

In summary, the gill is a sensitive indicator of environmental stress, including anthropogenic compounds in the water. However, it may be too sensitive, because fish can be killed with the complete absence of histologic lesions and because a variety of factors can result in the same suite of lesions.

SHELLFISH NEOPLASMS

Historical Information

For thorough coverage of neoplasia in shellfish, readers should consult various reviews (Pauley 1969; Dawe and Harshbarger 1969; Couch and Harshbarger 1985; Sparks 1985; Mix 1986a; 1986b; and Peters 1988). Four general types of neoplasms have been identified in feral shellfish: hematopoietic or sarcomatous neoplasms (softshell clams, blue mussels, and oysters), tumors of germ cell origin (hardshell and softshell clams), cutaneous neoplasms (Australian oysters), and gill carcinomas (the clam species *Macoma balthica*). With the exception of benzo[a]pyrene and the resultant leucocytic proliferative disorders in mussels from Yaquina Bay, Oregon (Mix et al. 1979), none of these neoplasms have been correlated with exposure to any xenobiotic chemical.

Laboratory Investigations

Proliferative disorders have been induced in bivalve molluscs in recent laboratory studies. A cause and effect relationship was established between neoplasms in oysters and exposure to sediment containing known carcinogens (Gardner et al. 1987; 1991a). Neoplasms developed after 30 days in oysters allowed to filter-feed a 20 mg/l concentration of contaminated BRH sediment particulate. Tumor incidence was 13.6% with organ/tissues ranked by frequency as renal excretory mucosa > gill > gonad > GI tract > heart > embryonic neural tissue. Renal cell carcinomas and GI tract adenocarcinomas in situ were observed. The same sediment (BRH) also induced myxomas in the hearts of blue mussels (Yevich et al. 1986). Khudoley and Syrenko (1978) described tumors of digestive diverticulae (composed of basophilic cells), hematopoietic (leukemic) tumors, and a renal carcinoma in freshwater mussels exposed to carcinogenic N-nitroso compounds. More recently, low grade renal cell tumors and gastrointestinal adenomas were induced in oysters by a mixture containing: aromatic hydrocarbons, an aromatic amine, polychlorinated biphenyls, chlorinated hydrocarbons, a nitrosamine and heavy metals (Gardner et al. 1991c).

Field Investigations

Recent field studies have confirmed the utility of bivalve mollusc neoplasms as a biomarker of exposure to carcinogens. Neoplasms in oysters transplanted to contaminated sites from clean reference sites and lesions in an indigenous population of softshell clams have recently been associated with environmental pollutants (Gardner and Pruell 1987; Gardner et al. 1987, 1991a). Neoplasms were found in oysters transplanted to BRH for 30 days or to Long Island Sound where sediment dredged from BRH was deposited (latency period = 36 days). Renal and GI neoplasms were also observed in oysters 40 days after transplantation to Quincy Bay (Boston Harbor, Massachusetts). Oysters transplanted to reference areas lacked neoplasms.

Malignant seminomas and dysgerminomas (germ cell neoplasms of male and female, respectively) have been observed in softshell clams collected from locations near Searsport, Dennysville and Machiasport, Michigan (Gardner et al. 1991b). Clams collected at Dennysville were also observed to have pericardial neoplasms (mesotheliomas) and teratoid siphons. Evidence suggests these clams were exposed to the teratogens 2,4-D and 2,4,5-T and the herbicide Tordon 101.

Recommendations for Shellfish Neoplastic Biomarkers

Based on a limited number of studies with bivalve molluscs, there is potential for development of biomarkers in these invertebrates. Their habitat and filter-feeding behavior places them in direct contact with contaminants in the sediment. Field application and verification are simplified with use of in situ exposures which can be controlled spatially and temporally. The brief time to endpoint as presented above is a distinct advantage. We encourage exposures with proven carcinogens to determine whether similar lesions to those presented above will result. For shellfish, studies to date indicate that neoplasms originating in (1) mucosal epithelium of digestive organs, (2) germinal epithelium, and (3) the cardiovascular system have the most promise.

The authors have communicated with Dr. John Harshbarger, Director, Registry of Tumors in Lower Animals, who has provided us with information regarding neoplasms in other invertebrate species. Convincing neoplasms have been reported in arthropods, molluscs, cnidari, shrimp, and insects. Lesions from other phyla including Echinodermata are under investigation. Much more work is necessary before histopathological biomarkers in the many other phyla and classes of invertebrates will be developed.

FIELD SAMPLING STRATEGIES

Site Selection and Organism Collection

The first choices required in any field study are the locations and organisms to be sampled. Usually a contaminated site is compared to a reference site. More information can often be obtained by comparing a series of locations with varying degrees of pollution or types of pollutants. If an adequate reference site (one using the same species and biomarker) exists in the literature, additional reference site collection may not be necessary. Adequate baseline or reference values for the majority of histopathologic biomarkers, however, have not yet been established and are a research need.

Benthic species are most often exposed to toxic and/or carcinogenic compounds in sediment and accumulations of such compounds in food organisms. Not surprisingly, much of the field validation of biomarkers have involved such species as sole, flounder, croakers, and bullhead. This preference, however, does not preclude the value of more pelagic species for detecting effects of water-soluble compounds, particularly fish in families known to be sensitive from laboratory or field studies, such as salmonids, percids, or cyprinodontids. If the locations to be sampled do not contain species of known sensitivity, then some sensitivity screening studies are appropriate prior to organism selection.

Fish should be collected by methods that avoid mortality and minimize trauma. Thus small seines, fyke nets, or short trawl runs are preferable to gill nets or lengthy trawls which could result in excessive injury or mortality. Electroshocking is a suitable technique for capturing fish for most biomarkers, but obviously not for vertebral deformities that can be caused by muscular contractions induced by electroshock capture. Captured fish should be kept in well-aerated holding facilities and processed rapidly to reduce the occurrence of autolytic lesions. Sample size (see Statistics, below) is critical.

Fish collection should ideally be stratified by age and sex. Older fish are more likely to have chronic lesions and neoplasms. The prevalence of liver neoplasms has been shown to increase with age in wild brown bullhead (Baumann et al. 1990) and English sole (Rhodes et al. 1987). In contrast, younger fish may show higher prevalences of acute toxicity biomarkers and preneoplastic lesions. Sex may affect the prevalence of some biomarkers because of differences in hormone metabolism and its effects on enzyme induction. Therefore, the reproductive status of the fish may influence selection of the appropriate collection time.

Timing of collections could also be influenced by seasonal variation in biomarkers, sometimes related to metabolic rate as influenced by water temperature for fish. Migration cycles must also be taken into account. Fish species that are either territorial or whose migrations are minimal and predictable make the most useful subjects for study.

Necropsy and Fixation

In those cases where known toxicants are the object of study, a particular biomarker or target organ might be selected to maximize sampling within funding and time limitations. Ideally, however, it is best to obtain as much data as possible from an animal that must be sacrificed. A complete necropsy should not only include a gross exam for obvious lesions such as ulcers, skeletal deformities, and cutaneous neoplasms, but should also involve a thorough exam of all internal organs, including the heart, liver, spleen, gonad, kidney, GI tract, and swim bladder. GI tract neoplasms can be found, if the tract is opened and rinsed prior to examination, and swim bladder tumors can be detected through the transparent bladder wall, if care is taken not to puncture it. As stated previously, skin scrapings and gill wet mounts will facilitate distinguishing parasitic etiologies from toxic ones.

While selection of organs for fixation and later histologic examination may be limited by time and budgetary constraints, the minimum complement of organs chosen should include liver, kidney, spleen, gonad, GI tract, heart, skin, and gill. Other organs possibly providing useful information include brain, olfactory organs, nares, lateral line, thymus, swim bladder, and endocrine organs (thyroid, pancreas, pituitary, and corpuscle of Stannius). If no gross lesions are visible, multiple random sections through the larger organs should be taken. Sectioning should be no thicker than 3 to 4 mm to ensure adequate fixative penetration.

Fixatives should be chosen according to need, Davidson's, 10% formalin, Dietrich's, or Bouin's are suitable for routine use. The latter two are superior for processing small fish, since they decalcify the specimen. Karnovsky's is the preferred preservative for a combination of LM and Em. Immunohistochemistry can be accomplished after a brief (1 to 2 h) fixation in 10% neutral-buffered formalin. Some techniques including enzyme histochemistry and immunocytochemistry may require freezing in liquid nitrogen-cooled freon or isopentane and storage in the former.

Statistics

Sample size should be sufficiently large to allow determination of significant differences (as measured in lesion prevalence) between different sites. The number of individuals needed increases as the differences in prevalence become smaller (Sokal and Rohlf 1981). Thus, documenting a low biomarker prevalence as significantly different from baseline requires many more specimens than proving a high prevalence site to be significantly different. Approximately 400 animals are needed to discriminate between a 1% prevalence and a 5% prevalence when the background prevalence is zero. Since segregation of samples by age and/or sex is often required, the total number of animals needed to distinguish among several age classes may soon outgrow the time or funding available for research. Such situations are not uncommon in field studies and require the judgement of the investigator to determine trade-offs between the level of discrimination attainable and the detail with which the study can be conducted. Funding agencies need to be aware of the costs necessary to complete adequate assessments of biomarkers.

RECENT ADVANCES STRENGTHENING HISTOPATHOLOGIC BIOMARKERS

Major technologic advances in histopathology include increased resolution through plastic embedment, quantitative assessment of specific cellular and tissue alterations, and the use of macromolecular markers such as lectins and immunochemical tags.

High resolution light microscopy with water soluble plastic resins, such as glycol methacrylate (GMA), increases resolution so that subtle alterations in specific cell types can be more readily detected. Application of morphometric methods to GMA sections permits analysis of quantitative changes in cells (hyperplasia, hypertrophy) (Zuchelkowski et al. 1981). Visualization of ligand-binding sites by fluorescein- or peroxidase-labeled lectins or antibodies allows analysis of very specific cell characteristics (Goldstein and Poretz 1986; Damjanov 1987; Goksyr et al. 1987; Droy et al. 1988; Miller et al. 1989; Stegeman et al. 1989).

In addition, fixatives suitable for light and electron microscopy (Ito and Karnovsky 1968; McDowell and Trump 1976) are available and have proven to be useful when light microscopy findings dictate ultrastructural confirmation in the same tissue or organ.

Collection, processing and archiving according to principles of morphometry

(Weibel 1979; 1980) enables the investigator to use these powerful quantitative approaches, should the study require them. Planning and processing, without complete, initial workup can be done at a minimal additional cost. The investigator controls the number of tiers of examination (Hampton et al. 1989) and may elect to proceed from the organ through tissue to cellular and organelle levels of organization. Meaningful data may only require the initial one or two levels of analysis (see Adams et al. 1989).

RESEARCH NEEDS

Perhaps the most pressing need for research with respect to histopathologic biomarkers is their application to field investigations. The methods and their utility have been known for years. Furthermore, analysis of alterations in tissues fits the biomarker approach since it affords an opportunity for the investigator to interpret a range of toxicant effects and host tissue responses.

The paucity of carefully executed descriptive studies, with the level of resolution necessary to describe the locale and morphology of individual cell types, is a drawback to future histopathologic studies of organs not commonly analyzed. The potential biomarkers previously listed have real promise and need to be applied in field conditions.

A limited number of species of freshwater and marine finfish have been subjected to detailed morphologic analysis. The normal morphologic database (at cell and tissue level) needs expansion in other promising indicator species. By using quantitative approaches to histopathology, collecting data amenable to statistical evaluation is an achievable goal. Such data may then be correlated with other more historically quantitative approaches such as chemical and physiological.

CONCLUSIONS

In summary, the present histopathologic alterations discussed in this section have immediate utility as biomarkers of effect, resulting from a variety of chemical contaminants in the environment. They are particularly appropriate since many have been validated in the laboratory and been found relevant to field investigations. As such, they provide higher level response signals, since morphologic alterations follow earlier biochemical and physiological alterations. The alterations present, in many instances, are an integration of physiological and chemical changes.

This chapter is not the first to extol the virtues of histopathology, yet these valuable approaches have not found common usage in field application. Histopathologic biomarkers are most valuable as indicators of general health of organisms and mirror effects of exposure to a variety of anthropogenic pollutants. Their utility as biomarkers is most studied in teleost fish, but changes in tissues and cells occur in all vertebrates and invertebrates. No geographic or ecosystem limitations are apparent. Acute changes are seen when contaminant levels are sufficiently high, while chronic duration is required to determine sublethal aspects of change. Many alterations persist even after exposure to a toxicant has ceased so that host responses to prior toxicity can also be used to determine effect.

Seasonal, physiologic, and sex related variation exist and must be taken into account, but should not prevent the immediate application of histopathologic biomarkers. Normal variation is at cellular and subcellular levels of organization while effective biomarkers primarily involve tissue components. Comparison of tissues from the same species at reference sites permits determination of change at contaminated sites. Extensive methodology for determining tissue, cellular, and subcellular responses exists. Newer plastic embedment procedures improve resolution without appreciably altering cost.

Responses are relatively easily recognized, provided proper reference and control data are available. Although somewhat subjective, user-oriented computer software programs for quantifying lesions also exist. When applied to estimate the magnitude of response, data amenable to statistical evaluation can be obtained.

Histopathologic biomarkers are ready for field use. Considerable testing has been completed, but inadequate application to field studies is the major cause of historical lack of data. The Status and Trends program of NOAA, "Mussel Watch," and limited monitoring efforts attest to the utility of these approaches.

REFERENCES

Abel, P.D., and J.F. Skidmore. "Toxic Effects of an Anionic Detergent on the Gills of Rainbow Trout," Water Res. 9:759–765 (1975).

Adams, S.M., K.L. Shepard, M.S. Greely Jr., B.D. Jimenez, M.G. Ryon, L.R. Shugart, J.F. McCarthy and D.E. Hinton. "The Use of Bioindicators for Assessing the Effects of Pollutant Stress on Fish," Mar. Environ. Res. 28:459–464 (1989).

Ahne, W., and I. Thomsen. "The Existence of Three Different Viral Agents in a Tumour Bearing European Eel (Anguilla anguilla)," Zentralbl Veterinarmed [B] 32(3):228–235 (1989).

Akberali, H.B., and E.R. Trueman. "Effects of Environmental Stress on Marine Bivalve Molluscs," Adv. Mar. Biol. 22:101–198 (1985).

Anderson, D.P. "Immunological Indicators: Effects of Environmental Stress on Immune Protection and Disease Outbreaks," Am. Fish. Soc. Symp. 8:38–50 (1990).

Anthony, A., E.L. Cooper, R.B. Mitchell, W.H. Neff and C.D. Thierren. "Histochemical and Cytophotometric Assay of Acid Stress in Freshwater Fish," Water Poll. Research Service 18050 D x J 05/71 (1971).

Arias, I.M., W.B. Jakoby and H. Popper, Eds. The Liver: Biology and Pathology, 2nd ed. (ew York: Raven Press,1988).

Avella, M., A. Masoni, M. Bornancin and N. Mayer-Gostan. "Gill Morphology and Sodium Influx in the Rainbow Trout (Salmo gairdneri) Acclimated to Artificial Freshwater Environments," J. Exp. Zoo. 241:159–169 (1987).

Bannasch, P., M. Bloch and H. Zerban. "Spongiosis Hepatis: Specific Changes of the Perisinusoidal Liver Cells Induced in Rats by N-nitrosomorpholine," Lab. Invest. 44:252–264 (1981).

Baumann, P.C., and S.J. Hamilton. "Vertebral Abnormalities in White Crappies, *Pomoxis annularis* Rafinesque, from Lake Decatur, Illinois, and an Investigation of Possible Causes," *J. Fish Biol.* 25:25–33 (1984).

Baumann, P.C., J.C. Harshbarger and K.J. Hartmann. "Relationship Between Liver Tumors and Age in Brown Bullhead Populations from Two Lake Erie Tributaries," *Sci. Tot. Environ.* 94:71–87 (1990).

Baumann, P.C., W.D. Smith and W.K. Parland. "Tumor Frequencies and Contaminant Concentrations in Brown Bullheads from an Industrialized River and a Recreational Lake," *Trans. Amer. Fish. Soc.* 116:79–86 (1987).

Baumann, P.C., and D.M. Whittle. "The Status of Selected Organics in the Laurentian Great Lakes: An Overview of DDT, PCBs, Dioxins, Furans, and Aromatic Hydrocarbons," *Aquat. Toxicol.* 11:241–257 (1988).

Bengtsson, B.E. "Vertebral Damage to Minnows (*Phoxinus phoxinus*) Exposed to Zinc," *Oikos* 25:134–139 (1974).

Bengtsson, B.E. "Vertebral Damage in Fish Induced by Pollutants," in *Sublethal Effects of Toxic Chemicals on Aquatic Animals*, J.H. Koeman, and J.J.T.W.A. Strik, Eds., (New York: Elsevier, 1975), pp. 23–30.

Benyi, S.J., G.R. Gardner, J.F. Heltshe and J. Rosen. "Pigment Localization in Spleen of the Winter Flounder (*Pseudopleuronectes americanus*) in Relation to Sediment Chemical Contamination," abstract presented at the American Fisheries Society, Fish Health Section, 1989.

Bernstein, J.J. "I. Anatomy and Physiology of the Nervous System," in *Fish Physiology Vol. 4, The Nervous System, Circulation, and Respiration*, W.S. Hoar, and D.J. Randall, Eds., (New York: Academic Press, 1970), pp. 1–90.

Berthou, F., G. Balouet, G. Bodennec and M. Marchand. "Occurrence of Hydrocarbons and Histopathological Abnormalities in Oysters for Seven Years Following the Wreck of the Amoco Cadiz in Brittany (France)," *Mar. Environ. Res.* 23:103–133 (1987).

Betzer, S.B., and P.P. Yevich. "Copper Toxicity in *Busycon canaliculatum* L.," *Biol. Bull.* 148:16–25 (1975).

Black, J.J. "Epidermal Hyperplasia and Neoplasia in Brown Bullheads (*Ictalurus nebulosus*) in Response to Repeated Applications of PAH Containing Extract of Polluted River Sediment," in *Polynuclear Aromatic Hydrocarbons: Formation, Metabolism, and Measurements*, M.W. Cook, and A.J. Dennis, Eds. (Columbus, OH: Battelle Press, 1983a).

Black, J.J. "Field and Laboratory Studies of Environmental Carcinogenesis in Niagara River Fish," *J. Great Lake Res.* 9(2):326–334 (1983b).

Black, J.J., H. Fox, P. Black and F. Bock. "Carcinogenic Effects of River Sediment Extracts in Fish and Mice," in *Water Chlorination Chemistry, Environmental Impact and Health Effects*, R.L. Jolley, R.J. Bull, W.P. Davis, S. Katz, M.H. Roberts, Jr., and V.A. Jacobs, Eds. (Chelsa, MI: Lewis Publishers, 1985), pp. 415–427.

Blazer, V.S., R.E. Wolke, J. Brown and C.A. Powell. "Piscine Macrophage Aggregate Parameters as Health Monitors: Effect of Age, Sex, Relative Weight, Season and Site Quality in Largemouth Bass (*Micropterus salmoides*).," *Aquat. Toxicol.* 10:199–215 (1987).

Bodammer, J.E., and R.A. Murchelano. "Cytological Study of Vacuolated Cells and Other Aberrant Hepatocytes in Winter Flounder from Boston Harbor," *Cancer Res.* 50:6744–6756 (1990).

Bolender, R.P. "Correlation of Morphometry and Stereology with Biochemical Analysis of Cell Fractions," *Int. Rev. of Cytol.* 55:247–289 (1978).

Bolender, R.P. "Morphometric Analysis in the Assessment of the Response of the Liver to Drugs," *Pharmacol. Rev.* 30(4):429–443 (1979).

Bolender, R.P. "Stereology: Applications to Pharmacology," *Annu. Rev. of Pharmacol. Toxicol.* 21:549–573 (1981).

Boyer, J.L., J. Swartz and N. Smith. "Biliary Secretion in Elasmobranchs. II. Hepatic Uptake and Biliary Excretion of Organic Anions," *Amer. J. Physiol.* 230:974–981 (1976).

Brooks, R.E., G.E. McArn and S.R. Wellings. "Ultrastructural Observations on an Unidentified Cell Type Found in Epidermal Tumors of Flounders," *J. Nat. Cancer Inst.* 43:97–109 (1969).

Brown, C.L., and C.J. George. "Age-dependent Accumulation of Macrophage Aggregates in the Yellow Perch, *Perca flavescens* (Mitchill)," *J. Fish Dis.* 8:135–138 (1985).

Brungs, W.A. "Chronic Effects of Constant Elevated Temperature on the Fathead Minnow (*Pimephales promelas* Rafinesque)," *Trans. Amer. Fish. Soc.* 100:659–664 (1971).

Budd, J.J., D. Schroder and K.D. Dukes. "Tumors of the Yellow Perch," in *Pathology of Fishes,* W.F. Ribelin, and G. Migaki, Eds. (Madison, WI: University of Wisconsin Press, 1975), pp. 895–906.

Calabrese, A., J.R. Macinnes, D.A. Nelson, R.A. Grieg and P.P. Yevich. "Effects of Long-term Exposure to Silver or Copper on Growth, Bioaccumulation and Histopathology in the Blue Mussel *Mytilus edulis, Mar. Environ. Res.* 11:253–274 (1984).

Chevalier, G., L. Gauthier and G. Moreau. "Histopathological and Electron Microscopic Studies of Gills of Brook Trout, *Salvelinus fontinalis*, from Acidified Lakes," Can. J. of Zoo. 63:2062–2070 (1985).

Cope, O. "Some Chronic Effects of 2,4-D on the Bluegill (*Lepomis macrochirus*)," *Trans. Amer. Fish. Soc.* 99:1–12 (1970).

Couch, J.A. "Ultrastructural Study of Lesions in Gills of a Marine Shrimp Exposed to Cadmium," *J. Invert. Pathol.* 29(3):267–288 (1977).

Couch, J.A. "Diseases, Parasites and Toxic Responses of Commercial Penaeid Shrimps of the Gulf of Mexico and South Atlantic Coasts of North America," *Fish. Bull.* 76(1):1–44 (1978).

Couch, J.A. "Atrophy of Diverticular Epithelium as an Indicator of Environmental Irritants in the Oyster, *Crassostrea virginica, Mar. Environ. Res.* 14:525–526 (1984).

Couch, J.A. "Prospective Study of Infectious and Noninfectious Diseases in Oysters and Fishes in Three Gulf of Mexico Estuaries," *Dis. Aquat. Organ.* 1:59–82 (1985).

Couch, J.A. "Pericyte of a Teleost Fish: Ultrastructure, Position, and Role in Neoplasm as Revealed by a Fish Model," *Anat. Rec.,* 228:7—14 (1990).

Couch, J.A. "Spongiosis Hepatis: Chemical Induction, Pathogenesis, and Possible Neoplastic Fate in a Teleost Fish Model," *Toxicol. Pathol.* 19(3):237—250 (1991).

Couch, J.A., and L.A. Courtney. "N-nitrosodiethylamine-induced Hepatocarcinogenesis in Estuarine Sheepshead Minnow (*Cyprinodon variegatus*): Neoplasms and Related Lesions Compared with Mammalian Lesions," *J. Nat. Cancer Inst.* 79:297–321 (1987).

Couch, J.A., L.A. Courtney, J.T. Winstead and S.S. Foss. "The American Oyster (*Crassostrea virginica*) as an Indicator of Carcinogens in the Aquatic Environment, in *Animals as Monitors of Environmental Pollutants,* National Academy of Sciences, (1979a), pp. 65–83.

Couch, J.A., and J.C. Harshbarger. "Effects of Carcinogenic Agents on Aquatic Animals: An Environmental and Experimental Overview," *Environ. Carcinogenesis Rev.* 3:63–105 (1985).

Couch, J.A., J.T. Winstead and L.R. Goodman. "Kepone-induced Scoliosis and Its Histological Consequences in Fish, *Science* 197:585–587 (1977).

Couch, J.A., J.T. Winstead, D.J. Hansen and L.R. Goodman. "Vertebral Dysplasia in Young Fish Exposed to the Herbicide Trifluralin," *J. Fish Dis.* 2:35–42 (1979b).

Cross, J.N., and J.E. Hose. "Reproductive Impairment in Two Species of Fish from Contaminated Areas Off Southern California," *Oceans '89* (1989).

Cross, J.N., and J.E. Hose. "Evidence for Impaired Reproduction in White Croaker (*Genyonemus lineatus*) from Contaminated Areas Off Southern California," Mar. Environ. Res. 24:185–188 (1988).

Damjanov, I. "Biology of Disease: Lectin Cytochemistry and Histochemistry," *Lab. Invest.* 57:5–20 (1987).

Daoust, P.Y., G. Wobeser and J.D. Newstead. "Acute Pathological Effects of Inorganic Mercury and Copper in Gills of Rainbow Trout," *Vet. Pathol.* 21:93–101 (1984).

Dawe, C.J. "Polyoma Tumors in Mice and X Cell Tumors in Fish, Viewed Through Telescope and Microscope," in *Phyletic Approaches to Cancer*, C.J. Dawe et al., Eds., (Tokyo: Japan Scientific Society Press, 1981), pp. 19–49.

Dawe, C.J., and J.C. Harshbarger, Eds. "Neoplasms and Related Disorders of Invertebrates and Lower Vertebrate Animals," *Nat. Cancer Inst. Monogr.* 31 (1969).

Dawe, C.J., M.F. Stanton and F. J. Schwartz. "Hepatic Neoplasms in Native Bottom-feeding Fish of Deep Creek Lake, Maryland," *Cancer Res.* 24:1194–1201 (1964).

Daye, P.G., and E.T. Garside. "Histopathologic Changes in Surficial Tissues of Brook Trout, *Salvelinus fontinalis* (Mitchill), Exposed to Acute and Chronic Levels of pH," *Can. J. Zool.* 54:2140–2155 (1976).

Dethlefsen, J. "Observations on Fish Diseases in the German Blight and Their Possible Relation to Pollution," *Rapp. P. V. Re'vn. Cons Int. Explor. Mer.* 179:110–117 (1980).

DiMichele, L., and M.H. Taylor. "Histopathological and Physiological Responses of *Fundulus heteroclitus* to Naphthalene Exposure," *J. Fish. Res. Bd. Can.* 35:1060–1066 (1978).

Dixon, K.C. *Cellular Defects in Disease* (Oxford, England: Blackwell Scientific Publications, 1982).

Doughtie, D.G., P.J. Conklin and R.K. Rao. "Cuticular Lesions Induced in Grass Shrimp Exposed to Hexavalent Chromium," *J. Invert. Pathol.* 42:249–258 (1983).

Doughtie, D.G., and K.R. Rao. "Histopathological and Ultrastructural Changes in the Antennal Gland, Midgut, Hepatopancreas and Gill of Grass Shrimp Following Exposure to Hexavalent Chromium," *J. Invert. Pathol.* 43:89–108 (1984).

Doyle, L.J., N.L. Blake, C.C. Woo and P.P. Yevich. "Recent Biogenic Phosphorite: Concentrations in Mollusk Kidneys," *Science* 199:1431–1433 (1978).

Droy, B.F., M.R. Miller, T. Freeland and D.E. Hinton. "Immuno-histochemical Detection of CCl_4-Induced, Mitosis-related DNA Synthesis in Livers of Trout and Rat," *Aquat. Toxicol.* 13:155–166(1988).

Egami, N., Y. Kyono-Hamaguchi, H. Mitani and A. Shima. "Characteristics of Hepatoma Produced by Treatment With Diethylnitrosamine in the Fish, *Oryzias latipes*, in *Phyletic Approaches to Cancer*, C.J. Dawe et al., Eds., (Tokyo: Japan Scientific Society Press, 1981), pp. 217–226.

Eller, L.L. "Pathology of Redear Sunfish Exposed to Hydrothol 191," *Trans. Amer. Fish. Soc.* 98:52–59 (1969).

Fabacher, D.L., C.J. Schmitt, J.M. Besser and M.J. Mac. "Chemical Characterization and Mutagenic Properties of Polycyclic Aromatic Compounds in Sediment from Tributaries of the Great Lakes," *Environ. Toxicol. and Chem.* 7:529–543 (1988).

Farber, E., and D.S.R. Sarma. "Hepatocarcinogenesis: A Dynamic Cellular Perspective," *Lab. Invest.* 56:4–22 (1987).

Ferguson, H.W. *Systemic Pathology of Fish*, (Ames, IA: Iowa State University Press, 1989).

Fujita, H., H. Tatsumi and T. Ban. "Fine-structural Characteristics of the Liver of the Cod (*Gadus macrocephalus*), with Special Regard to the Concept of a Hepatoskeletal System Formed by Ito Cells," *Cell Tissue Res.* 244:63–67 (1986).

Fujiya, M. "Studies on the Effects of Copper Dissolved in Sea Water on Oysters," *Bull. Jap. Soc. Sci. Fish.* 26:462–467 (1960).

Gardner, G.R. "Chemically Induced Lesions in Estuarine or Marine A Teleosts," in *The Pathology of Fishes*, W.C. Ribelin, and G. Migaki, Eds., Madison, WI: The University of Wisconsin Press, 1975), pp. 657–694.

Gardner, G.R., S.J. Benyi, J.F. Heltshe and J. Rosen. "Pigment Localization in Lymphoid Organs of the Winter Flounder (*Pseudopleuronectes americanus*) in Relation to Contaminated Sediment," in Society of Environmental Toxicology and Chemistry. Proceedings of 10th Annual Meeting, Toronto, November 1989a.

Gardner, G.R., and R.J. Pruell. "Quincy Bay Study, Boston Harbor: A Histopathological and Chemical Assessment of Winter Flounder, Lobster and Soft-shelled Clam Indigenous to Quincy Bay, Boston Harbor and an *in situ* Evaluation of Oysters Including Sediment (Surface and Cores) Chemistry," U.S. EPA Report, Region I, Boston, MA (1987).

Gardner, G.R., R.J. Pruell and L.C. Folmar. "A Comparison of Both Neoplastic and Non-neoplastic Disorders in Winter Flounder (*Pseudopleuronectes americanus*) from Eight Areas in New England," *Mar. Environ. Res.* 28:393–397 (1989b).

Gardner, G.R., R.J. Pruell and A.R. Malcolm. "Chemical Induction of Tumors in Oysters by a Mixture of Aromatic and Chlorinated Hydrocarbons, Amines and Metals," presented at Sixth International Symposium on Pollutant Responses in Marine Organisms, Woods Hole, MA., April 24–26 (1991c).

Gardner, G.R., and P.P. Yevich. "Comparative Histopathological Effects of Chemically Contaminated Sediment on Marine Organisms," *Mar. Environ. Res.* 24:311–316 (1988).

Gardner, G.R., P.P. Yevich, J.C. Harshbarger and A.R. Malcolm. "Carcinogenicity of Black Rock Harbor Sediment to the Eastern Oyster and Trophic Transfer of Black Rock Harbor Carcinogens from the Blue Mussel to the Winter Flounder," *Environ. Health Perspec.* 90:53–66 (1991a).

Gardner, G.R., P.P. Yevich, J. Hurst, P. Thayer, S. Benyi, J.C. Harshbarger and R.J. Pruell. "Germinomas and Teratoid Siphon Anomalies in Softshell Clam, *Mya arenaria*, Environmentally Exposed to Herbicides," *Environ. Health Perspect.* 90:43–51 (1991b).

Gardner, G.R., P.P. Yevich, A.R. Malcolm, R.J. Pruell, P.F. Rogerson, J. Heltshe, T.C. Lee and A. Senecal. "Carcinogenic Effects of Black Rock Harbor Sediment on Molluscs and Fish," Final Report to the National Cancer Institute NCI/EPA Collaborative Program on Environmental Cancer, December 31, 1987 p. 222.

Gardner, G.R., P.P. Yevich, A.R. Malcolm, P.F. Rogerson, L.J. Mills, A.G. Senecal, T.C. Lee, J.C. Harshbarger and T.P. Cameron. "Tumor Development in American Oysters and Winter Flounder Exposed to a Contaminated Marine Sediment Under Laboratory and Field Conditions," *Aquat. Toxicol.* 11:403–404 (1986).

Gardner, G.R., P.P. Yevich and P.F. Rogerson. "Morphological Anomalies in Adult Oysters, Scallop, and Atlantic Silversides Exposed to Waste Motor Oil," Proceedings of 1975 Conference Prevention and Control of Oil Pollution, San Francisco, CA, March 25–27, 1975 pp. 473–477.

Gerking, S.D. "Fish Reproduction and Stress," in *Environmental Physiology of Fishes*, Ali Ma, Ed. (New York: Plenum Press, 1980), pp. 569–587.

Gilderhaus, P.A. "Some Effects of Sublethal Concentrations of Sodium Arsenite on Bluegills and the Aquatic Environment," *Trans. Amer. Fish. Soc.* 95:289–296 (1966).

Gingerich, W.H. "Hepatic Toxicology of Fishes," in *Aquatic Toxicology*, L. Weber, Ed. (New York: Raven Press, 1982), pp. 55–105

Glaiser, J. *Principles of Toxicological Pathology* (London: Taylor and Francis, 1986).

Goksøyr, A., T. Andersson, T. Hanson, J. Klungsoyr, Y. Zhang and L. Förlın. "Species Characteristics of the Hepatic Xenobiotic and Steroid Biotransformation Systems of Two Teleost Fish, Atlantic Cod (*Gadus morhua*) and Rainbow Trout (*Salmo gairdneri*). *Toxicol. Appl. Pharmacol.* 89:347–360 (1987).

Goldstein, I.J., and R.D. Poretz. "Isolation, Physicochemical Characterization, and Carbohydrate-binding Specificity of Lectins," in *The Lectins: Properties, Functions and Applications in Biology and Medicine*, I.E. Liener, N. Sharon and I.J. Goldstein, Eds., (New York: Academic Press, 1986), pp. 33–248.

Greenwald, L., and L.B. Kirschner. "The Effect of Poly-L Sysine on Gill Ion Transport and Permeability in the Rainbow Trout," *J. Membr. Biol.* 26:371–383 (1976).

Grizzle, J.M., P. Melius and D.R. Strength. "Papillomas on Fish Exposed to Chlorinated Wastewater Effluent," *J. Natl. Cancer Inst.* 73(5):1133–1142 (1984).

Hampton, J.A., P.A. McCuskey, R.S. McCuskey and D.E. Hinton. "Functional Units in Rainbow Trout (*Salmo gairdneri*, Richardson) Liver: I. Arrangement and Histochemical Properties of Hepatocytes," *Anat. Rec.* 213:166–175 (1985).

Hampton, J.A., R.C. Lantz, P.J. Goldblatt, D.J. Laurén and D.E. Hinton. "Functional Units in Rainbow Trout (*Salmo gairdneri*, Richardson) Liver: II. The Biliary System," *Anat. Rec.* 221:619–634 (1988).

Hampton, J.A., R.C. Lantz and D.E. Hinton. "Functional Units in Rainbow Trout (*Salmo gairdneri*, Richardson) Liver: III. Morphometric Analysis of Parenchyma, Stroma, and Component Cell Types," *Am. J. Anat.* 185:58–73 (1989).

Harshbarger, J.C. "Pseudoneoplasms in Ectothermic Animals," Natl. Cancer Inst. Monogr. 65:251–273 (1984).

Harshbarger, J.C., and J.B. Clark. "Epizootiology of Neoplasms in Bony Fish of North America," *Sci. Total Environ.* 94:1–32 (1990).

Hawkins, W.E., J.W. Fournie, R.M. Overstreet and W.W. Walker. "Intraocular Neoplasms Induced by Methylazoxymethanol Acetate in Japanese Medaka (*Oryzias latipes*)," *J. Natl. Cancer Inst.* 76:453–465 (1986).

Hawkins, W.E., W.W. Walker, J.S. Lytle and R.M. Overstreet. "Carcinogenic Effects of 7,12-Dimethylbenz(a)anthracene on the Guppy (*Poecilia reticulata*)," *Aquat. Toxicol.* 15:63–82 (1989).

Hayes, M.A., I.R. Smith, T.H. Rushmore, T.L. Crane, C. Thorn, T.E. Kocal and H.W. Ferguson. "Pathogenesis of Skin and Liver Neoplasms in White Suckers from Industrially Polluted Areas in Lake Ontario," *Sci. Total Environ.* 94:105–123 (1990).

Hendricks, J.D., T.R. Meyers and D.W. Shelton. "Histological Progression of Hepatic Neoplasia in Rainbow Trout (*Salmo gairdneri*)," *Natl. Cancer Inst. Monogr.* 65:321–336 (1984).

Hendricks, J.D., R.O. Sinnhuber, M.C. Henderson and D.R. Buhler. "Liver and Kidney Pathology in Rainbow Trout (*Salmo gairdneri*) Exposed to Dietary Pyrrolizidine (Senecio) Alkaloids," Exper. Mol. Pathol. 35:170–183 (1981).

Herraez, M.P. and A.G. Zapata. "Structure and Function of the Melano-macrophage Centres of the Goldfish *Carassius auratus*," *Vet. Immunol. Immunopathol.* 12:117–126 (1986).

Hinton, D.E. "Environmental Contamination and Cancer in Fishes," *Mar. Environ. Res.* 28:411–416 (1989).

Hinton, D.E., J.A. Couch, S.J. Teh and L.A. Courtney. "Cytological Changes Durıng Progression of Neoplasia in Selected Fish Species," *Aquat. Toxicol.* 11:77–112 (1988).

Hinton, D.E., E.R. Walker, C.A. Pinkstoff and E.M. Zuchelkowski. "Morphological Survey of Teleost Organs Important in Carcinogenesis with Attention to Fixation," *Natl. Cancer Inst. Monogr.* 65:291–320 (1984a).

Hinton, D.E., R.C. Lantz and J.A. Hampton. "Effect of Age and Exposure to a Carcinogen on the Structure of the Medaka Liver: A Morphometric Study," *Natl. Cancer Inst. Monogr.* 65:239–249 (1984b).

Hinton, D.E., and D.J. Laurén. "Liver Structural Alterations Accompanying Chronic Toxicity in Fishes: Potential Biomarkers of Exposure," in *Biological Markers of Environmental Contamination*, J. McCarthy, and L.R. Shugart, Eds. (Boca Raton, FL: CRC Press, 1990a).

Hinton, D.E., and D.J. Laurén. "Integrative Histopathological Approaches for Detecting Effects of Environmental Stressors on Fishes," in *Biological Indicators of Fish Community Stress*, S.M. Adams, Ed., Amer. Fish. Soc. Special Pub (1990b).

Hinton, D.E., S.J. Teh, M.S. Okihiro, J.B. Cooke and L.M. Parker. "Phenotypically Altered Hepatocyte Populations in Diethylnitrosamine-induced Medaka Liver Carcinogenesis: Resistance, Growth, and Fate," *Mar. Environ. Res.*, in press.

Hoffman, C.L., C.E. Dunbar and A. Bradford. "Whirling Disease of Trout Caused by *Myxosoma cerebralis* in the United States," U.S. Bureau of Sport Fishing, Wildlife Special Scientific Report. Fisheries 427 (1962).

Holcombe, G.W., D.A. Benoit, E.N. Leonard and J.M. McKim. "Long-term Effects of Lead Exposure on Three Generations of Brook Trout (*Salvelinus fontinalis*)," *J. Fish. Res. Bd. Can.* 33:1731–1741 (1976).

Hose, J.E., J.B. Hannah, H.W. Puffer and M.L. Landolt. "Histologic and Skeletal Abnormalities in Benzo(a)pyrene-treated Rainbow Trout Alevins," *Arch. of Environ. Contam. and Toxicol.* 13:675–684 (1984).

Hughes, G.M., S.F. Perry and V.M. Brown. "A Morphometric Study of Effects of Nickel, Chromium and Cadmium on the Secondary Lamellae of Rainbow Trout Gills," *Water Res.* 13:665–679 (1979).

Ito, T., A. Watanabe and Y. Takahasaki. "Histological und Cytologische Unter-untersuchungen der Leber Bei Fisch und Cyclostomata, Nebst Bemerkrugen uber die Fettspeicherungszellen," *Arch. Histol. Japan* 22:429–463 (1962).

Ito, S., and M.J. Karnovsky. "Formaldehyde-glutaraldehyde Fixatives Containing Trinitro Compounds," *J. Cell Biol.* 39:168A–169A (1968).

Jenson, N.J., and J.L. Larsen. "The Ulcer Syndrome in Cod in Danish Coastal Waters," International Council Exploration of the Sea. C. M. 1978/E:28 (1978), p. 15.

Johnson, L.L., E. Casillas, T.K. Collier, B.B. McCain and U. Varanasi. "Contaminant Effects on Ovarian Development in English Sole (*Parophrys vetulus*) from Puget Sound, Washington," *Can. J. Fish. Aquat. Sci.* 45:2133–2146 (1988).

Kent, M.L., M.S. Myers, D.E. Hinton, W.D. Eaton and R.A. Elston. "Suspected Toxicopathic Hepatic Necrosis and Megalocytosis in Pen-reared Atlantic Salmon *Salmo salar* in Puget Sound, Washington, U.S.A.," *Dis. Aquat. Org.* 49:91–100 (1988).

Khudoley, V.V., and O.A. Syrenko. "Tumor Induction by N-nitroso Compounds in Bivalve Molluscs *Unio pictorum*," *Cancer Lett.* 4:349–354 (1978).

Kimura, I., N. Taniguchi, H. Kumai, I. Tomita, N. Kinae, K. Yoshizaki, M. Ito and T. Ishikawa. "Correlation of Epizootiological Observations with Experimental Data: Chemical Induction of Chromatophoromas in the Croaker, *Nibea mitsukurii*," *Natl. Cancer Inst. Monogr.* 65:139–154.(1984)

Kimura, T., M. Yoshimizu and M. Tanaka. "Fish Viruses; Tumor Induction in *Onchorhynchus keta* by the Herpes Virus," in *Phyletic Approaches to Cancer*, C.J. Dawe, J.C. Harshbarger and S. Kondo, Eds. (Tokyo: Japanese Scientific Society Press., 1981a), pp. 59–68.

Kimura, T., M. Yoshimizu and M. Tanaka. "Studies on a New Virus (OMV) from *Onchorhynchus masou* - II. Oncogenic Nature," *Fish Pathol.* 15:149–153 (1981b).

Kinae, N., M. Yamashita, I. Tomita, I. Kumura, H. Ishida, H. Kumai and G. Nakamura. "A Possible Correlation Between Environmental Chemicals and Pigment Cell Neoplasia in Fish," *Sci. Total Environ*. 94:143–153 (1990).

Klaunig, J.E., M.M. Lipsky, B.F. Trump and D.E. Hinton. "Biochemical and Ultrastructural Changes in Teleost Liver Following Subacute Exposure to PCB," *J. Environ. Pathol. Toxicol*. 2:953–963 (1979).

Koenig, C.C., and M.P. Chasar. "Usefulness of the Hermaphroditic Marine Fish *Rivulus marmoratus*, in Carcinogenicity Testing," *Natl. Cancer Inst. Monogr*. 65:15–33 (1984).

Kranz, H., and V. Dethlefsen. "Liver Anomalies in Dab (*Limanda limanda*) from the Southern North Sea with Special Consideration Given to Neoplastic Lesions," *Dis. Aquat. Org*. 9:171–185 (1990).

Kurubagaran, R., and K.P. Joy. "Toxic Effects of Mercuric Chloride, Methylmercuric Chloride, and Emisan 6 (an Organic Mercurial Fungicide) on Ovarian Recrudescence in the Catfish *Clarias batrachus* L," *Bull. Environ. Contam. Toxicol*. 41:902–909 (1988).

LaRoche, G., G.R. Gardner, R.E. Eisler, E.H. Jackim, P.P. Yevich and G.E. Zaroogian. "Analysis of Toxic Responses in Marine Poikilotherms," in *Bioassay Techniques and Environmental Chemistry*, G.E. Glass, Ed. (Michigan: Ann Arbor Science Publishers, Inc., 1973), pp 199–216..

Laurén, D.J., and D.G. McDonald. "The Effects of Copper on Branchial Ion Regulation in the Rainbow Trout, *Salmo gairdneri* Richardson — Modulation by Water Hardness and pH," *J. Comp. Physiol*. 155:635–644 (1985).

Lauren, D.J., Teh, S.J. and D.E. Hinton. "Cytotoxicity Phase of Diethylnitrosamine-induced Hepatic Neoplasia in Medaka," *Cancer Res*. 50:5504–5514 (1990).

Laurent, P., H. Hobe and S. Dunel-Erb. "The Role of Environmental Sodium Chloride Relative to Calcium in Gill Morphology of Freshwater Salmonid Fish," *Cell Tissue Res*. 240:675–692 (1985).

Leatherland, J.F., and R. Sonstegard. "Structure of Normal Testis and Testicular Tumors of Cyprinids from Lake Ontario," *Cancer Res*. 38:3164–3173 (1978).

Lesniak, J.A., and S.M. Ruby. "Histological and Quantitative Effects of Sublethal Cyanide Exposure on Oocyte Development in Rainbow Trout," *Arch. Environ. Contam. Toxicol*. 13:101–104 (1982).

Lightner, D.V. "Diseases of Cultured Penaeid Shrimp," in *CRC Handbook of Mariculture, Vol. 1 Crustacean Aquaculture*, J.P. McVey,Ed., (Boca Raton, FL: CRC Press, Inc., 1983), pp 289–320.

Lightner, D.V. "Examples of Lesions in the Crustacean Due to the Toxic Effects of Organic and Inorganic Toxins," presented at the *Soc. Invert. Pathol. Ann. Mtg*., Gainesville, FL. July 20–24, 1987.

Lock, R.A.C., and A.P. van Overbeeke. "Effects of Mercuric Chloride and Methylmercuric Chloride on Mucus Secretion in Rainbow Trout (*Salmo gairdneri*) Richardson," *Comp. Biochem. Physiol*. 69C:67–73 (1981).

Loud, A.V., and P. Anversa. "Biology of Disease: Morphometric Analysis of Biologic Processes," *Lab. Invest*. 50:250–261 (1984).

Lucke, B., and H.G. Schlumberger. "Transplantable Epitheliomas of the Lip and Mouth of Catfish," *J. Exper. Med*. 74:397–408 (1941).

Marine Technological Society. *Proceedings of the Marine Bioassays Workshop*, (Washington, D.C.: Marine Technological Society, Washington, 1974), p. 156–173.

Masahito, P., T. Ishikawa and H. Sugano. "Neural and Pigment Cell Neoplasms," in *Neoplasms and Related Disorders in Fishes*, C.J. Dawe, Ed., in press.

Mayer, F.L., P.M. Mehrle and R.A. Schoettger. "Collagen Metabolism in Fish Exposed to Organic Chemicals," in *Recent Advances in Fish Toxicology; A Symposium Held in Corvallis, Oregon.*," R.A. Tubb, Ed., U.S. EPA-600/3-77-085 (1977) pp. 31–54.

McAllister, P.E., T. Nagabayashi and K. Wolf. "Viruses of Eels with and without Stomatopapillomas," *Ann. N. Y. Acad. of Sci.* 298:233–244 (1977).

McCormick, J.H., G.N. Stokes and R.O. Hermanatz. "Oocyte Atresia and Reproductive Success in Fathead Minnows (*Pimephales promelas*) Exposed to Acidified Hardwater Environments," *Arch. Environ. Contam. and Toxicol.* 18:207–214 (1989).

McDonald, D.G. "The Effects of H^+ upon the Gills of Freshwater Fish," *Can. J. Zool.* 61:691–703 (1983).

McDowell. E.M., and B.F. Trump. "Histologic Fixatives Suitable for Diagnostic Light and Electron Microscopy," *Arch. Pathol. Lab. Med.* 100:405–414 (1976).

McIntyre, J.D. "Toxicity of Methylmercury for Steelhead Trout Sperm," *Bull. Environ Contam. Toxicol.* 9:98–99 (1973).

Mehrle, P.M., and F.L. Mayer. "Toxaphene Effects on Growth and Development of Brook Trout (*Salvelinus fontinalis*)," *J. Fish. Res. Board of Can.* 32:609–613 (1975).

Mehrle, P.M., and F.L. Mayer. "Bone Development and Growth of Fish as Affected by Toxaphene," in *Fate of Pollutants in Air and Water Environments. Part 2.* I.H. Suffet, Ed. (New York: Wiley Interscience Publishing, 1977), pp. 301–304.

Meyers, T.R., and J.D. Hendricks. "Histopathology," in *Fundamentals of Aquatic Toxicology*, G.M. Rand, and S.R. Petrocelli, Eds. (Washington, D.C.: Hemisphere Publishing Corp., 1985), pp. 283–331.

Miller, M.R., C. Heyneman, S. Walker and R.G. Ulrich. "Interaction of Monoclonal Antibodies Directed Against Bromodeoxyuridine with Pyrimidine Bases, Nucleosides, and DNA," *J. Immunol.* 136:1791–1795 (1986).

Miller, M.R., D.E. Hinton and J.J. Stegeman. "Cytochrome P-450 E Induction and Localization in Gill Pillar (endothelial) Cells of Scup and Rainbow Trout," *Aquat. Toxicol.* 14:307–322 (1989).

Miller, T.G., and W.C. Mackay. "Relationship of Secreted Mucus to Copper and Acid Toxicity in Rainbow Trout," *Bull. Environ. Contamin. Toxicol.* 28:68–74 (1982).

Mittal, A.K., and J.S.D. Munshi. "On the Regeneration and Repair of Superficial Wounds in the Chin of *Rita rita* (Ham.) (Bazridae, Pisces).," *Acta Anatom.* 88:424–442 (1974).

Mix, M.C. "Cancerous Diseases in Aquatic Animals and Their Sssociation with Environmental Pollutants: A Critical Literature Review," *Mar. Environ. Res.* 20:1–141 (1986a).

Mix, M.C. "Shellfish Diseases in Relation to Toxic Chemicals," *Aquat. Toxicol.* 11:29–42 (1986b).

Mix, M.C., Trenholm, S.R. and K.I. King. "Benzo(a)pyrene Body Burdens and the Prevalence of Proliferative Disorders in Mussels (*Mytilus edulis*) in Oregon," in *Animals as Monitors of Environmental Pollutants*, S.V. Nielsen, G. Migaki and D. Scarpelli, Eds. (Washington D.C.: National Academy Science, 1979), pp. 52–64.

Moon, T.W., P.J. Walsh and T.P. Mommsen. "Fish Hepatocytes: A Model Metabolic System," *Can. J. Fish. Aquat. Sci.* 42:1772–1782 (1985).

Moore, M.J., R. Smolowitz and J.J. Stegeman. "Cellular Alterations Preceding Neoplasia in *Pseudopleuronectes americanus* from Boston Harbor," *Mari. Environ. Res.* 28:425–429 (1989).

Morris, R.W. "Testing Statistical Hypotheses about Rat Liver Foci," *Toxicol. Pathol.* 17:569–578 (1989).

Mottet, N.K., and M.L. Landolt. "Advantages of Using Aquatic Animals for Biomedical Research on Reproductive Toxicology," *Environ. Health Perspect.* 71:69–75 (1987).

Mulcahy, M.F., and A. O'Leary. "Cell-free Transmission of Lymphosarcoma in the Northern Pike *Esox lucius* L. (Pisces: Esocidae)." *Experimentia* 26:891 (1970).

Murai, T., and J. Andrews. "Riboflavin Requirements of Channel Catfish Fingerlings," *J. Nutr.* 108:1512–1517 (1978).

Muramoto, S. "Vertebral Column Damage and Decrease of Calcium Concentration in Fish Exposed Experimentally to Cadmium," *Environ. Pollut.* 24:125–133 (1981).

Murchelano, R.A., and R.E. Wolke. "Epizootic Carcinoma in the Winter Founder (*Pseudopleuronectes americanus*). *Science* 228:587–589 (1985)

Murchelano, R.A., and J. Ziskowski. "Some Observations on an Ulcer Disease of Red Hake, *Urophycis chuss*, from the New York Bight," International Council on Exploration of the Sea, C. M. 1979/E:23, (1979), p 5.

Myers, M.S., L.D. Rhodes and B.B. McCain. "Pathologic Anatomy and Patterns of Occurrence of Hepatic Neoplasms, Putative Preneoplastic Lesions, and Other Idiopathic Hepatic Conditions in English Sole (*Parophrys vetulus*) from Puget Sound, Washington," *J. Natl. Cancer Inst.* 78(2):333–363 (1987).

Myers, M.S., J.T. Landahl, M.M. Krahn, L.L. Johnson and B.B. McCain. "Overview of Studies on Liver Carcinogenesis in English Sole from Puget Sound; Evidence for a Xenobiotic Chemical Etiology: Pathology and Epizootiology," *Sci. Total Environ.* 94:33–50 (1990).

Nakazawa, T., S. Hamaguchi and Y. Kyono-Hamaguchi. "Histochemistry of Liver Tumors Induced by Diethylnitrosamine and Differential Sex Susceptibility to Carcinogenesis in *Oryzias latipes*," *J. Natl. Cancer Inst.* 75:567–573 (1985).

Neff, J.M., R.E. Hillman, R.S. Carr, R.L. Buhl and J.I. Lahey. "Histopathologic and Biochemical Responses in Arctic Marine Bivalve Molluscs Exposed to Experimentally Spilled Oil," *J. Arctic* 40(1):220–229 (1987).

Nelson, D.A., A. Calabrese, R.A. Grieg, P.P. Yevich and S. Chang. "Long-term Silver Effects on the Marine Gastropod *Crepidula fornicata*," *Mar. Ecol.* 12:155–165 (1983).

Nunez, O., J.D. Hendricks and G.S. Bailey. "Enhancement of Aflatoxin B_1 and *N*-Methyl-*N'*Nitro-N-Nitrosoguanidine Hepatocarcinogenesis in Rainbow Trout *Salmo gairdneri* by 17-B-Estradiol and Other Organic Chemicals," *Dis. Aquat. Organis.* 5:185–196 (1988).

Nunez, O., J.D. Hendricks and J.R. Duimstra. "Ultrastructure of Hepatocellular Neoplasms in Aflatoxin B_1 (AFB_1)-Initiated Rainbow Trout (*Oncorhynchus mykiss*)." *Toxicol. Pathol.* 19:11–23 (1991).

Okihiro, M.S. "Chromatophoromas in Two Species of Hawaiian Butterflyfish, *Chaetodon multicinctus* and *C. miliaris*," *Vet. Pathol.* 25:422–431 (1988).

Okihiro, M.S., J. Whipple, J.M. Groff and D.E. Hinton. "Chromatophoromas and Related Hyperplastic Lesions in Pacific Rockfish (*Sebastes*)," *Mar. Environ. Res.*, in press 1991.

Packer, R.K., and W.A. Dunson. "Anoxia and Sodium Loss Associated with the Death of Brook Trout at Low pH," *Comp. Biochem. Physiol.* 41A:17–26 (1972).

Papas, T.S., J.E. Dahlberg and R.A. Sonstegard. "Type C Virus in Lymphosarcoma in Northern Pike (*Esox lucius*)," *Nature* 261:506–508 (1976).

Papas, T.S., T.W. Pry, M.P. Schafer and R.A. Sonstegard. "Presence of DNA Polymerase in Lymphosarcoma in Northern Pike (*Esox lucius*)," *Cancer Res.* 37:3214–3217 (1977).

Pauley, G.B. "A Critical Review of Neoplasia and Tumor-like Lesions in Mollusks," in *Neoplasms and Related Disorders in Invertebrates and Lower Vertebrate Animals, Natl. Cancer Inst. Monogr.* 31:509–539 (1969).

Perry, S.F., and C.M. Wood. "Kinetics of Branchial Calcium Uptake in the Rainbow Trout: Effects of Acclimation to Various External Calcium Levels," *J. Exper. Biol.* 116:411–433 (1985).

Peters, E.C. "Recent Investigations of the Disseminated Sarcomas of Marine Bivalve Molluscs," *Amer. Fish. Soc. Spec. Publ.* 18:74–92 (1988).

Pickering, A.D., T.G. Pottinger and P. Christie. "Recovery of the Brown Trout, *Salmo trutta* L., from Acute Handling Stress: A Time Course Study," *J. Fish Biol.* 20:229–244 (1982).

Pierce, K.V., B.B. McCain and S.R. Wellings. "Pathology of Hepatomas and Other Liver Abnormalities in English Sole (*Parophrys vetulus*) from the Duwamish River Estuary, Seattle, Washington," *J. Natl. Cancer Inst.* 60:1445–1453 (1978).

Pitot, H.C. "Hepatic Neoplasia: Chemical Induction," in *The Liver: Biology and Pathology*, I.M. Arias, W.B. Jakoby and H. Popper, Eds. (New York: Raven Press, 1988), pp. 1125–1146.

Pitot, H.C., H.A. Campbell, R. Maronpot, N. Bawa, T.A. Rizvi, Y. Xu, L. Sargent, Y. Dragan and M. Pyron. "Critical Parameters in the Quantitation of the States of Initiation, Promotion, and Progression in One Model of Hepatocarcinogenesis in the Rat," *Toxicol. Pathol.* 17:594–612 (1989).

Popp, J.A., and T.L. Goldsworthy. "Defining Foci of Cellular Alteration in Short-term and Medium-term Rat Liver Tumor Models," *Toxicol. Pathol.* 17:5661–5668 (1989).

Pritchard, J.B., and J.L. Renfro. "Interactions of Xenobiotics with Teleost Renal Function," in: *Aquatic Toxicology, Vol. 2*, L.J. Weber Ed. (New York: Raven Press, 1982), pp. 51–106.

Reid, S.D., and D.G. McDonald. "Effects of Cadmium, Copper, and Low pH on Ion Fluxes in Rainbow Trout, *Salmo gairdneri*," *Can. J. Fish. Aquat. Sci.* 45:244–253 (1988).

Reide, U.N., and A. Reith. *Morphometry in Pathology*, (New York:Gustav Fisher Verlag, 1980).

Reimschuessel, R., R.O. Bennett, E.B. May and M.M. Lipsky. "Development of Newly Formed Nephrons in the Goldfish Kidney Following Hexachlorobutadiene-induced Nephrotoxicity," *Toxicol. Pathol.* 18:32–38 (1990).

Rhodes, L.D., M.S. Myers, W.D. Gronlund and B.B. McCain. "Epizootic Characteristics of Hepatic and Renal Lesions in English Sole, *Parophrys vetulus*, from Puget Sound," *J. Fish Biol.* 31:395–407 (1987).

Roch, M., and E.J. Maly. "Relationship of Cadmium-induced Hypocalcemia with Mortality in Rainbow Trout (*Salmo gairdneri*) and the Influence of Temperature on Toxicity," *J. Fish. Res. Bd. Can.* 36:1297–1303 (1979).

Rogerson, P.F., S.C. Schimmel and G. Hoffman. "Chemical and Biological Characterization of Black Rock Harbor Dredged Material," U.S. EPA Report D-85-9 (1985), p. 110.

Rohr, H.P., M. Oberholzer, G. Bartsch and M. Keller. "Morphometry in Experimental Pathology: Methods, Baseline Data and Application," *Inter. Rev. Exper. Pathol.* 15:233–325 (1976).

Rose, F.L. "Tissue Lesions of Tiger Salamanders (*Ambystoma tigrinum*): Relationship to Sewage Effluents.," *Ann. N.Y. Acad. Sci.* 298:270–279 (1977).

Rose, F.L. "The Tiger Salamander (*Ambystoma tigrinum*): A Decade of Sewage Associated Neoplasia," in *Phyletic Approaches to Cancer*, C.J. Dawe, J.C. Harshbarger and S. Kondo, Eds. (Tokyo: Japanese Science Society Press, 1981), pp. 91–100.

Saiki, M., and T. Mori. "Studies on the Distribution of Administered Zinc in the Tissues of Fishes," *Bull. Jap. Soc. Sci. Fish.* 21:945–949 (1955).

Sangalang, G.B., H.C. Freeman and R. Crowell. "Testicular Abnormalities in Cod (*Gadus morhua*) Fed Aroclor 1254," *Arch. Environ. Contam. Toxicol.* 10:617–627 (1981).

Sangalang, G.B., and M.J. O'Halloran. "Cadmium-induced Testicular Injury and Alteration of Androgen Synthesis in Brook Trout," *Nature* (London) 240:470–471 (1972).

Sauer, G.R., and N. Watabe. "Zinc Uptake and Its Effect on Calcification in the Scales of the Mummichog, *Fundulus heteroclitus*," *Aquat. Toxicol.* 5:51–66 (1984).

Scarpelli, D.G., M.H. Grieder and W.J. Frajola. "Observations on Hepatic Cell Hyperplasia, Adenoma, and Hepatoma of Rainbow Trout (*Salmo gairdneri*)," *Cancer Res.* 23:848–857 (1963).

Schlumberger, H.C., and B. Lucke. "Tumors of Fishes, Amphibians, and Reptiles," *Cancer Res.* 8:657–760 (1948).

Schmidt, D.C., and L.J. Weber. "Metabolism and Biliary Excretion of Sulfobromophthalein by Rainbow Trout (*Salmo gairdneri*)," *J. Fish. Res. Bd. Can.* 30:1301–1308 (1973).

Schoor, W.P., and J.A. Couch. "Correlation of Mixed-function Oxidase Activity with Ultrastructural Changes in the Liver of a Marine Fish," *Cancer Biochem. Biophys.* 4:95–103 (1979).

Schwerdtfeger, W.K. "Morphometrical Studies of the Ultrastructure of the Epidermis of the Guppy, *Poecilia reticulata* Peters, Following Adaptation to Seawater and Treatment with Prolactin," *Gen. Comp. Endocrinol.* 38:476–483 (1979a).

Schwerdtfeger, W.K. "Qualitative and Quantitative Data on the Fine Structure of the Guppy (*Poecilia reticulata* Peters) Epidermis Following Treatment with Thyroxine and Testosterone," *Gen. Comp. Endocrinol.* 38:484–490 (1979b).

Segner, H., and T. Braunbeck. "Hepatocellular Adaptation to Extreme Nutritional Conditions in Ide, *Leuciscus idus melanotus* L. (Cyprinidae). A Morphofunctional Analysis," *Fish Physiol. Biochem.* 5(2):79–97 (1988).

Segner, H., and J.V. Juario. "Histological Observations on the Rearing of Milkfish, (*Chanos chanos*), by Using Different Diets," *J. Appl. Ichthyol.* 2:162–173 (1986).

Segner, H., and H. Möller. "Electron Microscopical Investigations on Starvation-induced Liver Pathology in Flounders *Platichthys flesus*," *Mar. Ecol. Progr. Ser.* 19:193–196 (1984).

Simon, K., and K. Lapis. "Carcinogenesis Studies on Guppies," *Natl. Cancer Inst. Monogr.* 65:71–81 (1984).

Sindermann, C.J. "*Principal Diseases of Marine Fish and Shellfish. Volume 1*," 2nd ed. (New York: Academic Press, 1990), pp. 521.

Skidmore, J.F., and P.W.A. Tovell. "Toxic Effects of Zinc Sulfate on the Gills of Rainbow Trout," *Water Res.* 6:217–230 (1972).

Sloof, W. "Skeletal Anomalies in Fish from Polluted Surface Waters," *Aquat. Toxicol.* 2:157–173 (1982).

Smith, C.E. "Hyperplastic Lesions of the Primitive Meninx of Fathead Minnows, *Pimephales promelas*, Induced by Amino: Species Potential for Carcinogen Testing," *Natl. Cancer Inst. Monogr.* 65:119–128 (1984).

Smith, I.R., K.W. Baker, M.A. Hayes and M.A. Ferguson. "Ultrastructure of Malpighian and Inflammatory Cells in Epidermal Papillomas of White Suckers *Catastomus commersoni*," *Dis. Aquat. Org.* 6:17–26 (1989).

Sokal, R.R., and F.J. Rohlf. *Biometry*, 2nd ed., (New York:Freeman, 1981), p. 733 .

Solangi, M.A., and R.M. Overstreet. "Histopathological Changes in Two Estuarine Fishes, *Menidia beryllina* (Cope) and *Trinectes maculatus* (Bloch and Schneider), Exposed to Crude Oil and Its Water Soluble Fractions," *J. Fish Dis.* 5:13–35 (1982).

Sonstegard, R.A. "Environmental Carcinogenesis Studies in Fishes of the Great Lakes of North America," *Ann. N.Y. Acad. Sci.* 298:261–269 (1977).

Sorenson, E.M.B., P.M. Cumbie, T.C. Bauer, J.S. Bell and C.W. Harlan. "Histopathological, Hematological, Condition Factor, and Organ Weight Changes Associated with Selenium Accumulation in Fish from Belews Lake, North Carolina," *Arch. Environ. Contam. Toxicol.* 13:153–162 (1984).

Sparks, A.K. *Synopsis of Invertebrate Pathology,* (New York: Elsevier, 1985), p. 424.

Spencer, S.L. "Internal Injuries of Largemouth Bass and Bluegills Caused by Electricity" *Prog. Fish Culturist* 29:168–169 (1967).

Spies, R.B., D.W. Rice and J. Felton. "Effects of Organic Contaminants on Reproduction of the Starry Flounder (*Platichthys stellatus*) in San Francisco Bay. I. Hepatic Contamination and Mixed-function Oxidase (MFO) Activity During the Reproductive Season," *Mar. Biol.* 98:181–189 (1988).

Stanton, M.F. "Diethylnitrosamine-induced Hepatic Degeneration and Neoplasia in the Aquarium Fish, *Brachydanio rerio*," *J. Natl. Cancer Inst.* 344:117–130 (1965).

Stegeman, J.J., R.L. Binder and A. Orren. "Hepatic and Extrahepatic Microsomal Electron Transport Components and Mixed-function Oxygenases in the Marine Fish *Stenotomus versicolor*," *Biochem. Pharmacol.* 28:3431–3439 (1979).

Stegeman, J.J., M.R. Miller and D.E. Hinton. "Cytochrome P-450 Induction and Localization in Endothelium of Vertebrate Heart," *Molec. Pharmacol.* 36:723–729 (1989).

Stehlik, L.L., and J.V. Merriner. "Effects of Accumulated Dietary Kepone on Spot (*Leiostomus xanthurus*)," *Aquat. Toxicol.* 3:345–358 (1983)

Stehr, C. "Ultrastructure of Vacuolated Cells in the Liver of Rock Sole and Winter Flounder Living in Contaminated Environments," Proceedings of the 12th International Congress for Electron Microscopy (1990), pp. 522–523.

Stott, G.G., N.H. McArthur, R. Tarpley, V. Jacobs and R.F. Sis. "Histopathologic Survey of Ovaries of Fish from Petroleum Production and Control Sites in the Gulf of Mexico," *J. Fish. Biol.* 18:261–269 (1981).

Stott, G.G., W.E. Haensly, J.M. Neff and J.R. Sharp. "Histopathologic Survey of Ovaries of Plaice, *Pleuronectes platessa* L., from Aber Wrac'h and Aber Benoit, Brittany, France: Long-term Effects of the Amoco Cadiz Crude Oil Spill," *J. Fish Dis.* 6:429–437 (1983).

Sunila, I. "Cystic Kidneys in Copper Exposed Mussels, *Mytilus edulis* L.," *Fourth International Colloquium of Invertebrate Pathology*, R.A. Samson, J.M. Vlak and D. Peters, Eds. (Veldhoven, The Netherlands: Society Invertebrate Pathologists, 1986).

Sunila, I. "Histopathological Effects of Environmental Pollutants on the Common Mussel, *Mytilus edulis* L. (Baltic Sea), and Their Application in Marine Monitoring," PhD Thesis, University of Helsinki (1987).

Sunila, I. "Pollution-related histopathological changes in the mussel, *Mytilus edulis* L. in the Baltic Sea," *Mar. Environ. Res.* 24:277–280 (1988).

Tafanelli, R., and R.C. Summerfelt. "Cadmium-induced Histopathological Changes in Goldfish," in *The Pathology of Fishes*, W.E. Ribelin, and G. Migaki, Eds. (Madison, WI: University of Wisconsin Press, 1975), pp. 613–645.

Tan, B., P. Melius and J. Grizzle. "Hepatic enzymes and Tumor Histopathology of Black Bullheads with Papillomas," in *Polynuclear Aromatic Hydrocarbons: Chemical Analysis and Biological Fate*, M. Cooke, and A.J. Dennis, Eds. (Columbus, OH: Battelle Press, 1981), pp. 377–386.

Thomas, P., M. Bally and J.M. Neff. "Ascorbic Acid Status of Mullet *Mugil cephalus* Linn., Exposed to Cadmium," *J. Fish Biol.* 20:183–196 (1982).

Trump, B.F., E.M. McDowell and A.U. Arstila. "Cellular Reaction to Injury," in *Principles of*

Pathobiology, 3rd ed., R.B. Hill, and M.F. LaVia, Eds. (New York: Oxford University Press, 1980), pp. 20–111.

Vaillant, C., C. Le Guellec and F. Padkel. "Vitellogenin Gene Expression in Primary Culture of Male Rainbow Trout Hepatocytes," *Gen. Comp. Endocrin.* 70:284–290 (1988).

Valentine, D.W. "Skeletal Anomalies in Marine Teleosts," in *The Pathology of Fishes*, W.R. Ribelin, and G. Migaki, Eds. (Madison, WI: University of Wisconsin Press, 1975).

van Bohemen, C.G., J.G.D. Lambert and J. Peute. "Annual Changes in Plasma and Liver in Relation to Vitellogenesis in the Female Rainbow Trout *Salmo gairdneri*," *Gen. Comp. Endocrinol.* 44:94–107 (1981).

Van deKamp, G. "Vertebral Deformities of Herring Around the British Isles and Their Usefulness for a Pollution Monitoring Programme," International Council of Exploration of the Sea, CM-E:5 (1977), p. 10

Varanasi, U. *Metabolism of Polycyclic Aromatic Hydrocarbons in the Aquatic Environment* (Boca Raton, FL: CRC Press Inc., 1989), pp. 341.

Varanasi, U., P.A. Robisch and D.C. Malins. "Structural Alterations in Fish Epidermal Mucus Produced by Water-borne Lead and Mercury," *Nature* (London) 258:431–432 (1975).

Varanasi, U., and D.J. Gmur. "Influence of Water-borne and Dietary Calcium on Uptake and Retention of Lead by Coho Salmon (*Oncorhynchus kisutch*)," *Toxicol. Applied Pharmacol.* 46:65–75 (1978).

Verbost, P.M., G. Flik, R.A.C. Lock and S.E. Wendelaar Bonga. "Cadmium Inhibition of Ca^{2+} Uptake in Rainbow Trout Gills," *Am. J. Physiol.* 253:R216–R221 (1987).

Vogelbein, W.K., J.W. Fournie, P.A. Van Veld and R.J. Huggett. "Hepatic Neoplasms in the Mummichog *Fundulus heteroclitus* from a Creosote-contaminated Site," *Cancer Res.* 50:5978–5986 (1990).

Walsh, A.H., and W.E. Ribelin. "The Pathology of Pesticide Poisoning," in *The Pathology of Fishes*, W.E. Ribelin, and G. Migaki, Eds. (Madison, WI: University of Wisconsin Press, 1975), pp. 515–577.

Walton, M.J., and C.B. Cowey. "Aspects of Intermediary Metabolism in Salmonid Fish," *Comp. Biochem. Physiol.* 73B:59–79 (1982).

Weibel, E.R. *Stereological Methods, Vol. I.* (New York: Academic Press, 1979), p. 161.

Weibel, E.R. *Stereological Methods, Vol. II.* (New York: Academic Press, 1980), p. 161.

Weis, J.S., and P. Weis. "Effects of Environmental Pollutants on Early Fish Development," *Aquat. Sci.* 1:45–73 (1989).

Weissenberg, R. "Fifty Years of Research on the Lymphocystic Virus Disease of Fishes (1914–1964)," *Ann. N.Y. Acad. Sci.* 126:362–374 (1965).

Wellings, S.R., H.A. Bern, R.S. Nishioka and J.W. Graham. "Epidermal Papillomas in the Flathead Sole," *Proc. Amer. Assoc. Cancer Res.* 4:71 (1963).

Wellings, S.R., R.G. Chuinard, R.T. Gourley and R.A. Cooper. "Epidermal Papillomas in the Flathead Sole, *Hippoglossoides elassodon*, with Notes on the Occurrence of Similar Neoplasms in Other Pleuronectids," *J. Natl. Cancer Inst.* 33:991–1004 (1964).

Wellings, S.R., B.B. McCain and B.S. Miller. "Epidermal Papillomas in Pleuronectidae of Puget Sound, Washington," *Prog. Exper. Tumor Res.* 20:55–74 (1976).

Wester, P.W., and J.H. Canton. "Histopathological Study of *Oryzias latipes* (medaka) After Long-term B-Hexachlorocyclohexane Exposure," *Aquat. Toxicol.* 9:21–45 (1986).

Westernhagen, H. von, H. Rosenthal and K.R. Sperling. "Combined Effects of Cadmium and Salinity on Development and Survival of Herring Eggs," *Helgolander wiss. Meeresunters* 26:416–433 (1974).

Wilcock, B.P. and T.W. Dukes. "The Eye" in *Systemic Pathology of Fish*, H.W. Ferguson, Ed. (Ames, IA: Iowa State University Press, 1989), pp. 156–173.

Wolf, K. *Fish Viruses and Fish Viral Diseases* (Ithaca, N.Y.: Cornell University Press, 1988).

Wolke, R.E., R.A. Murchelano, C.D. Dickstein and C.J. George. "Preliminary Evaluation of the Use of Macrophage Aggregates (MA) as Fish Health Monitors," *Bull. Environ Contam. Toxicol.* 35:222–227 (1985).

Wyllie, A.H., J.F.G. Kerr and A.R. Cumi. "Cell Death: The Significance of Apoptosis," *Int. Rev. Cytol.* 68:251–306 (1980).

Yamamoto, K., P.A. Sargent and M.M. Fisher. "Periductal Fibrosis and Lipocytes (Fat-storing Cells or Ito Cells) During Biliary Atresia in the Lamprey," *Hepatology* 6:54–59 (1986).

Yamamoto, Y., M. Sato and S. Ikeda. "Existence of L-Gulonolactone Oxidase in Some Teleosts," *Bull. Jpn. Soc. Sci. Fish.* 44:775–779 (1978).

Yamashita, M., N. Kinae, I. Kimura, H. Ishida, H. Kumai and G. Nakamura. "The Croaker (*Nibea mitsukurii*) and the Sea Catfish (*Plotosus anguillaris*): Useful Biomarkers of Coastal Pollution," in *Biomarkers of Environmental Contamination*, J.F. McCarthy, and L.R. Shugart Eds. (Boca Raton, FL.: Lewis Publishers, CRC Press, Inc., 1990), pp. 73–84.

Yevich, P.P. "Comparative Histopathological Effects of Metals on Marine Organisms," presented at Society for Invertebrate Pathology, 22nd Annual Meeting, August, 1989.

Yevich, P.P., C.A. Yevich, K.J. Scott, M. Redmond, D. Black, P.S. Schauer and C.E. Pesch. "Histopathological Effects of Black Rock Harbor Dredged Material on Marine Organisms," U. S. EPA Report D-86-1. (1986),.p. 72.

Yevich, C.A., and P.P. Yevich. "Histopathological Effects of Cadmium and Copper on the Sea Scallop, *Placopectin magellanicus*," Mar. Pollut. Physiol. 187–198 (1985).

Yoshimizu, M., M. Tanaka and T. Kimura. "*Onchorhynchus masou* Virus (OMV): Incidence of Tumor Development Among Experimentally Infected Representative Salmonid Species," *Fish Pathol.* 22(1):7–10 (1987).

Zuchelkowski, E.M., R.C. Lantz and D.E. Hinton. "Effects of Acid-stress on Epidermal Mucous Cells of the Brown Bullhead (*Ictalurus nebulosus* (Le Seur): A Morphometric Study," *Anat. Rec.* 200:33–39 (1981).

Zuchelkowski, E.M., R.C. Lantz and D.E. Hinton. "Skin Mucous Cell Response to Acid Stress in Male and Female Brown Bullhead Catfish, *Ictalurus nebulosus* (Le Sueur)," *Aquat. Toxicol.* 8:139–148 (1986).

CHAPTER 5

Immunological Biomarkers to Assess Environmental Stress

Beverly A. Weeks, Douglas P. Anderson, Alfred P. DuFour, Ann Fairbrother,
Arthur J. Goven, Garet P. Lahvis, and Gabriele Peters

INTRODUCTION

Numerous methods have been developed that are useful for screening chemicals and analyzing their modes of action on living organisms. Although many assays have revealed toxic effects in affected species, comparative immunologists have only recently become aware of the broad spectrum of xenobiotics that alter some of the immune functions (Vos 1977; Sharma 1981; Dean et al. 1985). Clearly the immune system has potential for assessing the toxic effects of chemicals. Since there is considerable information concerning the cellular, humoral, and molecular components of the immune system, immune responses are especially well-suited for comparative analyses which emphasize mechanisms of toxicity. Characteristics that render immunocompetent cells appropriate and sensitive for analyzing chemically induced toxic effects include:

1. Capacity to proliferate rapidly following activation with antigens, mitogens, or hormones. Several studies have shown that exposure to toxic chemicals interferes with early events of this process.

2. Potential to undergo terminal differentiation that results in production of mediators (e.g., antibodies and lymphokines) or effector functions (e.g., target cell killing).

3. Gene products that can be used as markers of maturation; for example, the Ia receptors are important markers of macrophage function. These markers can be detected with specific antisera, for instance, by immunofluorescent labeling to determine the state of maturation and functional capacity of these cells in the immune response. The Ia expression on macrophages is subject to modulation by a variety of internal signals (Zimmer and Jones 1990), which suggests that Ia expression might have potential use as a bioindicator.

Since immunocompetent cells are required for host resistance, measurement of increased susceptibility to infectious agents or tumor cells could provide insight into the biological significance of immune alteration induced by xenobiotics. The interaction of environmental chemicals with components of the immune system may suppress or enhance immune activity, and consequently may alter host defense against pathogens and neoplasia (Luster et al. 1988; Zeeman and Brindley 1981).

This chapter will provide a summary of the organization of the immune system and outline the state of knowledge of some of the applications of various immune parameters to assess the effects of exposure to toxic chemicals.

THE IMMUNE SYSTEM: STRUCTURE, FUNCTION, AND BASIC CONCEPTS

The immune system is a highly evolved system that functions to provide organisms with the ability to resist infectious agents, destroy neoplastic cells, and reject nonself components. The immune system has increased in complexity as animals have evolved. Immunocytes in one form or another are found in all phyla from the Porifera (sponges) to the Vertebrata. According to the present level of knowledge, differentiated lymphoid organs first appear in the primitive forms of fish, with an increase in structural definition through amphibians and reptiles to birds and mammals. Higher mammals possess five immunoglobulin classes, birds three, reptiles and amphibians two, and fish only a single class. No immunoglobulins are found in invertebrates, but functionally analogous proteins, mostly agglutinins, are present (Roitt et al. 1989). Although all vertebrate species synthesize immunoglobulins, selection of humoral immunological assays as biomarkers will be constrained by the sophistication of the immune system of the species under investigation. Similar assays have been developed in invertebrates using agglutinins. A more detailed description of the phylogeny of the immune system is shown in Table 1.

Several aspects of immunity amenable for use as panspecific biomarkers are conserved phylogenetically. Cell-mediated responses, found throughout the animal kingdom, provide mechanisms (e.g., phagocytosis) that can differentiate self from nonself and thus play a key role in the regulation of the immune response. Assays of these responses can provide useful measures of immunomodulation (Weeks et al. 1990).

Table 1. Phylogeny of the Immune System.

| Group | Immunoglobulins | Differentiated Macrophage | | | |
		Lymphoid Organ	Importance[a]	MHC	Temp.
Invertebrates	—	—	High	—	E[b]
Fish	IgM	Thymus Spleen Kidney Liver	High	+	E
Amphibians	IgG-like IgM	Thymus Spleen Kidney Lymph nodes[c] Bone marrow[c]	Moderate	+	E
Reptiles	IgG IgM	Thymus Spleen Bone marrow Lymph node (homologue) Kidney Liver	Moderate	+	E
Birds	IgG-like IgM IgA	Thymus Bursa Spleen Lymph nodes Bone marrow	Low	+	(40–42)
Mammals	IgG IgM IgA IgE IgD	Thymus Bone marrow Spleen Lymph nodes Fetal liver	Moderate	+	(37–40)

[a] Relative importance of macrophages in the immune response.
[b] Ectothermic.
[c] In the anuran amphibians.

The fundamental units of the immune system are the leukocytes, which develop from pluripotent stem cells and undergo differentiation, maturation, and proliferation into morphologically and functionally distinct cell populations; e.g., granulocytes, monocytes, macrophages, and lymphocytes in higher vertebrates (Figure 1).

Specific and Nonspecific Immunity

Leukocytes mediate immune function via two pathways:

1.   nonspecific immune responses mediated by mononuclear phagocytes (blood monocytes, tissue macrophages) and granulocytes, which can recognize foreign material nonspecifically

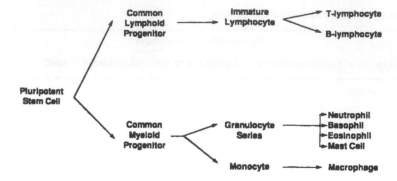

Figure 1.   Pluripotent stem cell differentiation in mammals.

2.   specific immune responses mediated by several effector leukocytes such as lymphocytes, which are directed against, and are specific for, an eliciting agent (antigen)

Nonspecific Immunity

Nonspecific immunity comprises two types of responses: phagocytosis and inflammation. Phagocytosis is the ingestion and destruction of foreign agents. Inflammation is accompanied by the infiltration of phagocytic cells into tissue at the site of injury or infection. Polymorphonuclear leukocytes (neutrophils) and mononuclear phagocytes (macrophages) mediate mainly nonspecific immune functions. Neutrophils are short-lived cells that are specialized for ingesting and destroying microorganisms in the circulation or in tissues following infiltration of an infected site. Macrophages may be fixed or wandering within an organ, or they may be infiltrative. Fixed cells such as liver and splenic macrophages clear the blood of foreign particles, while motile cells such as alveolar macrophages police the lungs. Infiltrative macrophages differentiate from monocytes as they leave the circulation and tissues and surrounding spaces (Roitt et al. 1989). Tissue macrophages are one component of the inflammatory response. The phagocytic and inflammatory responses can be enhanced by products of lymphocytes, such as specific opsonizing antibodies and soluble mediators (lymphokines), which are described in the next section.

Another important cell involved in immune surveillance in mammals is the natural killer (NK) cell. In humans NK cells are derived from large granular lymphocytes that comprise about 5% of the peripheral blood lymphoid cells. These cells can recognize changes in surface markers that occur on virally infected cells and on some tumor cells. They may be important in preventing metastases and in the prevention of virally-induced tumors. They can be activated by lymphokines such as interleukin-2 (IL-2).

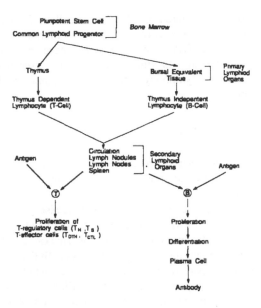

Figure 2.   Development of cells of the specific immune system in mammals.

Specific Immunity

The specific immune response has two major components: cell-mediated immunity (CMI) and humoral-mediated immunity (HMI) (Figure 2). CMI is primarily initiated by cells that mature in the thymus referred to as thymus dependent lymphocytes, T-lymphocytes, or T-cells. CMI responses mediate immunoregulation, delayed-type hypersensitivity, immunosurveillance, graft rejection, and resistance to infection by viral, bacterial, protozoan, and fungal agents. Lymphocytes that mature in the bursa of Fabricius in birds or bursal-equivalent tissues and organs in mammals (e.g., gut associated lymphoid tissue and bone marrow) produce antibodies that mediate HMI. These are also known as B-lymphocytes or B-cells. In anuran amphibians and reptiles, like other higher vertebrates, there is a functional bone marrow where B-cells are differentiated. However, in other ectothermic vertebrates (e.g., fish and nonanuran amphibians which lack bone marrow) B-cells or their precursors are generated in a variety of organs (Roitt et al. 1989).

After maturation in the primary lymphoid organs (thymus or bursal-equivalent tissue) mature T- and B-cells migrate via the circulation to secondary lymphoid organs such as lymph nodes and spleen. At these sites, antigen exposure may occur, provoking the lymphocytes to assume their genetically determined functional characteristics. Lymph nodes serve as filtering sites for lymphatic fluid, while the spleen provides the same function for blood. The filtered lymph or blood collected from distal sites may contain foreign antigens. The induction of a specific immune response by these antigens depends upon their physical and chemical nature (e.g., size or degree of foreignness), mode of presentation to lymphocytes, and their molecular configuration. Antigen stimulates only those lymphocytes that have receptors complementary to the antigen configuration.

Cell-mediated Immunity (CMI)

The CMI response is characterized by the sensitization of T-cells through antigen presentation and by the differentiation and proliferation of T-cells into effector, regulatory and memory cells.

Sensitization is assisted by macrophages and other accessory cells that concentrate antigenic determinants on their surfaces for presentation to T-cells. Macrophages also provide soluble factors called monokines (e.g., interleukin I). Macrophages are necessary for T-cells to function, specifically for T helper cells (T_H), T suppressor cells (T_S), and T delayed-type hypersensitivity cells (T_{DTH}). Some effector cells can lyse tumor target cells by releasing lymphotoxins and are called cytotoxic T-cells (T_C). T_C cells are believed to be important in specific defense against neoplastic cells or virally infected cells. Regulatory cells include T_H and T_S cells. T_H cells facilitate antibody responses of B-cells as well as other T-cell functions. T_S cells inhibit or suppress both T- and B-cell functions. The movement of inflammatory cells into sites of injury is directed by lymphokines, such as migration inhibition factor (MIF) and macrophage activating factor (MAF). Memory cells are produced for each T-cell type and are stimulated by antigen. This is the basis for secondary (anamnestic) immune responses. Figure 3 illustrates the interactions among these specialized lymphocytes in the cell-mediated immune system.

Humoral-mediated Immunity (HMI)

The HMI response is characterized by the production of antibody molecules that react specifically with antigen. Upon stimulation by antigen, B-cells undergo proliferation and differentiation into plasma cells, which synthesize and secrete antibody into the lymphatic and circulatory systems. In addition, some B-cells differentiate into memory cells, which provide a more rapid response to a second antigen exposure (Roitt et al. 1989). Based on structure and function, antibodies can be divided into five classes: IgM, IgG, IgA, IgE, and IgD. The major role of antibodies is to protect the host from infectious disease and can involve the following functions:

- Virus neutralization   A specific antibody binds to viral attachment sites on target cells, thus blocking attachment.

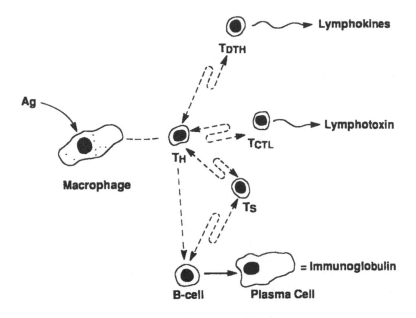

Figure 3.   Interaction among lymphocytes in the immune system.

- Opsonization   Antibody coating (opsonization) of viruses or bacteria
                                        renders the infectious agents more susceptible to
                                        phagocytosis by neutrophils and macrophages that possess
                                        antibody receptor sites (Fc receptors).
- Antibody-dependent cell   ADCC involves the binding of leukocytes to target
  cytotoxicity (ADCC)   cells via antibody bridges, after which the leukocyte can
                                        lyse the target cell.
- Complement-mediated lysis   Antibody binding to a target cell can activate the
                                        complement (C') system. The system consists of over 20
                                        serum proteins that interact with one another resulting in
                                        the lysis of target cells (e.g., neoplastic cells, transplanted
                                        cells, bacteria, or viruses).

USES OF IMMUNOLOGICAL BIOMARKERS: SELECTED ANIMAL STUDIES

Immunological parameters have been used in various laboratory and field experiments to analyze the effects of toxicants on the immune response and disease resistance of mammals (Sharma 1981). In the following examples, immune assays are broadly grouped according to the specificity of responses. Specific assays such as those that measure antibody production, the formation of antibody-producing cells (plaque assays), and secretory rosette (SR) formation, are activated cellular responses that result from exposure to specific antigen (Anderson 1974). Measurements of nonspecific responses, such as phagocytic or natural cytotoxic cell (NCC) activity and erythrocyte rosette (ER) formation, indicate the degree of immunocompetence without prior antigenic stimulation.

Some nonspecific assays such as the determination of leukocrit values (leucocyte volume), blood differential counts, or lysozyme levels have been used extensively in the past for analyzing fish health. Lysozyme is important in the initial destruction of invasive agents and in some cases can serve as an early biomarker indicating deterioration of some protective mechanism(s) (Peters et al., in press). Phagocytosis is an important parameter of nonspecific immunity as it is well conserved throughout phylogeny and important in immune surveillance in invertebrates as well as vertebrates. Therefore, assays of these immune functions have potential as bioindicators of toxic effects among several animal phyla.

In this respect, assays of fish macrophage functions have been proposed as sensitive indicators for the effects of certain toxic chemicals present in the aquatic environment (Weeks and Warinner 1984). For example, Weeks et al. (1986; 1987a; 1987b) and Warinner et al. (1988) have shown that fish exposed to polynuclear aromatic hydrocarbons (PAH), both in the Elizabeth River, Virginia and in the laboratory, show significant alterations of the kidney macrophage functions, such as phagocytosis of bacteria and yeast, chemotaxis, pinocytosis, chemiluminescence, and the accumulation of melanin. Responses were either suppressed or elevated depending upon the species and function tested. Figure 4 shows the results of the chemiluminescent, phagocytic, and chemotactic responses in spot (*Leiostomus xanthurus*) obtained from the PAH-contaminated Elizabeth River and from the nonpolluted York River. All alterations in fish obtained from the Elizabeth River, however, could be reversed when fish were held in relatively clean water.

Humoral antibody assays have been used in many studies to determine altered immune capacity which is caused by toxic chemicals. The advantages of measuring the humoral or circulating antibody concentrations include the ease of taking blood samples without killing the animal. Robohm (1986) showed that, at certain concentrations, cadmium stimulated antibody production in cunner (*Tautogalabrus adspersus*) but suppressed it in striped bass (*Morone saxatilis*). In field studies, equivocal results were sometimes obtained due to interference by background antibody levels. O'Neill (1981) also demonstrated immunosuppression of humoral antibody titers in brown trout (*Salmo trutta*) exposed to heavy metals. A reduction in antibody-producing cells following bath immunization was observed in rainbow trout that had been exposed to phenol, formalin, or detergent solutions (Anderson et al. 1984). Other pollutants in the environment may have similar effects in reducing the ability of fish to mount an effective immune response against disease-causing agents. Oxytetracycline, commonly used to treat bacterial diseases in fish, has also been implicated in the reduction of antibody-producing cells (Siwicki et al. 1989).

Humoral-mediated immunity has also been evaluated in chickens following exposure

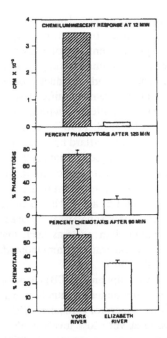

Figure 4.   Macrophage functions in spot (*Leiostomus xanthurus*) from polluted (Elizabeth River) and nonpolluted (York River) waters (a: Warinner et al. 1988; b: Weeks and Warinner 1984; c: Weeks et al. 1986).

to pentachlorophenol (PCP) (Prescott et al. 1982). Although there was no effect on antibody production, PCP decreased the *in vitro* lymphoproliferative response to Concanavalin A. Another study (Zeakes et al. 1981) examined the integrated functioning of the entire immune system of bobwhites (*Colimus virginianus*) exposed to the insecticide Sevin (1-napthyl N-methyl carbamate) by challenging them with *Histomonas meleagridis*. Birds not treated with chemicals exhibited no pathological responses or mortality due to histomonad infection whereas Sevin-treated birds developed histomoniasis and had significant mortality rates. Friend and Trainer (1974) also conducted a challenge experiment exposing mallards (*Anas platyrhynchos*) to dieldrin and challenging them with duck hepatitis virus. Dieldrin exposure significantly increased mortality.

Copper has been shown to be an immunosuppressive agent in aquatic environments. Exposing oysters *in vivo* to copper resulted in a reduction in the percentage of hyalinocytes (Cheng 1988). In fish, *in vitro* laboratory studies have shown copper to reduce the numbers of antibody-producing cells (Anderson et al. 1989). On the other hand, heavy metals appear to have less effect on the avian immune responses. Chickens exposed to aqueous lead acetate per os had no alterations in interferon or antibody production capabilities (Vengris and Mare 1974) and mallards given oral doses of lead shot had normal numbers of splenic plaque-forming cells (Rocke, T., personal communication).

Fairbrother and Fowles (in press) exposed mallards to selenium as sodium selenite or as the more bioavailable selenomethionine in drinking water for 83 d and conducted several immune function assays representative of many aspects of the immune system. The delayed-type hypersensitivity response was significantly impaired in birds exposed to selenomethionine. A companion study conducted on mallard pairs and offspring kept in cages placed on selenium-treated artificial streams showed an increased mortality in 15-day-old ducklings challenged with duck hepatitis virus (Fairbrother, personal communication). Rocke et al. (1984) examined the effects of petroleum hydrocarbons and dispersants on the mallard humoral immune response and showed a reduction in numbers of splenic plaque-forming cells.

The earthworms, *Lumbricus terrestris* and *Eisenia foetida,* are proving to be valuable organisms for environmental monitoring using immunological biomarkers (Goven et al. 1988). These animals are easy and inexpensive to maintain, and their size facilitates harvesting leukocytes for immunoassays. The earthworm immune system is sufficiently analogous to mammalian systems such that established mammalian immunoassays can be used to demonstrate the immunomodulatory effects of exposure to toxic materials (Goven et al. 1988).

The effects of polychlorinated biphenyl (PCB) (Aroclor 1254) and refuse-derived fuel (RDF) fly ash on humoral-mediated immunity in earthworms were assayed by examining rosette formation. Secretory rosette (SR) formation by earthworm coelomocytes was suppressed at very low levels of PCBs (180 µg/g dry weight), suggesting that the SR immunoassay is sensitive to PCB levels within an order of magnitude similar to those reported for a wide variety of wildlife from contaminated areas (Figure 5) (Rodriguez et al. in press). In addition, earthworms exposed to PCB for up to 122 h showed significant reductions in phagocytic activity of coelomocytes (Goven et al. 1988). Exposure of earthworm coelomocytes to RDF fly ash leachate suppressed both phagocytosis and SR formation but had no effect on erythrocyte rosette (ER) formation (Figure 6; Venables et al. 1989).

APPLICATION OF IMMUNE BIOMARKERS TO ASSESS ENVIRONMENTAL IMPACT

Currently Available Tests

Due to the complexity of the immune system, many assays have been developed in numerous species for use in examining either the integrated functioning of the entire system or its various component parts. Several authors have organized these tests into a tiered system for the most effective approach to assessing the effects of chemicals on the immune system (Vos 1980; Kohler and Exon 1985; Kerkvliet 1986; Luster et al. 1988). Table 2 is an integration of the previously reported tier approaches and the immunological assays applicable to several classes of animals.

Tier I provides a general screening of immune function. It comprises tests that are relatively easy and inexpensive, and requires very little specialized equipment.

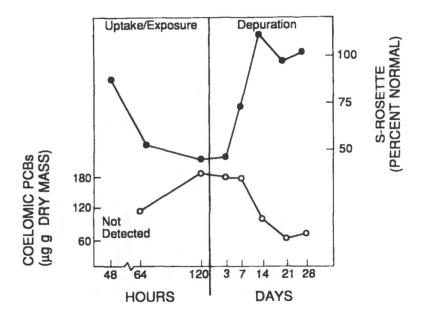

Figure 5.   S-rosette formation (0) and PCB levels (0) in coelomocytes of *Lumbricus terrestris* (Rodriguez et al., in press).

Complete and differential blood cell counts as well as weights and histological integrity of lymphoid organs provide a general overview of the adequacy of the structural parts of the immune system. Wound healing provides information about the inflammatory response and cell division. Natural killer cell activity (as measured by a chromium release assay), macrophage phagocytosis of fluorescent beads or stained yeast cells, killing of bacteria, and measurement of lysozyme activity by agar gel immunodiffusion can be performed with almost any animal and provides a great deal of information about nonspecific immune responses.

Humoral-mediated immunity can be assessed easily by quantitating the amount of circulating antibodies or factors using tests such as hemagglutination of sheep red blood cells or rosette formation with sheep or rabbit red blood cells. Graft rejection responses such as tissue transplantation in earthworms (Goven et al. 1988) and mice (Luster et al. 1988), scale transplantation in fish (Botham et al. 1980; Nakanishi 1986), or pock formation in embryonated eggs of birds (Fairbrother and Fowles, in press) provide an assessment of cell-mediated immune function. Macrophage chemiluminescence and melanomacrophage center formation in fish also provide additional measures of cell-mediated immunity (Weeks et al. 1990).

Tier II tests comprise a comprehensive evaluation of all components of the immune response. Although Tier II tests are specific and sensitive assays, extensive studies using mice have shown that no compound has had an effect in Tier II without also

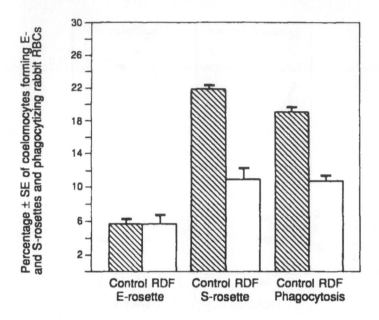

Figure 6.   Immune functions of coelomocytes of *Lumbricus terrestris* exposed to refuse-derived fuel (RDF) extract (Venables et al. 1988).

demonstrating some effect(s) in Tier I (Luster et al. 1988). Therefore, Tier II tests should be used when a thorough understanding of the mechanism of action of immunotoxicants or other stressors is needed.

Additionally, some Tier II tests that have been developed in mice may not be applicable or have not yet been adapted to other species. The lack of readily available standardized reagents may also increase the difficulty of conducting these tests in nonmammalian species. B- and T-cell surface markers, used to quantitate lymphocytes and differentiate them into subgroups, are commercially available for mice and have been produced by several researchers for chickens (Bucy et al. 1988; van de Water et al. 1989), but are not yet available for other animals. Flow cytometer separation of cell populations using light scatter and granularity criteria may be a viable alternative. Quantitation of native immunoglobulins (e.g., IgG, IgM) is easily conducted by radial immunodiffusion or by ELISA tests (provided antiimmunoglobulins are available, as is the case for many mammals, several species of birds, and fish). Measurement of numbers of plaque-forming cells provides information about the number of immunocytes capable of producing antibodies.

Lymphocyte blastogenesis, mixed-lymphocyte responses, and cytotoxic T-lymphocyte assays all provide information about various aspects of the cell-mediated immune system. The lymphocyte blastogenesis assay (measurement of ^{3}H-thymidine

Table 2. Immune Function Assays for Use in Screening or Comprehensive Analysis of Immunomodulatory Effects of Chemicals.

		Species Applicability			
			Vertebrates		
Test	Immune Component	Invertebrates (Earthworm)	Fish	Birds	Mammals

Screen (Tier I)

Test	Immune Component	Invertebrates (Earthworm)	Fish	Birds	Mammals
Complete blood count	General	+[a]	+	+	+
Cell differential	General	+	+	+	+
Hematocrit	General	−	+	+	+
Leukocrit	General	−	+	−	−
Organ weights (spleen, thymus, bursa, etc.)	General	−	+	+	+
Histology (spleen, thymus, bursa, lymph nodes, somatopleure pronephros etc.)	General	+	+	+	+
Wound healing	Inflammation	+	+	−	−
NK cell activity	Nonspecific	+	+	+	+
Macrophage phagocytosis and killing	Nonspecific	+	+	+	+
Lysozyme activity	Nonspecific	+	+	+	+
Agglutination assay	Humoral	+	+	+	+
Chemiluminescence	CMI	−	+	−	+
Graft rejection	CMI	+	+	+	+
Melanomacrophage centers	CMI	−	+	−	−

Comprehensive (Tier II)

Test	Immune Component	Invertebrates (Earthworm)	Fish	Birds	Mammals
Immune cell quantitation	General				
Surface markers		+[b]	+	+ (Chicken)	+
Flow cytometry		+	+	−	+
Native immunoglobulin quantitation	Humoral	−	+	+ (Chicken, mallard)	+
Plaque-forming cell assay	Humoral	+	+	+	+
Lymphocyte blastogenesis	CMI	+	+	+	+
Mixed leukocyte response	CMI	+	+	−	+
Cytotoxic T-cell (leukocyte) activity	CMI	+	+	−	+
Delayed-type hypersensitivity response	CMI	−	−	+	+
Macrophage responses	Nonspecific		+	−	+
Melanin accumulation		−	+	−	−
Chemotaxis		−	+	−	+
Pinocytosis		−			
NBT reduction	Nonspecif	+	+	+	+
Lymphokine quantitation	Soluble mediators	−	(±)	+ (Chicken)	+

Table 2. Cont'd.

Test	Immune Component	Invertebrates (Earthworm)	Vertebrates		
			Fish	Birds	Mammals
		Host Resistance Challenge (Tier III)			
Mortality	Comprehensive	+	+	+	+
Bacteremia/viremia/ parasitemia/tumor quantitation and duration	Comprehensive	+	+	+	+
Specific antibody quantitation	Humoral	−	+	+	+

a   Complete coelomocyte counts.
b   Coelomocyte cell surface markers.

uptake) is applicable across animal species and measures one of the earlier steps of cell responses to stimuli. The cytotoxic T-lymphocyte assay (another chromium-release assay) requires species-specific reagents against target cells, which may limit its applicability to only a few species within each class. A delayed-type hypersensitivity response is an *in vivo* assay that measures the integration of several components of the cell-mediated immune system, including cell migration and subsequent migration inhibition in response to lymphokine production. This response can be generated by several mitogens (e.g., phytohemagglutinin) or antigens (e.g., *Mycobacterium tuberculosis*).

Several *in vitro* macrophage function tests (chemiluminescence, chemotaxis, phagocytosis, pinocytosis, nitroblue tetrazolium dye reduction, and melanin accumulation) provide information about the integrity of the nonspecific immune response. Lymphokine quantitation (e.g., measurement of circulating interferons; Graham and Secombs 1990; de Kinkelin et al. 1982) and interleukins (Caspi and Avtalion 1984) or the ability of macrophages and lymphocytes to produce these compounds (Elsasser and Clem 1986), has been extensively used in mammalian immunology and to a lesser extent in chickens and fish.

The final step in immune function assessment requires the use of a host resistance challenge study (Tier III). Animals are exposed to appropriate bacterial, viral, or parasitic pathogens, or to tumor cells. Subsequent survivorship, amount and duration of pathogen replication, and specific protective immune responses are measured. Challenge agents differ across animal species, but should be selected to give information about both HMI and CMI responses (Luster et al. 1988). Timing and route of challenge in relation to chemical exposure are important considerations as they can affect results (Kerkvliet 1986).

Additional Considerations

The battery of immune function tests previously discussed have been used primarily in controlled laboratory exposures. These types of experiments are necessary for standardization and development of assays, and for the screening of numerous chemicals. Laboratory tests have also been used to determine the mechanism of action of various individual chemicals on the immune system. Such findings have been extrapolated to general classes of compounds; for example, various isomers and mixtures of PCBs and chlorinated dioxins.

Difficulties arise when applying tests to animals exposed to environmental stresses under natural conditions. Several of the Tier I and all of the Tier II and III assays require laboratory facilities difficult to provide at a field site. Therefore, the animal or its organs, tissues, or cells must be brought to the laboratory for complete analysis. Furthermore, capture and handling stress, especially if the animal is maintained in captivity for several days, can cause immunosuppression through the central nervous or endocrine systems discussed in the next section. Various techniques used in fish immunology have been published in a book entitled *Techniques in Fish Immunology* (Stolen et al. 1990). Techniques for maintaining viability and activity of various immune cells in vitro have been well-developed for only a few species (e.g., mammalian lymphocytes). Therefore, assays of immune function may vary because of species differences in immune responses, available laboratory facilities and species reaction to handling and captivity stress (Stolen et al. 1990).

An alternative approach is to reproduce environmental conditions in the laboratory. For example, fish can be held in aquaria containing contaminated sediments or water from the natural environment; earthworms can be placed in enclosures filled with soil from contaminated sites. However, these "microcosms" reduce the complexity of the environment compared with natural conditions. Ambient temperature, solar radiation, wind, currents, interactions with predator/prey species, and foraging activities are all stress factors that are significantly reduced in the laboratory.

Immune function screening of some wild species can be limited by legal or aesthetic restrictions on animal sacrifice. This could preclude the use of some animals for immunoassays such as organ weight determination, splenic plaque-forming and natural killer cell assays, lymphocyte blastogenesis, or host resistance to challenge. However, most of these tests have alternatives that can provide similar information on the same aspect of the immune system without harming the target organism.

Agglutination assays provide information about the total amount of circulating antibodies (or their functional equivalent in the lower classes of animals) in lieu of a plaque-forming cell assay. Natural killer cell activity and lymphocyte blastogenesis tests can be performed on circulating lymphocytes as well as on spleen cells, provided sufficient numbers of cells can be harvested.

INTERACTIONS OF THE IMMUNE, NERVOUS, AND ENDOCRINE SYSTEMS

According to emerging views, homeostasis is achieved in all vertebrates by the coordinated activities of the nervous and endocrine systems. Selye (1936) described the General Adaptation Syndrome (GAS) as a basic strategy by which all vertebrates adapt to altered conditions in the environment, thereby regaining and maintaining homeostasis. The adaptation process, stabilized by a network of biofeedback mechanisms, is accompanied by well known physiological and cellular alterations (Pickering 1981). These alterations are mediated via three main categories of hormones (i.e., corticoids, catecholamines, and opioid peptides). All three groups of hormones have direct and indirect effects on various aspects of the immune system.

The neuroendocrine reaction to exogenous challenges is nonspecific; i.e., independent of the nature of the stimulus. If an unfavorable state persists for a long period, or if the demands of the challenge exceed the capability of the organism to respond, various functional and structural "maladaptations" occur. Experimental studies have provided evidence that stressors primarily cause an overproduction of the previously mentioned hormones, which have modulatory effects on humoral and cellular defense mechanisms.

The effect of cortisol on leukocytes and other immune responses is well-documented in vertebrates. *In vitro* incubation with cortisone suppressed the chemiluminescence response of striped bass phagocytes (Stave and Roberson 1985) and the proliferative response of lymphocytes to mitogens (Tripp et al. 1987). Moreover, rainbow trout exposed to cortisone showed lower immune responses to a *Yersinia ruckeri* bacterin (Anderson et al. 1982). Recently, the receptors for cortisone were discovered on the surface of rainbow trout lymphocytes (Maule et al. 1989). Experimentally applied cortisol has been found to reduce the number of circulating lymphocytes in salmonids (Ralph et al. 1987).

Increased production of catecholamines related to stress leads to a mobilization of blood granulocytes and macrophages, which are released in large numbers from the hemopoietic tissue. They show enhanced chemotactic and phagocytic activity. It is generally assumed that this activation is an expression of an adaptive attempt by the cells to maintain basic functions. In the stressed organism and in their ischemic microenvironment, leukocytes eventually lose their functional capabilities.

Stressors that induce the classical General Adaptation Syndrome and well-documented maladaptations play an important role in field and laboratory studies. In fish (often used for biomarker assays), stressors such as handling, transportation, or social interaction among individuals are especially important and may interfere with the experimental design. Immunological disturbances have been shown to occur in fish after a few minutes of transport (Elsasser and Clem 1987), among individuals kept in groups in experimental tanks (Peters et al. 1980), or in fish that had been threatened by a predator or a dominant partner (Peters and Schwarzer 1985; Faisal et al. 1989). The stressed fish showed no external sign of physiological imbalance. An additional "stress" can be attributed to fish exposed to chemical toxicants as they initially sense an altered environment.

SUMMARY

Immunological biomarkers in fish that have been observed to react to experimental stress include:

1. decreased lymphocyte counts (Pickford et al. 1971; Esch and Hazen 1980; Peters 1982)
2. decreased mitogenic response of lymphocytes (Elsasser and Clem 1986)
3. increased leucocrits as a result of hypertrophy of granulocytes and macrophages (McLeay and Gordon 1977; Tomasso et al. 1983; Peters and Schwarzer 1985; Peters and Hong 1985)
4. hypertrophy of granulocytes and macrophages (Peters and Schwarzer 1985)
5. increased phagocytosis and pinocytosis (Peters et al. in press) or decreased phagocytosis (Weeks and Warinner 1984)
6. increased macrophage counts and degenerated macrophage structures (Peters and Schwarzer 1985)
7. increased susceptibility to bacterial infections (Angelidis et al. 1987; Peters et al. 1988)
8. decreased lysozyme activity (Peters et al., in press; Figure 7)
9. decreased natural cytotoxic activity (Ghoneum et al. 1988; Figure 8)
10. increased respiratory burst (Herrmann, personal communication) and increased autophagocytosis (Peters et al. 1990)

POTENTIAL VALUE AND RESEARCH RECOMMENDATIONS

The immune system comprises a network of cells capable of rapid proliferation and differentiation regulated by a variety of soluble factors and is tightly integrated with other organ systems and functions. As such, it is extremely vulnerable to insult from exogenous chemicals. Although severe immune suppression quickly results in morbidity and death, subtle changes in specific components of the immune system often occur first and provide sensitive early warning indicators of toxicity. Such effects frequently occur at chemical doses much lower than those that cause acute toxicity (Kohler and Exon 1985). Due to the ability of memory cells to proliferate rapidly, the immune

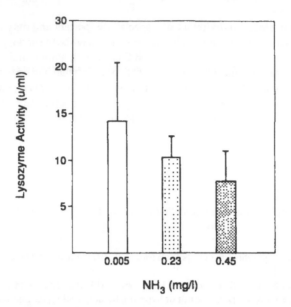

Figure 7.   Lysozyme activity of rainbow trout exposed to ammonia in the water. n=10 (Mock and Peters 1990).

system can respond quickly to secondary exposures. Therefore, chronic exposure or repeated short exposures to toxicants are much more likely to produce immune suppression than would single acute exposure.

As with other biomarkers, immune responses provide an integrated measure of exposure over time and may reflect the combined results of simultaneous exposure to several chemicals. It is, however, not possible to determine which chemical has caused the observed effect as no one change in immune function has been shown to be pathognomic for a specific compound or class of chemicals.

The literature contains many conflicting reports about the immunologic effects of environmental chemicals (Rice and Weeks 1989; Robohm 1986; Weeks et al. 1990). For a particular chemical, different investigators have shown immune suppression, immune augmentation, or no effect depending on the species studied, routes and duration of exposure to the chemical, time of antigenic stimulus in relation to chemical exposure, and specific protocol for conducting similar assays (Kohler and Exon 1985; Sharma 1981). This variability reflects species-related differences and the complexity of the immune system itself, although standardization of immune function assays within species could help to reduce some of the variability. However, researchers must keep in mind the particular hypotheses being tested by the experimental system and avoid the tendency to extrapolate the data further than the constraints imposed by species and assay choices. In particular, screening (Tier I) tests should be used to

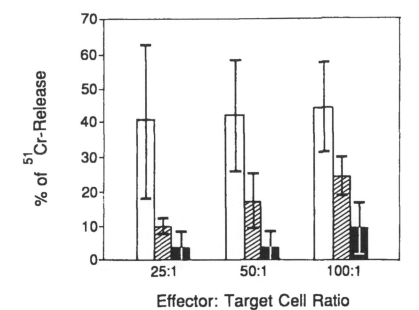

Figure 8.   Effect of stress on nonspecific cytotoxicity in *Tilapia*. Unstressed (open); slightly stressed (striped); and severely stressed (solid) (Ghoneum et al. 1988).

determine if immune dysfunction has occurred, but determination of mechanism(s) of the observed effect requires one or more additional Tier II tests. Additionally, Tier II assays have been used to determine mechanisms of action of relatively few environmental toxicants. Research is needed in this area if scientists are to predict the potential immunotoxic properties of classes of compounds and thereby avoid extensive testing of individual chemicals.

In addition to assay standardization, an important area of future research concerns the question of how to conduct immunological screening tests of organisms exposed to environmental chemicals in the field. Collecting samples or individuals in field situations poses difficulties. Capture, handling, and maintenance of animals in captivity introduce exogenous stress that can impose added immunomodulatory effects. Holding animals in captivity for a considerable duration may also provide the immune system time to recover if the stress of the environmental chemical has been removed. Methods are needed for collecting and preserving organ, tissue, cell, or blood samples from field-exposed animals; such methods should provide a minimum amount of stress and allow for the most realistic and precise assessment of the immune function of the animal at the time of capture.

Other areas of fruitful research include identification of appropriate challenge pathogens for use in Tier III immunological tests for various species. Basic disease parameters of these organisms must be defined and exposure methods standardized prior to their acceptance as indicators of immune dysfunction. In addition, known positive immunosuppressant chemicals need to be identified and validated for the various classes of animal species. These chemicals could then be used to develop and validate new immunological tests and to provide positive controls in all biomarker assays for quality assurance purposes.

In summary, measurement of the various components or integrated functioning of the immune system is a sensitive indicator that reflects exposure to low concentrations of environmental toxicants in any aquatic or terrestrial animal species. Highly reliable and reproducible tests of immune function exist for many classes of animals and can be used as screening tools or as a means of investigating mechanisms of effect. As standardized reagents become more readily available, tests will become more reproducible and cost effective. Immunological biomarkers have an important role to play in monitoring the health of animals prior to the occurrence of devastating disease outbreaks and as early warning indicators of the potential harm of environmental chemicals. Since many parameters of the immune system are similar in different species, animals may serve as sentinels of potential environmental hazards for humans.

REFERENCES

Anderson, D.P. *Fish Immunology* (Neptune, N.J.: T.F.H. Publications, 1974), p. 239.

Anderson, D.P., O.W. Dixon, J.E. Bodammer and E.F. Lizzio. "Suppression of the Antibody-producing Cells in Rainbow Trout Spleen Sections Exposed to Copper In Vitro," *J. Aquat. Anim. Health* 1:57–61 (1989).

Anderson, D.P., O.W. Dixon and F.W. van Ginkel. "Suppression of Bath Immunization in Rainbow Trout by Contaminant Bath Pretreatments," in *Chemical Regulation of Immunity in Veterinary Medicine*, M. Kende, J. Gainer and M. Chirigos, Eds. (New York: Alan R. Liss, Inc., 1984), pp. 289–293.

Anderson, D.P., B.S. Roberson and O.W. Dixon. "Immunosuppression Induced by a Corticosteroid or an Alkylating Agent in Rainbow Trout (*Salmo gairdneri*) Administered a *Yersinia ruckerri* Bacterin," *Dev. Comp. Immunol. (Suppl.)* 2:197–204 (1982).

Angelidis, P., F. Baudin-Laurencin and P. Youinou. "Stress in Rainbow Trout, *Salmo gairdneri*: Effects Upon Phagocyte Chemiluminescence, Circulating Leucocytes and Susceptibility to *Aeromonas salmonicida*," *J. Fish Biol. (Suppl. A)* 31:113–122 (1987).

Botham, J.W., M.F. Grade and M.J. Manning. "Ontogeny of First Set and Second Set Autoimmune Reactivity in Fishes," in *Phylogeny of Immunological Memory*, M.J. Manning, Ed. (North-Holland, Amsterdam: Elsevier, 1980), pp. 83–92.

Bucy, R.P., C.L. Chen, J. Cihak, U. Losch and M.D. Cooper. "Avian T Cells Expressing Gamma Delta Receptors Localize in the Splenic Sinusoids and the Intestinal Epithelium," *J. Immunol.* 141:2200–2205 (1988).

Caspi, R.R., and R.R. Avtalion. "Evidence for the Existence of an IL-2 Like Lymphocyte Growth Promoting Factor in a Bony Fish, *Cyprinus carpio*," *Dev. Comp. Immunol.* 8:51–60 (1984).

Cheng, T.C. "In Vivo Effects of Heavy Metals on Cellular Defense Mechanisms of *Crassostrea virginica*: Total and Differential Cell Counts," *J. Invert. Pathol.* 51:207–214 (1988).

Dean, J.H., M.I. Luster, A.E. Munson and H. Amos, Eds. *Immunotoxicology and Immunopharmacology* (New York: Raven Press, 1985).

de Kinkelin, P., M. Dorson and A.M. Hattenberger-Baudouy. "Interferon Synthesis in Trout and Carp After Viral Infection," *Dev. Comp. Immunol.* 8:51–60 (1982).

Elsasser, C.F., and L.W. Clem. "Haematological and Immunological Changes in Channel Catfish Stressed by Handling and Transport," *J. Fish Biol.* 28:511–521 (1986).

Elsasser, C.F., and L.W. Clem. "Cortisol Induced Hematologic and Immunologic Changes in Channel Catfish *Ictalurus punctatus*," *Comp. Biochem. Physiol. (Suppl. A)* 87:405–408 (1987).

Esch, G.W., and T.C. Hazen. "Stress and Body Condition in a Population of Largemouth Bass: Implication for Red-sore-disease," *Trans. Am. Fish. Soc.* 109:532–536 (1980).

Fairbrother, A., and Fowles. "Subchronic Effects of Sodium Selenite and Selenomethionine on Several Immune Functions in Mallards," *Arch. Environ. Chem. Toxicol.*, in press.

Faisal, M., F. Chiappelli, I.I. Ahmed, E.L. Cooper and H. Weiner. "Social Confrontation Stress in Aggressive Fish Is Associated with an Endogenous Opioid-mediated Suppression of Proliferative Responses to Mitogens and Nonspecific Cytotoxicity," *Brain Behav. Immun.*, 3:223–233 (1989).

Friend, M., and D.O. Trainer. "Experimental Dieldrin-duck Hepatitis Virus Interaction Studies, *J. Wild. Manage.* 38:896–900 (1974).

Ghoneum, M., M. Faisal, G. Peters, I.I. Ahmed and E.L. Cooper. "Suppression of Natural Cytotoxic Cell Activity by Social Aggressiveness in *Tilapia*," *Dev. Comp. Immunol.* 12:595–602 (1988).

Goven, A.J., B.J. Venables, L.C. Fitzpatrick and E.L. Cooper. "An Invertebrate Model for Analyzing Effects of Environmental Xenobiotics on Immunity," *Clin. Ecol.* 4:150–154 (1988).

Graham, S., and C.J. Secombs. "Do Fish Lymphocytes Secrete Interferon-y?" *J. Fish Biol.* 36:563–573 (1990).

Kerkvliet, N. "Measurements of Immunity and Modifications by Toxicants," in *Safety Evaluation of Drugs and Chemicals*, W.E. Lloyd, Ed. (Washington, D.C.: Hemisphere Publ. Corp., 1986), pp. 235–256.

Kohler, L.D., and J.H. Exon. "The Rat as a Model for Immunotoxicity Assessment," in *Immunotoxicology and Immunopharmacology*, J.H. Dean, M.I. Luster, A.E. Munson and H. Amos, Eds. (New York: Raven Press, 1985), pp. 99–112.

Lang, T., G. Peters, R. Hoffman and E. Meyer. "Experimental Investigations on the Toxicity of Ammonia: Effects on Ventilation Frequency, Growth, Epidermal, Mucous Cells, and Gill Structure of Rainbow Trout *Salmo gairdneri*," *Dis. Aquat. Org.* 3:159–165 (1987).

Luster, M.I., A.E. Munson, P.T. Thomas, M.P. Holsapple, J.D. Fenters, K.L. White, L.D. Lauer, D.R. Germolec, G.J. Rosenthal and J.H. Dean. "Development of a Testing Battery to Assess Chemical-induced Immunotoxicity: National Toxicology Program's Guidelines for Immunotoxicity Evaluation in Mice," *Fundam. Appl. Toxicol.* 10:2–19 (1988).

Maule, A.G., C.B. Schreck and S.L. Kaatari. "Changes in Number and Affinity of Cortisol Receptors in Coho Salmon Leukocytes Appear to Mediate Organ-specific Immunosuppression by Cortisol In Vitro." in *Fish Health Section/American Fisheries Society and Eastern Fish Health Workshop*, (Abstract), Annapolis, MD, July 17–21, 1989.

McLeay, D.J., and M.R. Gordon. "Leucocrit: A Simple Hematological Technique for Measuring Acute Stress in Salmonid Fish, Including Stressful Concentration of Pulpmill Effluent," *J. Fish. Res. Bd. Can.* 34:2164–2175 (1977).

Mock, A., and G. Peters. "Lysozyme Activity in Rainbow Trout (*Oncorhynchus mykiss*) Stressed by Handling and Transport and Water Pollution," *J. Fish Biol.*, 37:873–885 (1990).

Nakanishi, T. "Ontogenic Development of the Immune Response in the Marine Teleost *Sebastiscus marmoratus*," *Bull. Jpn. Sci. Soc. Sci. Fish* 52(3):473–478 (1986).

O'Neill, J.G. "Effects of Intraperitoneal Lead and Cadmium on the Humoral Immune Response of *Salmo trutta*," *Bull. Environ. Contam. Toxicol.* 27:42–48 (1981).

Peters, G. "The Effect of Stress on the Stomach of the European Eel *Anguilla anguilla* L," *J. Fish Biol.* 21:497–512 (1982).

Peters, G., H. Delventhal and H. Klinger. "Physiological and Morphological Effects of Social Stress in the Eel (*Anguilla anguilla* L.)." *Arch. Fisch. Wiss.* 30:157–180 (1980).

Peters, G., M. Faisal, T. Lang and I. Ahmed. "Stress Caused by Social Interaction and Its Effect on Susceptibility to *Aeromonas hydrophila* Infection in Rainbow Trout *Salmo gairdneri*," *Dis. Aquat. Org.* 4:83–89 (1988).

Peters, G., and L.Q. Hong. "Gill Structure and Blood Electrolyte Levels of European Eels Under Stress," in *Fish and Shellfish Pathology*, A.E. Ellis, Ed. (London: Academic Press, 1985), pp. 183–196.

Peters, G., T. Lang, M. Faisal and I. Ahmed. "The Relationship of Stress Produced in Culture to the Natural Defense Mechanisms Against Disease," in *Proceeding of the International Symposium on Icthyopathology in Aquaculture*, N. Fijan, Ed., Yugoslavia Academy of Science, in press (1990).

Peters, G., and R. Schwarzer. "Changes in the Hemopoietic Tissue of Rainbow Trout Under Influence of Stress," *Dis. Aquat. Org.* 1:1–10 (1985).

Pickering, A.D. "Introduction: The Concept of Biological Stress," in *Stress and Fish*, A.D. Pickering, Ed. (London: Academic Press, 1981), pp. 1–9.

Pickford, G.E., A.K. Srivastava, A.M. Slicher and P.K.T. Pang. "The Stress Response in the Abundance of Circulating Leucocytes in the Killifish *Fundulus heteroclitus*. III. The Role of the Adrenal Cortex and a Concluding Discussion of the Leucocyte Stress Syndrome," *J. Exp. Zool.* 177:109–117 (1971).

Prescott, C.A., B.N. Wilkie, B. Hunter and R.J. Julian. "Influence of a Purified Grade of Pentachlorophenol on the Immune Response of Chickens," *Am. J. Vet. Res.* 43:481–487 (1982).

Ralph, A.T., A.G. Maule, C.B. Schreck and S.L. Kaatari. "Cortisol Mediated Suppression of Salmonid Lymphocyte Responses In Vitro," *Dev. Comp. Immunol.* 11:565–576 (1987).

Rice, C.D., and B.A. Weeks. "Influence of Tributyltin on In Vitro Macrophage Activation of Oyster Toadfish Macrophages," *J. Aquat. Anim. Health* 1:62–68 (1989).

Robohm, R.A. "Paradoxical Effects of Cadmium Exposure on Antibacterial Antibody Responses in Two Fish Species: Inhibition in Cunners (*Tautogolaburs adspersus*) and Enhancement in Striped Bass (*Morone saxatilis*)," *Vet. Immunol. Immunopathol.* 12:251–262 (1986).

Rocke, T.E., T.M. Yuill and R.E. Hinsdill. "Oil and Related Toxicant Effects on Mallard Immune Defenses," *Environ. Res.* 133:343–352 (1984).

Rodriguez, J.B., B.J. Venables, L.C. Fitzpatrick, A.J. Goven and E.L. Cooper. "Suppression of Secretory Rosette Formation by PCBs in *Lumbricus terrestris*: An Earthworm

Immunoassay for Humoral Immunotoxicity of Xenobiotics," *J. Environ. Toxicol. Chem.*, in press (1990).

Roitt, I.M., J. Brostoff and D.K. Male. *Immunology* (London: Gower Medical Publishing, 1989), pp. 312.

Selye, H. "A Syndrome Produced by Diverse Nocuous Agents," *Nature* (London) 138:32 (1936).

Sharma, R.P., Ed. *Immunologic Considerations in Toxicology*, Vols. I and II. (Boca Raton, FL: CRC Press, Inc., 1981).

Siwicki, A.K., D.P. Anderson and O.W. Dixon. "Comparisons of Nonspecific and Specific Immunomodulation by Oxolinic Acid, Oxytetracycline, and Levamisol in Salmonids," *Vet. Immunol. Immunopathol.*, 23:195–200 (1989).

Stave, J.W., and B.S. Roberson. "Hydrocortisone Suppresses the Chemiluminescent Response of Striped Bass Phagocytes," *Dev. Comp. Immunol.* 9:77–84 (1985).

Stolen, J.S., T.C. Fletcher, D.P. Anderson, B.S. Roberson and W.B. van Muiswinkel, Eds. *Techniques in Fish Immunology* (Fair Haven, NJ: SOS Publications, 1990).

Tomasso, J.R., B.A. Simco and K.B. Davis. "Circulating Corticosteroid and Leucocyte Dynamics in Channel Catfish During Net Confinement," *Tex. J. Sci.* 35:83–88 (1983).

Tripp, R.A., A.G. Maule, C.B. Schreck and S.L. Kaatari. "Cortisol Mediated Suppression of Salmonid Lymphocyte Responses In Vitro," *Dev. Comp. Immunol.* 11:565–576 (1987).

van de Water, J., L. Haapanen, R. Boyd, H. Abplanalp and M.E. Gershwin. "Identification of T Cells in Early Dermal Lymphocytic Infiltrates in Avian Scleroderma," *Arthritis Rheum.* 32:1031–1040 (1989).

Venables, B.J., K.E. Daugherty, A.J. Goven and O.O. Ohlsson. "Characterization of Toxicity of Fly Ash from the Combustion of Refuse Derived Fuel," *81st Annual Air Pollution Control Association* Dallas, Texas (1988).

Vengris, V.E., and C.J. Mare. "Lead Poisoning in Chickens and the Effect of Lead on Interferon and Antibody Production," *Can. J. Comp. Med.* 38:328–335 (1974).

Vos, J.G. "Immune Suppression as Related to Toxicology," *CRC Crit. Rev. Toxicol.* 5:67–101 (1977).

Vos, J.G. "Immunotoxicity Assessment: Screening and Function Studies," *Arch. Appl. Toxicol. Suppl.* 4:95–108 (1980).

Warinner, J.E., E.S. Mathews and B.A. Weeks. "Preliminary Investigations of the Chemiluminescent Response in Normal and Pollutant-exposed Fish," *Mar. Environ. Res.* 24:281–284 (1988).

Weeks, B.A., R.J. Huggett, J.E. Warinner and E.S. Mathews. "Macrophage Responses of Estuarine Fish as Bioindicators of Toxic Contamination," in *Biomarkers of Environmental Contamination*, J.F. McCarthy, and L.R. Shugart, Eds. (Boca Raton, FL: Lewis Publishers, 1990), pp. 193–201.

Weeks, B.A., A.S. Keisler, Q.N. Myrvik and J.E. Warinner. "Differential Uptake of Neutral Red by Macrophages from Three Species of Estuarine Fish," *Dev. Comp. Immunol.* 11:117–124 (1987a).

Weeks, B.A., A.S. Keisler, J.E. Warinner and E.S. Mathews. "Preliminary Evaluation of Macrophage Pinocytosis as a Fish Health Monitor," *Mar. Environ. Res.* 22:205–213 (1987b).

Weeks, B.A., and J.E. Warinner. "Effects of Toxic Chemicals on Macrophage Phagocytosis in Two Estuarine Fishes," *Mar. Environ. Res.* 14(1–4):327–335 (1984).

Weeks, B.A., J.E. Warinner, P.L. Mason and D.S. McGinnis. "Influence of Toxic Chemicals on the Chemotactic Response of Fish Macrophages," *J. Fish Biol.* 28(6):653–658 (1986).

Zeakes, S.J., M.F. Hansen and R.J. Robel. "Increased Susceptibility of Bobwhites (*Colinus virginianus*) to *Histomonas meleagridis* After Exposure to Sevin Insecticide," *Avian Dis.* 25:981–987 (1981).

Zeeman, M.G., and W.A. Brindley. "Effects of Toxic Agents on Fish Immune Systems: A Review," in *Immunologic Considerations in Toxicology*, Vol II, R.P. Sharma, Ed. (Boca Raton, FL: CRC Press, Inc., 1981).

Zimmer, T., and P.P. Jones. "Combined Effects of Tumor Necrosis Factor-alpha, Prostaglandin E-2, and Corticosterone on Induced Ia Expression on Murine Macrophages," *J. Immunol.* 145:1167–1175 (1990).

CHAPTER 6

Molecular Responses to Environmental Contamination: Enzyme and Protein Systems as Indicators of Chemical Exposure and Effect

John J. Stegeman, Marius Brouwer, Richard T. Di Giullo, Lars Förlin, Bruce A. Fowler, Brenda M. Sanders and Peter A. Van Veld

INTRODUCTION

Detecting and evaluating biological changes that result from chemical contamination are essential steps to determining the significance of such contamination, and could help to identify the active pollutant compounds and their source. Biological changes resulting from exposure to chemicals can be used in biomonitoring. Specific changes can serve as markers of exposure to specific chemicals and/or markers of adverse effects of exposures.

The effects of chemical contaminants can be viewed as occurring at different levels of biological organization, extending from the molecular or biochemical level, to the physiology of the individual, and ultimately to the levels of population and ecosystem (Figure 1). Changes at the biochemical level offer distinct advantages as biomarkers for two major reasons. First, biochemical or molecular alterations are usually the first detectable, quantifiable responses to environmental change, including changes in the

- ecosystem

- community

- population

- individual

- organ

- tissue

- cell

- organelle

- pathway

- molecules

Figure 1.   Levels of biological organization.

chemical environment. Second, biochemical alterations can serve as markers of both exposure and effect. By definition, a chemically induced change in biochemical systems represents an effect of the chemical.

Alterations in biochemical systems are often more sensitive indicators than those at higher levels of biological organization. Indeed, changes at the molecular level will underlie the effects at higher levels of organization. Depending upon the function of the systems affected and the nature of the response, biochemical perturbations can indicate whether additional effects (at the organ or individual level) are likely to occur. We emphasize that changes detected in individuals are of primary importance in assessing contaminant exposure, underscoring the intrinsic value of biomarkers.

In this chapter, we focus on a number of enzymes and other proteins that show responses to pollutants. We consider their biological roles, evaluate their use as markers, and discuss their potential for application in monitoring programs. We also point out areas where research is needed to enhance our understanding of these systems, thereby enhancing their potential utility as biomarkers. The enzymes and other proteins we discuss are involved in specific responses of organisms to toxic chemicals. In many cases these responses are *adaptive*, however, the same systems are involved in reactions *leading to* toxic effects. These systems are

1.   *Cytochrome P450 monooxygenases* are a protein family involved in the biotransformation of organic chemicals, resulting in molecular changes and either their activation to toxic metabolites or their inactivation. Cytochrome P450

induction by specific classes of hydrocarbon compounds has been well documented experimentally, and has been linked to such chemicals in the environment.

2. *Metallothioneins* represent a class of low molecular weight, metal-binding proteins involved in the sequestration and the metabolism of heavy metals. Their synthesis can be induced by a wide variety of nutritionally-required and nonessential toxic heavy metals, both in the laboratory and in the field.

3. *Stress proteins or heat shock proteins* are a group of proteins whose synthesis is induced by a wide variety of physical conditions and chemical agents. Some of these proteins are believed to play a role in protecting the cell from damage that can result from environmental perturbations. Others are involved in the regulation of various genes.

4. *Phase II (conjugating) enzymes* aid in detoxification and excretion of foreign compounds, including reactive metabolites formed by cytochrome P450, by linking them to water-soluble endogenous compounds.

5. *Oxidant-mediated responses* enhance the production of oxyradicals in the cell. Increased fluxes of oxyradicals can induce a number of antioxidant enzymes, alter concentrations of other antioxidants, and/or produce biochemical lesions associated with oxidative damage.

6. *Heme and porphyrin* synthesis is essential to life. Though little is known about this pathway in nonmammalian species, extensive mammalian literature documents its sensitivity to organic and inorganic chemicals. Disturbances in this pathway correlate with overt cell injury.

Several of these systems, and also processes in other chapters, have been discussed together in previous compendia dealing with pollution effects and biomonitoring (McIntyre and Pierce 1980; Cairns et al. 1984; Giam and Ray 1987; Haux and Förlin 1988). The present effort expands upon those earlier works, benefiting from the advances made in biochemistry and molecular biology in recent years. This chapter is meant to be substantive, but not exhaustive; each section could be a chapter in itself.

In the sections that follow, we provide background information on the nature, function, and biological importance of the protein or metabolic systems. We consider the mechanisms by which the systems respond to chemicals, the sensitivity and specificity of the response, and the biological importance of changes in the content or activity of these proteins. We include an appraisal of whether changes can be causally linked or physiologically linked with subsequent biological effects. We also consider the potential for using these proteins (i.e., changes in their levels or functions) as biomarkers and evaluate the responsiveness of systems to concentrations of chemicals that actually occur in contaminated systems. Experimental and field studies validating the use of these biochemical changes as markers are discussed, as are other biological and environmental variables influencing the responses to chemical stimuli. Each of the sections includes a specific set of research needs. Finally, we summarize the present understanding and evaluate how these biochemical systems may be judged against the attributes of an "ideal" biomarker, and consider how they may be employed in a tiered approach to biomonitoring.

We reemphasize that many of the systems we discuss play vital dual roles in protecting against chemical injury and in mediating chemical injury. They are crucial to the health effects associated with chemical insult, and can be linked to adverse chemical effects at higher levels of organization. Accordingly, basic research on these systems is crucial to understanding toxic mechanisms and to fulfilling the very real potential for using these systems as biomarkers.

CYTOCHROME P450

Cytochrome P450 (or simply P450) refers to a family of enzymes that transform the structure of organic chemicals (for reviews, see Ortiz de Montellano 1986.) More than 100 P450 genes or proteins from procaryotes and eucaryotes have been sequenced, and organized into 27 gene families (Nebert et al. 1991.) The reactions catalyzed by these proteins are of a type generally referred to as mixed-function oxidase (MFO) or monooxygenase reactions. The toxicity of organic chemicals can be drastically altered by structural transformation. By affecting chemical structures, cytochrome P450 enzymes may render a given compound nontoxic or, by contrast, drastically increase its toxicity. The amounts of some types of P450 can be induced (increased) in response to an organism's exposure to many types of chemical, including many substrates. As a result, the rate of chemical transformation catalyzed by these enzymes is altered. P450 induction can also serve as a highly sensitive indicator of an organism's toxic burden, or the extent to which it has been exposed to chemical inducers in the environment.

Cytochrome P450 enzymes transform both endogenous and exogenous (foreign) compounds. In endogenous pathways of animals, P450 synthesizes and degrades steroids, prostaglandins, fatty acids, and other biological molecules. In transforming foreign compounds, P450 plays key roles in the toxicology, action, and excretion of pollutant chemicals, drugs, and many chemical carcinogens. Some compounds newly formed by P450 are carcinogenic. The presence of different P450 proteins will determine which of the various pathways occur in different cells, organs, individuals, or species. The catalytic function and regulation of P450 therefore can define the susceptibility of those different cells, organs, individuals, or species of animals to foreign compounds, particularly those compounds whose toxicity may depend upon biotransformation. Knowledge of the biochemical details of regulation of P450 is essential to understanding the toxicity of chemicals and to using P450 as a biomarker.

In this section, we consider P450 as a biomarker in aquatic systems, primarily fish. However, we also consider P450 action and induction in birds, mammals, and invertebrates.

Cytochrome P450 Systems in Fish

The structure and function of microsomal cytochrome P450 proteins that metabolize foreign compounds have been studied most extensively in mammalian systems, primarily in liver. Fish and invertebrate species possess microsomal P450 systems which are in many ways similar to those present in mammals. The basic biochemistry of the microsomal P450 systems in aquatic species has been described in considerable detail in several earlier reviews (Bend and James 1978; Stegeman 1981; Lech et al. 1982; Lee 1982).

Induction

One of the hallmark features of P450 systems in fish as well as in mammals, is their inducibility by substrates, and by compounds structurally related to substrates (Stegeman 1981; Kleinow et al. 1987). Induction is a process by which a chemical increases the amount of P450; this generally involves synthesis of new messenger RNA and, subsequently, new enzyme protein. Either P450 enzyme activity, protein, or mRNA might be measured to detect induction. These are considered separately as follows.

Catalytic (MFO) activity. Studies over the past 15 years have shown induction of P450 enzyme activity (MFO activity) in fish liver (Payne et al. 1987). Liver microsomes of animals treated with aromatic or halogenated hydrocarbons show enhanced rates of MFO activity. Early studies (Statham et al. 1978; Elcombe et al. 1979) showed that various polynuclear aromatic hydrocarbons (PAH) could induce MFO activity. Table 1 lists several other compounds active as inducers. PAHs and halogenated hydrocarbons (such as polychlorinated biphenyls, dibenzofurans, and dibenzodioxins) include many important and potentially hazardous environmental contaminants. Two activities catalyzed by P450, ethoxyresorufin O-deethylase (EROD) activity and aryl hydrocarbon (benzo[a]pyrene) hydroxylase (AHH) activity, are largely specific in their response to these compounds. These activities occur at very low, often undetectable levels in many control or untreated animals, but are highly induced by treatment with the hydrocarbon compounds. Many other monooxygenase activities are not increased by such treatment, substantiating that EROD and AHH are selectively induced.

Antibodies to P450. Studies in several fish species have demonstrated the presence of multiple cytochrome P450 proteins that are characterized by different properties. These include multiple forms purified from rainbow trout (Williams and Buhler 1983a; 1984; Miranda et al. 1989), scup (Klotz et al. 1983; 1986), cod (Goksøyr 1985); and other species reviewed elsewhere (Stegeman and Kloepper-Sams 1987; Stegeman 1989). One type of P450 purified from liver of several fish species has been identified as the single P450 primarily induced by PAH and planar chlorinated aromatic hydrocarbons. The three fish proteins of this type most well-studied to date are scup P450E, trout P450LM$_{4b}$, and cod P450c (Table 2). These three P450s are structurally similar and catalyze similar reactions.

Polyclonal and monoclonal antibodies have been prepared to the PAH-inducible P450s from several species. Inhibitory antibodies to the fish PAH-inducible P450s confirm that these are the enzymes responsible for the EROD and AHH catalytic activities; i.e., the MFO activities selectively induced by hydrocarbons. The fish P4501A enzymes are also responsible for metabolism of several aromatic hydrocarbon

Table 1.   Representative Aromatic and Chlorinated Hydrocarbons that Induce Microsomal
Monooxygenase Activity in Fish.

Active as Inducers	Inactive as Inducers
Benzo[a]pyrene	DDT
Dibenzanthracene	2,2´,4,4´-Tetrachlorobiphenyl
Methylcholanthrene	2,2´4,4´5,5´-Hexachlorobiphenyl
3,3´,4,4´-Tetrachlorobiphenyl	
2,3,7,8-Tetrachlorodibenzodioxin	

Source: Information reviewed in Stegeman 1981.

carcinogens in teleost microsomal systems. Antibodies to these fish P450s cross-react with proteins specifically induced by PAH or PCB compounds in many vertebrate species. For example, using Western blot analysis, monoclonal antibody (MAb) 1-12-3 to scup P450E has been shown to strongly recognize single proteins induced by PAH or PCB compounds in vertebrates including fish, birds, and mammals (Goksøyr et al. 1991; Stegeman 1989). This indicates that such antibodies can be used as reagents for detecting induction of P450. Table 3 lists a selection of fish species in which antibody studies have demonstrated immunological similarities in PAH-inducible P450s.

The PAH-inducible fish proteins appear to be members of the PAH-inducible P450 gene family, the P4501 family (Nebert et al. 1987; Stegeman 1989). Based on earlier arguments (Stegeman 1989), we might refer to this PAH-induced type of P450 in fish as fish P4501A1. While structural features including cDNA or protein sequence of the hydrocarbon-inducible P450LM$_4$ from trout and P450E from scup indicate that these are P4501A1 proteins, the sequence of related proteins in most species is unknown, and the designation as P4501A1 must be tentative. It is more appropriate to refer to the PAH-inducible fish proteins as P4501A proteins.

DNA probes.   Recently, a cDNA for fish PAH-inducible P450 was derived from 3-MC-treated rainbow trout liver. This complementary DNA has been cloned and sequenced (Heilmann et al. 1988). It is this sequence which confirmed that the PAH-inducible P450 from teleosts is a member of the cytochrome P4501A family. The cDNA hybridizes to rainbow trout genomic DNA and to a messenger RNA from induced rainbow trout liver (Heilmann et al. 1988). The probe also has been shown to hybridize with genomic DNA and messenger RNAs induced by PAH in various vertebrate species, such as brook trout, scup, catfish, *Fundulus*, garter snake, turtle, bullfrog, quail, and rat (Haasch et al. 1989; Kloepper-Sams and Stegeman 1989). These results demonstrate that similarities in gene sequence occur in PAH-inducible P4501A genes in many vertebrates. This structural similarity corroborates the antigenic similarities of the proteins. The results indicate that cDNA probes may be used to detect P450 mRNA induction in many vertebrate species, just as the cross-reactive antibodies to the teleost P450 forms can be used to detect the proteins.

Table 2.   PAH-Inducible Cytochrome P450 Forms from Fish.

Species	Cytochrome P450 Designation	Major Catalytic Functions[a]	Reference
Scup	P450E	EROD, AHH	Klotz et al., 1983
Rainbow trout	P450 LM$_4$	EROD, AHH	Williams and Buhler, 1984
Cod	P450c	EROD, AHH	Goksøyr, 1985

[a] EROD = ethoxyresorufin O-deethylase, AHH = aryl hydrocarbon hydroxylase.

Source: Table adapted from Stegeman and Kloepper-Sams 1987.

Table 3.   A Selection of Fish Species Showing Induction of P450 Recognized by MAb to Scup P450IA.

Family	Species Name	Common Name
Squalidae	*Squalus acanthus*	Dogfish
Rajidae	*Raja erinacea*	Little skate
Salmonidae	*Salmo gairdneri*	Rainbow trout
	Salvelinus fontalis	Brook trout
Cyprinidae	*Cyprinus carpio*	Carp
Gadidae	*Gadus morhua*	Cod
Cyprinidontidae	*Fundulus heteroclitus*	Killifish
Pleuronectidae	*Platichthys stellatus*	Starry flounder
	Pseudopleuronectes americanus	Winter flounder
Sparidae	*Stenotomus chrysops*	Scup

Source: Stegeman 1989.

Methods of Analysis

As just discussed, induction of P4501A proteins in fish can be evaluated by measuring increases in catalytic activity (EROD or AHH activity), protein detected immunochemically, and messenger RNA detected with cDNA probes. These catalytic activities, the antibodies to P450, and the DNA probes for P450 mRNA are highly specific to the type of P450 induced by PAH, and by selected PCBs, dioxins, and others. However, some such probes also appear to be generic, recognizing that specific form of P450 in all vertebrates examined. The probes can therefore be used as reagents for analysis of induction in vertebrates. A summary of the methods for measuring induction using the various probes is given in Table 4.

The measurement of P4501A activity, protein, and mRNA can supplement each

Table 4.   Methods for Analysis of Cytochrome P4501A Induction.

Process	Measurement	Specific Probes	Common Assays
Enzyme reaction	Catalytic rate	Substrates: Ethoxyresorufin Benzo[a]pyrene	EROD — fluorometric, spectrophotometric AHH — fluorometric, radiometric
Translation	Protein levels	Antibodies: monoclonal or polyclonal	Western Blot, Dot or Slot Blot, ELISA
Transcription	mRNA levels	DNA probes: cDNA, oligonucleotides	Immunohistochemistry, In vitro translation, Dot Blot, Slot Blot, Northern Blot

other in the analysis of the induction process. Detailed studies showing this have now been accomplished in several fish species including rainbow trout, scup, and the killifish (*Fundulus heteroclitus*) (Haasch et al. 1989; Kloepper-Sams and Stegeman 1989). Under some circumstances, one type of measurement may have distinct advantages over another; i.e., one assay might reveal induction not detected by another assay. For example, some inducers also inhibit or "kill" the catalytic activity of P450 induced. In such cases, analysis of catalytic activity alone might show no response, but strong induction can still be seen by analysis of the P4501A protein or mRNA (Gooch et al. 1989).

Immunohistochemistry (Stegeman et al. 1991) may be particularly useful for analysis of embryos or some organs for which other biochemical assays would be very difficult. Analysis of induction by immunohistochemical assay has further advantage in that it can reveal the cell types where induction occurs. This could be important to evaluating the effects of induction in a given organ, and could show the route of exposure.

The various methods described in Table 4 can all reveal induction, and can be applied to studies of experimental or environmental samples. At present these methods require an increasing effort and cost as one goes from catalytic assay to mRNA analysis. Analysis of induction by these methods also requires laboratories that are properly equipped, and personnel experienced in biochemistry. Some of the methods, particularly those with DNA probes, are still under study as applied to fish systems. However, each of the methods has usefulness in studying the basic features of the induction process, features important in interpreting induction in the context of biomonitoring.

Mechanisms of P4501A Induction in Fish

If the mechanisms of induction in different groups of animals are similar, it may lead to predictive capability regarding the nature of inducing compounds. Understanding those mechanisms, therefore, could be important in interpreting results of biomarker analysis. Representatives of teleost P4501A are induced by many of the same compounds that elicit induction of P4501A1 proteins in mammalian species, where mechanisms are better known. The results suggest that the induction of P4501A in teleost and mammalian species occurs by similar mechanisms. Thus, the analysis of mRNA has demonstrated that fish treated with compounds such as β-naphthoflavone show increased amounts of mRNA for P4501A prior to increases in the P4501A protein itself (Haasch et al. 1989; Kloepper-Sams and Stegeman 1989). This pattern is consistent with the operation of a receptor-mediated system in teleosts similar to that found in mammals, in which the inducing chemical binds first to a receptor (see Figure 2). The properties of such receptors can determine the specificity of response to different chemicals. The nature of these receptors in nonmammalian species are only now beginning to come to light (Hahn et al. 1991).

Specificity

The earlier studies on induction of catalytic activity indicated that EROD and AHH activity could be induced by a wide variety of chlorinated and aromatic hydrocarbons (Stegeman 1981; Kleinow et al. 1987; Payne et al. 1987). More recent studies are comparing the responses as detected not only by catalytic activity but also by antibody analysis of protein levels and analysis of messenger RNA levels. These studies show that many, *but not all* aromatic and chlorinated hydrocarbons that can strongly induce P4501A in mammals can strongly induce P4501A forms in fish. Thus, several orthosubstituted chlorobiphenyl congeners that are potent inducers of P4501A forms in mammalian systems have little capacity to induce cytochrome P4501A in scup (Gooch et al. 1989) or in trout (Kleinow et al. 1990).

The inducing capacity of Aroclor (PCB) mixtures stems almost exclusively from the coplanar fraction of chlorobiphenyls (Gooch et al., unpublished) and may reside in as few as six of the 209 chlorobiphenyl compounds. However, structure-activity relationships need to be further studied with chlorobiphenyls and other groups of compounds, including aromatic hydrocarbons. It will then be possible to state with greater certainty which compounds could act to cause induction in the environment. The possibility that natural products may elicit hydrocarbon type induction in fish has been discussed (Stegeman 1981) but not yet carefully addressed. Nevertheless, at present, environmental induction in fish appears to be linked to relatively few organic pollutant chemicals, most of which are usually toxic.

Sensitivity

As yet, few studies have carefully evaluated the sensitivity of the induction response in fish. Most studies to date have used single doses of pure compounds, often very high doses. Many studies with chemical mixtures (e.g., Aroclors, PCBs) have also used high doses, as high as 500 mg/kg. The few dose-response studies carried out have shown that strong induction by single chlorobiphenyls, aromatic hydrocarbons, and β-naphthoflavone occurs at much lower doses, in the range of 0.1 to 1.0 mg/kg (Pesonen et al. 1987; Zhang et al. 1990; Gooch et al. 1989). Even lower doses of dibenzofurans are highly potent inducers (Hahn et al. 1989).

Most dose-response studies have involved intraperitoneal injection. Although this is an artificial route, the doses used suggest that contaminant levels in the environment would be active in fish. Moreover, the induction of cytochrome P4501A in the gut of the marine fish spot was demonstrated using benzo[a]pyrene at environmentally realistic doses in a dietary regimen (Van Veld et al. 1988). Such indications of the activity of environmental concentrations of known inducers are substantiated by comparing tissue chemical residue contents and levels of induction in treated fish to the residue contents and induction in fish sampled from the wild (Gooch et al. 1989; Melancon et al. 1989). Elevated content of P4501A in many fish in the environment confirms that environmental concentrations of some chemicals are causing induction. Nevertheless, detailed studies are still needed to establish dose-response relationships, at environmental concentrations and by environmental routes, for the environmental compounds thought to be capable of induction and therefore potentially responsible for effects in the wild.

Temporal Patterns of Induction

Knowledge of the turnover of induced proteins (i.e., rates of synthesis and degradation) is important to the interpretations one can make regarding when an animal may have been exposed. Studies have analyzed the sequence of changes that occur following treatment with a single dose of inducer in the killifish and in rainbow trout. In both species, the induction of P450 protein followed the induction of P450 mRNA, but with a considerable lag time of 12 h or more. The *Fundulus* and rainbow trout studies both show that induction in fish is evident as early as 6 h after treatment when mRNA is analyzed, and by 24 h when evaluating P4501A protein or EROD activity (Haasch et al. 1989; Kloepper-Sams and Stegeman 1989).

The levels of mRNA in β-naphthoflavone-treated fish peak at about 2 d and fully decay by 5 d. However, the *Fundulus* work shows that P4501A protein induced by β-naphthoflavone is maintained at elevated levels for weeks after the levels of mRNA have returned to control values. The mechanism for this prolonged maintenance of P4501A protein levels is unknown. In contrast, studies in scup treated with chlorinated hydrocarbons rather than aromatic hydrocarbons have indicated that elevated mRNA levels also persist following that treatment. The distinction probably results from slower metabolism and removal of the chlorinated hydrocarbon inducer, which can thereby continue to stimulate mRNA synthesis. The measurement of catalytic activity, protein, and mRNA may therefore yield different results depending upon the time of analysis after exposure or treatment, and the nature of the inducing compound. Clearly, further studies on the turnover of P450 protein and mRNA are needed before generalizations can be made.

Monooxygenases in Invertebrates

The properties of MFO systems in marine invertebrates have been detailed in a number of recent reviews (Lee 1982; Livingstone 1989; James 1989; Livingstone 1991). Some features of these systems have been reported consistently in molluscs and crustacea. Thus, the levels of total cytochrome P450 in some species are comparable to those seen in some teleost fish. Molluscan and crustacean microsomal enzymes catalyze transformation of a diverse suite of xenobiotic substrates including aromatic hydrocarbons, but the rates of these processes in invertebrates are substantially lower than those in most fish.

At present, there seems to be little potential for using monooxygenase activity or P450 levels in molluscs or crustaceans to assess their exposure to pollutants such as aromatic and chlorinated hydrocarbons. Induction has been suggested to occur after weeks of treatment (Schlenk and Buhler 1989), but there is no convincing evidence that hydrocarbons can rapidly induce P450 forms in molluscs or crustaceans. We have not seen an Ah-type receptor in any of 10 invertebrate species (Hahn et al. 1991).

Some P450 forms have been partially purified from crustaceans (James and Shiverick 1984), but the relationship of these crustacean P450s to those in fish or mammals is not yet known. Recent studies have shown that chiton tissue microsomes contain a protein immunodetected by an antiP4501A (Schlenk and Buhler 1989), but, the degree of relationship to vertebrate P4501A proteins is not clear. Whether other molluscs contain a P450 related to PAH-inducible vertebrate P450 is presently unknown. However, DNA and RNA hybridizing to cDNA for another type of inducible mammalian P450 (the clofibrate and phthalate-inducible P4504A1 form) have been identified in mussels (Goldfarb et al. 1989).

The lack of demonstrable strong induction of P4501A in marine invertebrates does *not* imply that studies of invertebrate P450 are of little importance. Such research is needed to evaluate the role of these enzymes in toxic mechanisms and could reveal specific effects that may be useful in monitoring.

Mammals and Birds

As stated previously, most detailed information on molecular aspects of the P450 system stems from studies in mammals. There is extensive knowledge on P450 forms and their regulation in laboratory animals, and to a lesser extent in humans. However, mammalian P450 has been explored little in the context of environmental biomonitoring. Recent studies involving whales (Goksøyr et al. 1989b) indicate a high content of protein in adult and fetal liver that cross-reacts with anti-P4501A. This analysis with antibodies to P4501A strongly suggests that liver samples from whales caught in the Norwegian Sea contained large amounts of a PAH-inducible P4501A protein, presumably resulting from environmental chemicals. There are also studies showing

P4501A1 induction in human placenta, related to the PCB poisoning in Yusho disease (Lucier et al. 1987).

There have been some studies to evaluate P450 induction as a biomarker in birds (Knight and Walker 1982; Rattner 1989a). Birds have P4501A forms induced by hydrocarbon structures similar to those active in mammals (Sinclair et al. 1989). There are also reports indicating environmental induction in wild birds, as demonstrated by higher rates of MFO activity and the presence of P450 detected with antibodies to P4501A proteins (Ronis et al. 1989).

Studies such as those in whales, seabirds, and humans further support the use of P450 induction as a biomarker in vertebrate species. As with fish, studies on the specificity, sensitivity, and temporal aspects of induction are needed in other wildlife species.

Induction in Extrahepatic Organs of Fish

In addition to induction in liver, monooxygenases occur and are induced in several extrahepatic organs of fish (Lindström-Seppä et al. 1981; Stegeman et al. 1979; Koivusaari et al. 1981a). There is regional distribution of P450 induction in some organs, including kidney (Pesonen et al. 1990). Analysis of scup gill, kidney, gut, and heart with MAb 1-12-3 to P450E showed that β-naphthoflavone strongly induced P4501A in each of these organs (Stegeman 1989). The induction in gut and heart are particularly strong, with P4501A content increasing from less than 0.05 to 0.5 nmol/mg protein in cardiac microsomes (Stegeman et al. 1989), and from 0.01 to 0.2 nmol/mg in the gut (Van Veld et al. 1988). Furthermore, the induction in many extrahepatic organs has been found to occur in specific cell types, identified in immunohistochemical analysis. The pillar cells (endothelial cells) of gill (Miller et al. 1989) and the endothelial cells in the heart (Stegeman et al. 1989) were identified as the primary sites of P4501A induction in those organs. More recently, detailed studies in 10 or more organs (Smolowitz et al. 1991) have shown that endothelial cells are a common site of induction in most and probably all organs. P450 induction in extrahepatic organs and its cellular specificity could be useful as biomarkers. Organ-specific induction and cellular localization require much additional study.

Environmental Induction and Monitoring

Many studies of environmental induction have involved measurement of catalytic activity. Studies carried out over the last 15 years have clearly demonstrated that rates of PAH-inducible P450 monooxygenase activities (EROD and AHH) are elevated in liver of fish from polluted waters. The earliest field study to observe increased hepatic AHH activity in fish was that of Payne and Penrose (1975). Subsequent studies showed that fish caught in waters polluted with oil (Lindström-Seppä 1988) or industrial municipal wastes exhibit increased cytochrome P450 monooxygenase activity (see reviews by Förlin et al. 1986; Payne et al. 1987). Recent comparisons between the levels of chemical residues (polychlorinated biphenyls) and the rates of monooxygenase activities support the interpretation that the elevated rates stem from chemical pollutants (Addison and Edwards 1988). Studies in lake trout larvae have also shown a correlation between rates of MFO activity and PCB content (Binder and Lech 1984).

During recent years, antibodies raised against the PAH-inducible P450 in fish have been used to detect environmental induction (Stegeman et al. 1987; Schoor et al. 1988). Levels of putative P4501A forms have also been related to the content of contaminants in organisms or their environment (Stegeman et al. 1986; Varanasi et al. 1986; Stegeman et al. 1987; 1988; Melancon et al. 1989; Stegeman et al. 1990). In some studies, the levels of P4501A proteins in liver were associated with PCB content, with an $r^2 > 0.95$. Correlation of induction with content of rapidly metabolized compounds (PAH) can be difficult (Foureman et al. 1983). Yet, in some studies, the levels of liver microsomal P4501A protein have also been correlated with content of PAH residues; for example, in the killifish (Elskus and Stegeman 1989) and English sole (Varanasi et al. 1986). Induction has been seen also in extrahepatic organs (e.g., heart and gut) of environmentally sampled fish (Stegeman et al. 1989; Van Veld et al. 1990). The growing number of such studies provides a consistent picture, confirming the idea that the levels of a specific P450 form can reflect the levels of contaminants in the environment and/or in the organisms themselves.

The specificity of the induction response allows informative sampling strategies, such as along a gradient; e.g., outside the discharge point of a waste effluent. This strategy was successfully employed in recent investigations showing that bleached kraft mill effluents induce EROD activity in fish (Andersson et al. 1987; Andersson et al. 1988). The interpretation of these results was facilitated by two significant observations. First, exposure of fish to the effluents, under laboratory conditions, resulted in elevated EROD activity (Andersson et al. 1987). Second, an inverse relationship was found between the distance from the discharge point of the effluents and the EROD activity in the fish caught along the gradient in the receiving water (Andersson et al. 1988). Full interpretation of field studies generally requires (1) reference station(s) located in an unpolluted area, (2) some knowledge that the pollutants of concern include structures similar to chemicals known to induce the enzyme system, and (3) some experimental measure of the capacity of the species to respond, for comparison with field results. Analysis of monooxygenase activity or P450 in fish, however, can provide useful information even if these conditions are not fully met.

Induction of P4501A proteins in fish clearly offers a valuable approach for sensitive and early detection of chemical exposure and effect. There is a great increase in the number of studies supporting the measurement of induction as an environmental monitor. We can expect many additional papers supporting the use of induction in monitoring various effluents, oil spills, dredge spoils, urban harbors, rivers, etc.

Sources of Variability

In natural populations, the combined influences of biological and environmental factors are known to cause background and seasonal variations in P450 content and activity (Koivusaari 1981b; Lindström-Seppä 1985). The content of P450 and induction of monooxygenase activities in teleosts can be suppressed by estradiol (Förlin et al. 1984; Pajor et al. 1990), contributing to the large sex differences in such activity seen during reproduction in some species. The biochemical site and mechanism of this action involving estradiol are not yet known. Similarly, low temperature can cause attenuation of the induction response (Stegeman 1979), but the mechanism of that temperature effect is not known. Environmental factors (e.g., temperature and photoperiod) and biotic factors (e.g., sex, reproduction, developmental status, and nutrition; Kleinow et al. 1987) have to be taken into account in the use of P450 induction as a biomarker for environmental contamination. Understanding the mechanism by which these factors affect P450 levels, as well as the magnitude and timing of changes, can allow one to account for these influences.

Relationship of P450 to Health and Disease

The induction of cytochrome P4501A forms by hydrocarbons, PCBs, or other compounds can have profound significance for the organism. In fish, P4501A proteins appear to be primarily responsible for hydrocarbon metabolism. This includes metabolism leading to detoxification as well as metabolism leading to toxic injury from reactive metabolites. The PAH-inducible fish P450s appear to be counterparts of the mammalian P4501A1 forms such as rat P450c and mouse P_1450, which are proven to activate PAH procarcinogens. The relationships between these teleost and mammalian P450 forms further indicate a role for these fish P450 forms in activation of such toxic compounds.

Given its dominant role in PAH activation, the induction and action of P4501A could be a prerequisite for the formation and presence of PAH metabolites, whether in bile or bound to DNA or protein. The induction of P450 in specific cells in extrahepatic organs could have important consequences for the normal physiological function of those organs. Induction in brain or gonad, for example, could affect reproduction. Induction also occurs in fish embryos and larvae (Binder and Stegeman 1980; Goksøyr et al. 1988), and could influence chemical effects on development. P450 induction can alter the balance of oxidative and conjugation metabolism. P4501A or the NADPH-cytochrome P450 reductase can contribute to free radical formation and oxidative damage, including DNA damage.

In addition to direct involvement in toxic injury, P450 induction could act to signal coincident toxic effects. Such coincident effects might involve other roles of the receptors that bind inducing compounds and that cause induction to proceed (Poland and Knutson 1982). Many of the effects of 2,3,7,8-tetrachlorodibenzo-p-dioxin (TCDD), for example, are now believed to result from the binding of this compound to the Ah receptor, independent of its role in induction (see Figure 2). In such cases, induction would be a surrogate for toxicity, rather than a direct participant. This is an important area for further research.

Many consequences of P450 induction, whether in fish or mammals, are only now coming to light. In fish, the relationship between induction of P450 activity and altered steroid metabolism in vitro (Hansson et al. 1980) and in vivo (Förlin and Haux 1985) and fertilization success (Spies et al. 1984) deserves further study. We can expect an ever-increasing recognition of the role of P450 in health, disease, reproduction, and behavior. Ultimately, "typing" a species for P450 profiles and inducibility could define the sensitivity of that species for different chemical effects.

Summary and Conclusions

Laboratory and field studies have shown that cytochrome P450 monooxygenase enzymes in fish can be induced by several types of hydrocarbons. These compounds include many that are known environmental threats, such as PAH, certain dioxins, certain congeners of chlorobiphenyls, other halogenated aromatics, and mixed petroleum products, as well as chemical formulations containing these compounds. For many toxic organic compounds, the induction of P450 may be the most sensitive early indicator of exposure. Induction in fish in the laboratory can be readily seen 12–24 h after exposure.

Induction can now be reliably detected by use of a set of specific probes. Selected catalytic activities (particularly ethoxyresorufin O-deethylase activity), specific antibodies, and cDNA probes, can reveal an explicit effect of PAH-type compounds. At minimum, the induction of P450 can be interpreted as a clear sign that exposure to biochemically significant levels of inducer has occurred. PCB compounds have been linked to environmental induction in several cases. Relationships between the degree of induction in liver and the concentration of contaminants in the organism or its environment are now being established. Relationships between organ dose and induction in organs other than liver are not known. Additional studies on time-course and dose-response relationships are needed.

Presently, there is great potential for detecting exposure and early effects through use of P4501A protein induction. However, the ability to analyze xenobiotic effects on other isozymes of P450 could be extremely important. Knowledge of the function and regulation of specific isozymes of P450 in addition to the major PAH-inducible form, could be used to define species and/or organ susceptibility to other specific compounds. Such knowledge could possibly disclose a fingerprint of responses to a variety of compounds.

Defining mechanistic origin of other effects that might occur subsequent to induction or other modulation of P450 forms would be of obvious value in assessing the significance of that induction or modulation. Different forms of P450 play key roles in activation and deactivation of many toxic chemicals, in synthesis and breakdown of steroid hormones, and other functions. There is a crucial need to understand these

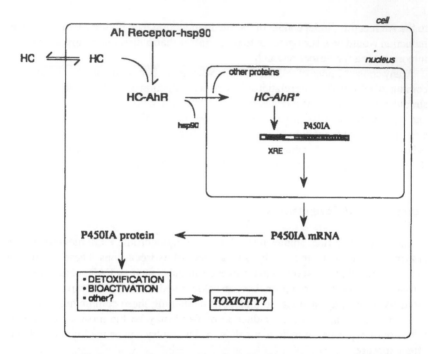

Figure 2.   Scheme showing mechanism of cytochrome P4501A induction. Inducer (HC) enters cell, binds to receptor (AhR) which is activated and then enters the nucleus. The receptor-inducer complex binds to regulatory elements (XRE) in the DNA, stimulating transcription of P4501A mRNA. The message is translated to apoenzyme; insertion of heme results in active enzyme. (Courtesy of M. Hahn.)

systems in fish and other nontraditional species, and their roles in endocrinology, toxicology, and chemically-induced disease. The return of understanding toxicology in aquatic species that can be expected from basic research on this system is enormous.

The objective of immediately employing or implementing methods involving P450 induction in analyses of exposure has strong positive features. The analysis of natural ecosystems may be the best biomonitoring use for P450 induction at present. It is likely that studies of induction using available probes could also delineate broad geographic regions where biochemical effects are occurring. Analysis of aquatic or terrestrial wildlife for chemical effects or biochemical responses involving P450 could eventually be linked to population or ecosystem effects, conferring predictive significance to such responses. However, P450 induction could also be used in monitoring effluents and screening compounds for effect. Basic research as a component of efforts in monitoring or ecological assessment could provide the definitive probes for analysis and interpretation, diagnosis, and possible prognosis. In the long-term view, the need for such tools may extend well into the next century. Thus, the most sensible approach is to invest in basic research on these systems, phasing in their applications when mechanistic understanding is achieved.

Recommendations

Induction of P450, as evaluated by MFO (EROD or AHH) activity, P450 protein content, and/or P450 mRNA content, is a powerful signal for detecting exposure to and effects induced by certain classes of chemicals. Research in a number of areas will improve the usefulness of the P4501A response in biomonitoring programs.

1. Identifying compounds responsible for environmental induction may benefit from studies to:

    a. determine experimentally the dose-response relationships for different compounds that are known or suspected environmental inducers
    b. establish clear relationships between the magnitude of P450 induction and the organ and environmental concentrations of inducers in field studies
    c. establish interactive effects for combinations of inducers and noninducers
    d. perform chemical analyses aimed at the identification of as yet unknown chemical residues that may be inducers
    e. examine usefulness of isolated cell cultures (e.g., isolated fish liver cells) in studying relative potency of inducers, identifying inducing components fractionated from complex mixtures (e.g., industrial effluents), and analyzing interactive effects and other aspects of regulation

2. Defining regulation of cytochrome P450s by xenobiotics needs studies to:

    a. analyze in detail the temporal patterns of transcription and translation of P4501A by various inducers (evaluated by specific cDNA probes, antibodies, and catalytic activities), in different species
    b. establish whether P450 genes additional to the PAH-inducible P4501A forms are regulated by these or other classes of chemicals (These studies must be performed in the various organisms of concern, e.g., fish and birds.)
    c. conclusively determine whether P4501A genes, or genes for other P450s, are regulated by xenobiotics in invertebrates
    d. establish the phylogeny and role of Ah receptors

3. Further characterization of P4501A forms in diverse species, and of P450 forms in other gene families is a very important research area. Aspects needing study include:

    a. biological and toxicological functions of various P450s
    b. tissue and cell specificity of various P450s

c.   regulation of various P450s by endogenous (e.g., hormones) and exogenous factors

4.   Further study is required on the catalytic roles of various P450 proteins in the metabolism of endogenous compounds (such as steroids and fatty acids) and of foreign compounds, and how these processes are affected by P450 induction or inactivation by chemicals in various cells, organs, and species.

5.   The most reliable approaches for measuring P4501A induction in monitoring should be compared. Present assays have been used successfully, and will continue to be used in future "biomarker" programs. However, such methods need to be adapted or modified to obtain simple and inexpensive assays for routine use and particularly use in the field. Standardization may be necessary.

6.   The nature of relationships between effects of environmental chemicals at the biochemical level (mRNA, protein, and activity) and effects at higher levels of biological organization (e.g., organ or individual) needs to be determined.

METALLOTHIONEINS

Metallothioneins constitute a family of low-molecular weight, cysteine-rich, metal-binding proteins and oligopeptides. The cysteinyl residues serve as ligands for metal chelation. These metalloproteins/peptides occur throughout the animal kingdom, as well as in plants and eukaryotic microorganisms. Their synthesis can be induced by a wide variety of metal ions, such as Cd, Cu, Zn, Hg, Co, Ni, Bi, and Ag. Given the extensive scientific information base and available methods for monitoring both changes in metallothionein synthesis and metal composition, these proteins have much to offer as potential biomarkers of excess exposure to common toxic metals in aquatic environments. Analysis of the metal composition of the protein confers a degree of chemical specificity with regard to the probable inductive agent.

The biological function of metallothionein is by no means fully understood. It has therefore not been possible, to date, to clearly link changes in the levels of metallothionein to injury at the cellular or organismal level. In order to exploit metallothionein as a predictive indicator of toxicity it will be necessary to expand our knowledge base concerning the normal physiological function of this protein.

Potential Biomarkers for Heavy Metal Exposure

Metallothioneins (MT) are small cysteine-rich proteins/peptides that can combine with a variety of metal ions. Metallothioneins have been reported to occur throughout the animal kingdom, in vertebrates (fish, amphibians, reptiles, birds, mammals) and invertebrates (echinoderms, annelids, molluscs, crustaceans, insects), as well as in plants, several eukaryotic microorganisms, and in some prokaryotes (Engel and Brouwer 1989; Hamer 1986; Kagi and Kojima 1987; Rauser 1990). Since the synthesis of MT can be induced by exposure to metal ions, such as Cd, Hg, Zn, and Cu, MTs have been considered potential biomarkers for metal exposure in the environment. In order to assess the feasibility of this proposal it will be necessary to review metallothioneins' chemical characteristics, proposed physiological functions, and the regulation of their biosynthesis and biodegradation. This information will provide the background against which use of MT as a biomarker should be evaluated.

Structure of Metallothioneins

Mammalian MTs usually contain 61 amino acids, 20 of which are cysteinyl residues, which serve as ligands for metal chelation. The protein does not contain aromatic amino acids. NMR spectroscopy (Otvos and Armitage 1980; Worgotter et al. 1988) and x-ray crystallography (Furey et al. 1986) have demonstrated that mammalian CdMT is a pseudosymmetrical, dumbbell-shaped molecule, composed of two domains enfolding a 3-cadmium/9-cysteine and a 4-cadmium/11-cysteine cluster. Each metal (cadmium and/or zinc) is tetrahedrally coordinated by the cysteine-residue sulfur atoms. The copper-containing form of MT appears to contain 12 copper atoms, which are proposed to be tetrahedrally (Freedman et al. 1986) or trigonally (George et al. 1988) coordinated to sulfur. Native MTs often contain mixed-metal clusters containing both Zn and Cu. After exposure to cadmium the protein may contain Cd, Cu, and Zn. Metallothioneins from marine crustacea consist of two 9-cysteine domains that each bind three metals (Otvos and Armitage 1982; Overnell et al. 1988; Brouwer et al. 1989).

Many of the proteins isolated from nonmammalian vertebrates and invertebrates are different from the typical mammalian MT. Because of the wide variation in chemical structure, it has become desirable to subdivide MTs into three classes. According to the new recommendations (Fowler et al. 1987), Class I MTs are defined to include all polypeptides related in primary structure to equine renal MT. Primary structure data have been reported for 38 Class I MTs, including the amino acid sequence of MTs isolated from the fish *Pleuronectes platessa* (Kagi et al. 1984), the crab *Scylla serrata* (Lerch et al. 1982), the lobster *Homarus americanus* (Brouwer et al. 1989), and the oyster *Crassostrea virginica* (Roesijadi et al. 1989). Class II MTs comprise forms displaying none, or only a very distant evolutional correspondence to the mammalian forms, such as the MTs from the sea urchin (Nemer et al. 1985) and yeast (Winge et al. 1985). Class III comprises the γ-glutamyl-cysteinyl oligopeptides isolated from plants (Grill et al. 1985), and the microorganisms *Euglena gracilis* (Weber et al. 1987) and *Schizosaccharomyces pombe* (Reese et al. 1988). In the following we will limit ourselves to the discussion of the properties of Class I and Class II MTs, and hence will refer to MTs as proteins.

Metal-Binding Properties

Virtually all metal ions tested can bind to apo-MT in vitro (Nielson et al. 1985). In vivo, under unstressed conditions, only Cu and Zn, and sometimes Cd, are found to be associated with MT. In addition, when exposed to metal ions, mammalian MTs can combine in vivo with Hg, Ag, Pt, Au, and Bi.

Despite our understanding of MT structure, relatively little is known about the affinities of metal ions for the multiple binding sites in this protein. On a semiquantitative level, pH titrations have shown that the metal-MT bond strength decreases in the order of Hg(II) > Cu(I) > Cd(II) > Zn (Rupp and Weser 1978). Based on spectrophotometric titrations the following values have been estimated: $K_{Cd} = 10^{22} M^{-1}$ and $K_{Zn} = 10^{18} M^{-1}$ (Vasak and Kagi 1983). However the assumptions underlying the calculations of the binding constants are somewhat doubtful (Engel and Brouwer 1989).

Even less is known about the metal-binding affinities of nonmammalian MTs. Unpublished observations, discussed during a recent workshop on "High Affinity Metal-Binding Proteins in Non-Mammalian Species" (National Institutes of Environmental Health Sciences, Research Triangle Park, NC, 1986) strongly suggested that invertebrate MTs bind cadmium less tightly than mammalian MT (Petering and Fowler 1986). Kinetic studies of the EDTA-mediated removal of Cd and Zn from MT from horse kidney (Li et al. 1980) and the crab *Cancer pagurus* (Higham et al. 1986) underscore these unpublished results. In addition to the common six metal-binding sites, the CdMT from *Cancer pagurus* contains a seventh low-affinity site for cadmium with an estimated dissociation constant of 3.6×10^{-7} M (Sparla and Overnell 1990).

Aspects of Metallothionein Function

MTs are thought to be involved in interrelated processes associated with metal metabolism, such as metal detoxification, regulation of zinc and copper, and donation of metals to metalloproteins (Foulkes 1982; Cousins 1985; Hamer 1986; Engel and Brouwer 1987; Brouwer et al. 1989). Evidence presented by several investigators (Ohtake et al. 1978; Brady and Webb 1981; Danielson et al. 1982; Cousins 1985) is consistent with the hypothesis that MTs supply zinc, and possibly copper, for nucleic acid metabolism, protein synthesis, and other metabolic processes, in growing, injured or regenerating tissues. MT also plays a role in the response to stresses such as cold, heat, exercise, and injection of chemicals (Oh et al. 1978). The stress-related increase of MT synthesis seems to be mediated by glucocorticoid hormones (Karin 1985). MT has also been shown to be an effective free-radical scavenger, which can be important both in normal cellular metabolism and in modifying radiation sensitivity (Thornalley and Vasak 1985; Bakka et al. 1982).

Studies on three marine species have shown how natural environmental and physiological factors can affect the mobilization and partitioning of metals by organisms. The following processes affected trace metal partitioning and accumulation: in oysters, the reproductive cycle/season of the year; in blue crabs, growth and the molt cycle; and in the beaked whale, the type of food and habitat (Engel 1988). It is thus evident that measurement of MT in mammals or nonmammalian organisms does not necessarily reflect the degree of exposure to metals.

Induction and Biodegradation of Metallothionein

Metallothionein synthesis in mammals can be induced directly by a wide variety of agents. Best known are of course metal ions such as Cd, Cu, Zn, Hg, Co, Ni, Bi, and Ag (Bracken and Klaassen 1987). Other agents are glucocorticoids (Karin 1985; Cousins 1985), progesterone (Bremner et al. 1981), catecholamines (Brady and Helvig 1984), glucagon (DiSilvestro and Cousins 1984), interleukin I (Cousins and Leinart 1988), and interferon (Friedman and Stark 1985). Indirect inducers, such as carbon tetrachloride (Oh et al. 1978) and ethanol (Bracken and Klaassen 1987), require mediation by the organism such as the stress response. With the exception of humans where MTs are represented by a single MT-II and at least five MT-I variant proteins, most animals possess primarily two major isoforms of MT that can be resolved by ion-exchange chromatography and have closely related but unique amino acid sequences. Genes coding for MT isoforms in mammalian cell lines have been shown to be inducible and differentially regulated at the level of transcription by heavy metals (i.e., copper, cadmium, and zinc), glucocorticoid hormones, and cytokines (Waalkes and Goering 1990; Sadhu and Gedamu 1988; Hamer 1986; Richards et al. 1984). Differential regulation of MT gene expression has also been demonstrated in rainbow trout cell lines (Olsson et al. 1990; Zafarullah et al. 1990).

Much less is known about the induction of MT in nonmammalian species. A wide variety of organisms (sea urchins, polychaetes, mussels, crabs, oysters, insects, fish, frogs, turtles, birds, and lizards) synthesize MT upon exposure to cadmium (for a review, see Engel and Brouwer 1989). Hormonal control of MT synthesis in invertebrates has not been studied. However, intriguing relationships between levels of MT and the ecdysteroid molting hormone have been observed during the molt cycle of the blue crab *Callinectes sapidus* (Engel and Brouwer 1987). Whether differential regulation of Class I MTs occurs in invertebrates is unknown. However, we have recently demonstrated that the cadmium- and copper-inducible MTs in blue crabs are distinct proteins (Brouwer and Brouwer-Hoexum 1991). We also have demonstrated that CuMT isoforms in the American lobster may be structurally and functionally distinct (Brouwer et al. 1989). One form may be involved in the detoxication of copper, whereas the other may act as a donor of Cu(I), via glutathione, for the biosynthesis of hemocyanin (M. Brouwer, unpublished results). Such observations clearly suggest that MT gene expression may be metal specific. This opens the possibility of making MT isoform-specific nucleotide probes, which can be used to quantitate expression of each metal-specific MT isoform.

The rates of biodegradation of MT are dependent on the metal bound. MT induced in the rat liver by Cu, Zn, or Cd has a half time of 12 to 16, 20, and 70 to 90 h, respectively (Feldman and Cousins 1976; Bremner et al. 1978). Both Cu and Zn are lost when MT biodegrades. Cadmium however, remains bound to MT in the steady state of biosynthesis and biodegradation of MT (Ridlington et al. 1981). Similar observations have been made for the blue crab, where cadmium concentrations in the gill and digestive gland remained constant for a period of 12 d after cadmium exposure was stopped (Brouwer et al. 1984).

Methods for Detection and Quantification of Metallothionein

Since MT has no known catalytic function, measurements of its concentration are based on a quantitative assay of the protein itself. Several existing methods have been reviewed recently (Engel and Roesijadi 1987). A brief overview follows.

Isolation of MT by size-exclusion and ion-exchange chromatography, coupled with atomic absorption spectroscopy. MT purification by means of gel filtration on Sephadex G-75, followed by ion-exchange chromatography on DEAE-Cellulose constitutes a standard procedure for the purification of both mammalian and nonmammalian MTs, but is by no means trivial, especially for the copper-containing form of MT (Brouwer et al. 1986). This method determines the actual amount of metals bound to MT. For well characterized MTs the metal/protein stoichiometry is known, and hence the amount of protein can be estimated from the metal measurements. For less well characterized MTs, assumptions concerning the stoichiometry have to be made, or alternatively, more detailed information on the composition of the MTs has to be determined. In the latter case further purification of the MT by reverse phase HPLC, prior to amino acid composition analysis, may be necessary (Brouwer et al. 1989).

The method described above is rather slow and requires a substantial amount of protein. Moreover, this procedure will not detect any apoMT that may be present. The latter problem may be overcome by repeating the purification procedure after treating the supernatant (cytosolic fraction) with Cd, to saturate empty binding sites.

Metal substitution assays. This method measures MT protein concentration in situ without isolation, by taking advantage of the natural order of stability constants for metals that bind to MT, i.e., Hg > Ag > Cu > Cd > Zn. In the "Cd-hemoglobin" method supernatant is treated with radiolabeled Cd. All non-MT Cd is bound by the addition of hemoglobin, which is precipitated from solution by heating (Eaton and Toal 1982). MT concentration, which is taken to be one seventh of the Cd content, is determined by measurement of Cd in the supernatant. Detection limits are in the microgram range. However, protein concentration can be underestimated in samples that contain Hg- or Cu-thioneins due to incomplete metal displacement. Moreover, this method does not provide information on the actual metal content of MT, and assumes that the measured cadmium is only bound to MT. The latter assumption needs to be validated by chromatographic procedures outlined above, whenever the technique is applied for the first time to estimate MT concentration in an organism.

Polarographic analysis of protein thiol content. In the differential pulse polarographic technique, MT is quantitated by electrochemical measurement of its thiol content (Olafson and Sim 1979; Olafson 1981). Current techniques provide nanogram sensitivity (Thompson and Cosson 1984). Low molecular weight, SH-containing compounds, such as glutathione and cysteine produce weak signals, which do not interfere significantly with the quantification of MT. However, this technique suffers from the same shortcomings as the metal-displacement assay, in that no information on the metal content of MT is obtained and that it is assumed, without confirmation, that the measured response is solely due to MT.

Immunochemical techniques. A sensitive radioimmunoassay and colorimetric ELISA for the detection of mammalian MTs have been developed (Vandermallie and Garvey 1979; Garvey 1984). Detection levels using the ELISA range from 500 to 5000 pg (Garvey 1984). Unfortunately, antibodies raised against mammalian MT have not shown significant cross-reactivity with proteins obtained from invertebrates and fish (Garvey and Kay, personal communication in Engel and Roesijadi 1987). Antibodies raised against MTs from fish and invertebrate species have recently been developed into quantitative assays (Hogstrand and Haux 1990; Roesijadi et al. 1988). However, as with methods two and three, this quantitation does not provide information about the metals bound to the protein. Due to its great sensitivity immunochemical assays will be the method of choice when the supply of protein is limited. A disadvantage that is common to metal substitution, polarographic, and most immunoassays is their inability to distinguish between MT isoforms. As discussed above this distinction is important because the induction of MT isoforms may be metal-specific.

Sources of Variability

It is now clear that, at least in mammals, insects, and crustaceans, metallothionein is induced under many other conditions besides metal exposure. Both glucocorticoid hormones (progesterone, glucagon) and peptide hormones (interleukin I and interferon) have been found to induce metallothionein synthesis. As a consequence levels of metallothionein, and the amounts of copper and zinc bound to it, are affected by processes such as growth, reproduction, tissue regeneration, and, in the case of insects and crustaceans, molting. Recently it has been demonstrated that factors, such as temperature and/or nutritional status may have profound effects on the Cu/Zn ratios bound to metallothionein in marine invertebrates. This phenomenon could be correlated to the synthesis of the copper-containing respiratory protein, hemocyanin (Engel and Brouwer 1987). Finally, as discussed before, synthesis of metallothionein in mammals can be increased in response to several stresses, as mediated by glucocorticoid hormones. The effect of stress on metallothionein levels in aquatic species is still poorly known. From the foregoing it seems to be clear that measurement of copper and zinc bound to metallothionein does not necessarily reflect exposure to these metals.

Measurement of nonessential trace metals such as Hg, Ag, and Cd will be less likely to be confounded by the sources of variability discussed previously.

Field Studies

There have been many reports that elevated metal concentrations in the aquatic environment are correlated with the presence or concentrations of metals bound to MT in fish (Hamilton and Mehrle 1986) and marine invertebrates (Roesijadi 1981). In an extensive study of lakes along the Campbell River, British Columbia, that were contaminated with metals (Zn, Cu, and Cd), it was found that the concentration of MT is a useful quantitative measure of the degree of exposure of fish to heavy metals (Roch et al. 1982; Roch and McCarter 1984). Similarly it was found that increased hepatic MT content correlates with cadmium accumulation in environmentally exposed perch (*Perca fuviatilis*) (Olsson and Haux 1986). It is interesting that liver metal concentrations in fish collected around municipal wastewater outfalls, despite large accumulations of metals in the sediments and in water borne particulates, tend to be low. This phenomenon may be associated with reduction of MT synthesis due to increased demands for cysteine residues for glutathione synthesis during detoxication of organic contaminants (Brown et al. 1987).

The presence of CdMT in marine molluscs from cadmium-contaminated areas has been demonstrated (Talbot and Magee 1978; Noel-Lambot 1980; Langston and Zhou 1986). Tissue concentrations of nonessential metals (Ag, Cd, Hg) reflect environmental levels. Cu and Zn appear to be regulated. In blue crabs from polluted areas (from Hudson River, Foundry Cove and Haverstraw Bay; and from Baltimore Harbor), the digestive glands contained the highest concentrations of trace metals and had MTs that contained Cd, Cu, and Zn. The gill MT contained mostly Cd (Engel and Brouwer 1984a,b). A CdMT has been isolated from the crab *Cancer pagurus* collected from an area remote from any form of industrial pollution (Shetland Islands, North of Scotland). In this case it was postulated that the cadmium originated from underwater outcrops of cadmium-rich minerals (Overnell and Trewhella 1979). The foregoing summary shows that MT may be indeed a useful indicator of exposure to toxic metals in the aquatic environment. Establishing a link between observed elevated levels of MT and injury at the cellular or organismal level constitutes a major challenge.

Positive Features of Metallothioneins as Biomarkers

Given the extensive basic scientific information base and available methods for monitoring both changes in metallothionein synthesis and metal composition, these proteins have much to offer as possible biomarkers of excess exposure to some common toxic metals in aquatic environments. Assay procedures for these proteins are sensitive and permit evidence of induction at relatively low metal dosage levels. Application of metal composition studies to analyses of this protein also confers a degree of chemical specificity with regard to the probable inductive agent, provided interactive factors such as those discussed below do not obfuscate this interpretation.

A positive scientific reason exists for studying metal-binding to metallothionein compared to simply measuring total tissue concentrations of metals that induce/bind to metallothionein: it is increasingly clear, that knowledge of intracellular metal compartmentation is essential to understanding mechanisms of metal-induced cell injury (Fowler 1987). This is particularly true in marine invertebrates where there may be several intracellular mechanisms for sequestering metals (Fowler 1987; George et al. 1979; 1980; Nolan and Duke 1983; Roesijadi and Klerks 1989). Since the normal physiological function of metallothionein is presently unknown (Kagi and Kojima 1987) there is no way to determine if metallothionein itself plays a direct role in the pathophysiology of cell injury. The vast majority of the data from the scientific literature suggests that the reverse is in fact the case: the nonmetallothionein bound fractions of these metals participate in the cell injury process, and metallothionein induction appears to be a protective cellular response. In other words, metallothionein induction and attendant metal-binding appear to be a cellular defense mechanism against cell injury. Metal toxicity seems to occur only after this capacity is exceeded (Squib et al. 1984; Engel and Fowler 1979).

Feasibility of Metallothioneins as Biomarkers for Metal Exposure or Metal-Induced Stress

A biomarker is a biochemical or physiological response to anthropogenic contaminants, one that can provide a sensitive index of exposure or sublethal stress at the organismal level. Ideally this response should have a predictive value for the effects of the contaminant at higher levels of biological organization (populations, communities, ecosystems). For a protein to be used as a biomarker, the following properties and information base should be available or be established: (1) the assay to quantify biomarkers should be sensitive, reliable, and "relatively" easy; (2) baseline data for the concentration/activity of the biomarker should be known in order to be able to distinguish between natural variability (noise) and contaminant-induced stress (signal); (3) the basic biology/physiology of the test organism should be known so that sources of uncontrolled variation (growth and development, reproduction, food sources) can be minimized; (4) all the factors, intrinsic as well extrinsic, that affect the biomarker should be known; (5) it should be established whether changes in biomarker concentration are due to physiological acclimation or to genetic adaptation; and finally (6) increased levels of the biomarker should be correlated with the "health" or "fitness" of the organism.

In the following section we will discuss the feasibility of metallothionein as a biomarker for metal exposure or metal-induced stress, using the six criteria outlined above as a guideline.

Criterion 1. As described in the section on "Methods for the detection and quantification of metallothionein", the procedure for quantifying metallothionein and its bound metals is rather elaborate. This raises the obvious question of how much is gained by measuring the distribution of metals among intracellular binding sites such as MT, instead of total metal concentrations in the tissues or in the environment. Techniques, such as flameless atomic absorption spectroscopy or inductively coupled plasma spectroscopy, allow the determination of metal ions in the nM range. Such measurements will provide significant information about environmental contamination and its effect on metal accumulation by animals. However, such measurements will not provide any detail about possible molecular mechanisms that may be responsible for metal-induced toxicity. For example, MT induction is associated with the acquisition of increased tolerance to metal toxicity in organisms, including several marine species (Bouquegneau 1979; Roesijadi et al. 1982; Young and Roesijadi 1983; Roesijadi and Fellingham 1987). Elevated levels of MT may therefore reflect acclimatization or adaptation to low-level metal contamination of the environment. On the other hand, mobilization of MT in response to environmental stress may occur at the expense of other physiological processes, resulting in reduced fitness of the organism. In addition, perturbation of constitutive MT levels by metal exposure, may interfere with MT's regulatory role in "normal" metal metabolism, which, in turn, may interfere with the biosynthesis of metal-requiring enzymes. Increased levels of MT may represent both compensatory and toxic responses to metal exposure.

Another potentially important consideration that may influence the utility of metallothionein as a biomarker in marine invertebrates is the fact that, in contrast with mammals or fish, this compartment frequently does not constitute the major intracellular depot for metals in target tissues. Other cellular storage sites such as lysosomes/granules or concretions frequently contain the majority of the intracellular metal burden (Carmichael and Fowler 1981; Carmichael et al. 1979; 1980; George et al. 1980; Nolan and Duke 1983).

One possible reason for this phenomenon may be the generally lower affinities of nonmammalian metallothioneins for metals such as cadmium (see Petering and Fowler 1986 for discussion). Another potentially important consideration is the fact that metallothioneins are a large gene family of proteins that sometimes show marked differences in metal-binding capability even among closely related marine species living in the same environment (Nordberg et al. 1986). In conclusion we may say that for a better understanding of the correlation between metal exposure and organismal stress, refined analysis of the cellular metal distribution will be necessary (see also Criterion 6).

Criteria 2 and 3. As indicated above, the interpretation of the meaning of elevated MT levels in cells is difficult. Such an interpretation may be confounded by natural changes in MT concentrations, which are independent of exposure to metals (Engel 1988; Engel and Brouwer 1989; Petering and Fowler 1986). It is therefore mandatory to demonstrate first MT's role in normal trace metal metabolism. Such studies will help in establishing the natural levels and variability of MT levels, and may serve as a basis for choosing indicator species. Use of MT as a monitoring tool, without understanding its involvement in normal physiological processes, will undoubtedly lead to errors in the interpretation of the monitored data.

Criterion 4. As described before, MT synthesis in mammals may be induced by a wide variety of agents. Several forms of stress can induce MT synthesis, mediated through glucocorticoid hormones (Cousins 1985). It needs to be established whether analogous processes occur in marine/aquatic organisms as well. Studies in sea scallops (Fowler and Gould 1988) using combined exposures to Cu and Cd have shown that Cu markedly influences the binding of Cd and Zn to the metallothionein-like protein from this species. These data suggest that multielement exposure situations may be more complicated to interpret than cases where only a single metal is present in excess concentrations. The presence of such interactive factors may hence limit the field use of metallothionein for identifying the metal responsible for initiating the inductive process.

Criterion 5. Another complexity in the interpretation of the meaning of elevated MT levels concerns the question of the evolution of resistance to metals. Exposure to metals can lead to physiological acclimation within the life-span of the exposed animal, but can also lead to natural selection of an increased resistance, resulting in genetic adaptation to the metals. It has been demonstrated, for example, that strong selection has resulted in the evolution of resistance to cadmium in an aquatic oligochaete. Increased levels of MT are, at least partly, responsible for this resistance (Klerks and Levinton 1988). Similarly, MT gene duplications have been linked to increased metal tolerance in natural populations of *Drosophila melanogaster* (Maroni et al. 1987). This phenomenon may lead to errors in the understanding of the response of species to toxic substances and the fate of toxic substances in food webs. As an example, a cadmium exposure experiment on native oligochaetes might lead to the incorrect conclusion that the introduction of cadmium would result in the demise of population. However, with the evolution of resistance, the population may actually maintain densities comparable to those of pristine areas. In addition, the protective mechanism may result in enhanced uptake of cadmium, with the resultant danger that it may be transferred through the food web (Klerks and Levinton 1988).

Criterion 6. A biomarker should provide a sensitive index of exposure or, more importantly, of sublethal stress. Sanders and coworkers (Sanders et al. 1983; Sanders and Jenkins 1984) have shown that growth of the larvae of the crab *Rhithropanopeus harrisii* can be inhibited by exposure to copper. At free cupric ion concentrations within the ambient range, cytosolic copper was associated with both MT and higher molecular weight (HMW) ligands, and was independent of the external free Cu(II) concentration. At higher copper concentrations, copper accumulated in the MT and "very low molecular weight" (VLMW) ligand pool. Reductions in larval growth could be correlated with copper accumulation in the MT and VLMW pools, suggesting that MT did not protect against copper-induced stress. No correlation between copper associated with HMW ligands and either external Cu(II) or inhibition of larval growth was found.

Conclusions and Recommendations

Given the extensive scientific information base and available methods for measuring changes in metallothionein synthesis and its metal composition, these proteins show a strong potential for use as biomarkers of exposure to toxic metals in the environment.

Analysis of the metals bound to metallothionein isoforms also provides a degree of chemical specificity with regard to the probable inductive agents. Simplified assay procedures such as immunoaffinity chromatography coupled with atomic absorption spectrophotometry need to be developed. Since MT isoforms seem to be differentially induced by various heavy metals, MT isoform-specific nucleotide probes for use in quantitation of expression of each MT isoform need to be developed.

It is our view that, if MTs are to be exploited for assessment of organismal health or fitness, an extensive knowledge base concerning their normal physiological function and the factors that control the levels of MT in selected marine/aquatic organisms needs to be established. Based on this information it may be possible to select species that may serve as indicator organisms in a monitoring program.

STRESS PROTEINS

Recently, it has become apparent that all cells undergo alterations in gene expression in response to environmental stressors (for reviews see Schlesinger et al. 1982; Atkinson and Walden 1985; Moromoto et al. 1990). This response was initially referred to as the heat shock response because it was discovered upon exposure to elevated temperatures. It is now referred to as the stress protein response since it can be elicited by a variety of physical and chemical stressors including anoxia (Spector et al. 1986), metals (Hammond et al. 1982; Caltabiano et al. 1986a), and xenobiotics (Sanders 1990). Changes in gene expression associated with this response are extremely rapid, and result in the induced synthesis and accumulation of stress proteins.

Many of the stress proteins and the genes that code for them have been sequenced in a wide range of organisms and found to be remarkably conserved (Schlesinger et al. 1982). Although a few stress proteins are found only in cells responding to environmental stressors, most are also present at much lower concentrations under normal conditions where they play a role in normal cellular function (Moromoto et al. 1990). Two stress proteins, hsp60 and hsp70, appear to be involved in protein homeostasis under normal conditions, taking on protective and repair roles upon exposure to adverse environmental conditions (Rothman 1989; Welch 1990). For example, members of the large hsp70 family are found in several subcellular compartments where they are catalysts of protein folding and are involved in intercompartmental transport under normal conditions (Rothman 1989; Beckmann et al. 1990; Chirico et al. 1988; Deshaies et al. 1988; Craig 1989). In contrast, members of the hsp60 family are found in the mitochondria, chloroplasts, and in procaryotes where they are involved in the folding and assembly of the numerous large enzyme-protein complexes associated with the inner membrane (Cheng et al. 1989; Ostermann et al. 1989). Under adverse environmental conditions, hsp60 and hsp70 may also perform the related functions of renaturing damaged peptides and resolubilizing protein aggregates (Rothman 1989). It is through this renaturing capability that stress proteins facilitate the repair of proteins and protein complexes associated with critical physiological processes and protect cells from stress induced damage (Gaitanaris et al. 1990; Skowyra et al. 1990).

In this section, which deals with the use of stress proteins as biomarkers, we focus on only two closely related protein groups, the heat shock proteins and the glucose-regulated proteins. Several proteins induced by specific classes of stressors, including heme oxygenase (Keyse and Tyrrell 1989) and metallothionein (Kagi and Nordberg 1979), can also be included under the broad category of stress proteins. These proteins are discussed in separate sections of this chapter.

Although bacteria, yeast, *Drosophila*, and mammalian cell lines have been used frequently as model systems, the highly conserved nature of this response allows for broad extrapolation to other organisms. Stress proteins make ideal candidates as biomarkers for environmental contamination since they are (1) part of the cellular protective response; (2) induced by a wide variety of environmental stressors; and (3) highly conserved in all organisms from bacteria to man (Schlesinger et al. 1982).

The Stress Protein Response

The stress protein response includes two major groups of gene products: the heat shock protein (hsp) group and the glucose-regulated protein (grp) group (Welch 1990). Synthesis of hsp is dramatically increased by exposure to heat and a variety of other physical and chemical stressors, while synthesis of grp is increased in cells by such factors as deprivation of glucose or oxygen. These two protein groups are closely related, having similar biochemical and immunological characteristics; considerable homology exists between families.

Each stress protein comprises a multigene family in which some proteins, often called cognates, are constitutively expressed while others are highly inducible in response to environmental stressors. The cognates play a role in basic cellular physiology, and are present in cells under normal conditions.

The Stress Proteins Induced by Heat: Heat Shock Proteins

The term heat shock proteins derives from the original description of the "heat shock response", although such proteins are now known to be induced by a variety of other environmental perturbations including heavy metals (Hammond et al. 1982; Caltabiano et al. 1986a), xenobiotics (Sanders 1990), oxidative conditions (Kapoor and Lewis 1987), anoxia (Spector et al. 1986), salinity stress (Ramagopal 1987), teratogens (Bournias-Vardiabasis et al. 1983; Bournias-Vardiabasis and Buzin 1986), and hepatocarcinogens (Carr et al. 1986). The number of heat shock proteins and their exact size are specific to both tissue and species; however, five "universal" stress proteins are found in all eukaryotes. Four of these are referred to by their apparent molecular weight on SDS-polyacrylamide gels: hsp90, hsp70, hsp60, and the low molecular weight hsp20-30. The fifth heat shock protein is an 8 kDa protein called ubiquitin. Except for the 72 kDa protein, a highly inducible member of the hsp70 family, all of these proteins are present in low concentrations under normal conditions in most organisms studied and play a role in normal cellular function (Welch 1990).

hsp90

The hsp90 protein is also referred to as hsp83 or hsp89 depending upon the species under study. It is found in association with several cellular proteins, including steroid receptors (Catelli et al. 1985; Ziemiecki et al. 1986), several kinases (Rose et al. 1988), and retrovirus encoded tyrosine kinases (Welch 1990). Depending upon the protein involved, the association of hsp90 with these proteins will either prevent them from carrying out their normal functions or enhance those functions. In mammals, hsp90 is abundant in cells under normal conditions, and its synthesis increases about three- to fivefold upon exposure to environmental insults (Welch 1990). In light of what is known about its function, this protein may participate in redirecting cellular metabolism in perturbed cells through such mechanisms as the alteration of signal transduction. Given the normal abundance of hsp90 and its limited induced synthesis upon exposure to environmental stressors, this stress protein alone may not have a great deal of potential as a biomarker. However, it may have more potential when used in conjunction with other stress proteins (e.g., hsp70).

hsp70

Two stress proteins, hsp73 and hsp72, are members of the highly conserved hsp70 family. The larger protein of the two, hsp73, is often referred to as the hsp cognate because its synthesis is constitutive. It also exhibits a marked increase in synthesis upon exposure to environmental perturbations. It appears to act to either stabilize or unfold target proteins (Welch 1990). Under normal conditions, such binding may serve a "chaperon" function for newly synthesized secretory and organellular proteins by helping them to translocate across a membrane (Chirico et al. 1988; Craig et al. 1987).

The smaller protein, hsp72, is only synthesized in response to environmental stressors and is not found in most cells under normal conditions. This highly inducible protein, with other members of the hsp70 family, may perform a similar role in cells subjected to environmental perturbations. Under such conditions, hsp72 rapidly migrates to the nucleolus where it is speculated to resolubilize denatured preribosomal complexes and help restore nucleolar function during recovery. During recovery, it migrates to the cytoplasm and associates with ribosomes and polyribosomes where, speculatively, it may bind to denatured proteins and, in an ATP-dependent manner, facilitate their resolubilization (Welch and Feramisco 1984; Pelham 1988). Since the hsp70 family accounts for much of the translational activity in cells responding to environmental perturbation and since it is one of the most highly conserved proteins known in biology, it is an excellent candidate as a biomarker for chemical contamination.

hsp60 or Chaperonin-60

Members of the hsp60 family in eucaryotes are located in the mitochondria and chloroplasts, and facilitate the translocation and assembly of oligomeric proteins in those compartments (Deshaies et al. 1988; Cheng et al. 1989). Because hsp60 is highly conserved and its rate of synthesis is increased in stressed cells, it is also a good candidate as a biomarker.

hsp20-30

These stress proteins are highly specific to species and considerable variation exists even within the same class of organisms. This protein family shows homology to the alpha crystalline lens protein, and also shares with that protein the tendency to form higher ordered structures (Arrigo et al. 1988). These proteins are regulated during development and differentiation, and by the hormones estrogen, progesterone, and ecdysone; otherwise, we know very little about their function in cells under normal conditions (Edwards et al. 1980; Ireland and Burger 1982; Lindquist 1986; Welch 1990). We do know that they are found in either a nuclear or perinuclear location after heat shock and return to the cytoplasm upon recovery (Arrigo et al. 1988). Since they are highly specific to species, regulated by several factors besides exposure to environmental stressors, and little is known about their function, it is difficult to evaluate their utility as biomarkers at this time.

Ubiquitin

Ubiquitin is a small molecular weight (7 kDa) protein. It is involved in the nonlysosomal degradation of intracellular proteins (Schlesinger 1988). These proteins become conjugated by a ubiquitin-protein ligase system and are selectively degraded. Ubiquitin is found in all eukaryotes. Its synthesis increases with exposure to heat and contaminants (Finley et al. 1987). In cells responding to environmental perturbation, the role of this protein could complement the resolubilization and stabilization function of hsp70 by targeting denatured proteins for degradation and removal. Little data are available that would aid in evaluating this stress protein as a biomarker. Since the generation of denatured proteins is a common problem in stressed cells, however, increased synthesis of this protein could be an excellent candidate as a biomarker.

The Glucose-Regulated Proteins

The glucose-regulated protein (grp) group comprises 100 and 78 kDa proteins, and are structurally and functionally related to the hsp group. The grp78 protein, also known as BiP, is a member of the hsp70 family and is located in the endoplasmic reticulum (ER) (Munro and Pelham 1986). The grp100, is also located in the ER and is homologous to hsp90 (Sargan et al. 1986). These stress proteins are present under normal conditions and show increased synthesis in cell cultures deprived of glucose or oxygen, exposed to elevated lead concentrations, or subjected to agents that perturb calcium homeostasis or protein homeostasis in the ER (Welch 1990; Shelton et al. 1986). Very little is known about the induced synthesis of these stress proteins in whole organisms, so their potential as biomarkers is unclear. Since their induced synthesis appears specific to a few types of stressors, the grp group may have potential as biomarkers to identify anoxic or suboptimal nutritional conditions.

The Protective Role of Stress Proteins

Collectively, heat shock proteins and glucose-regulated proteins appear to be involved in the protection, enhanced survival, and restoration of normal cellular activities in cells responding to environmental perturbation (Subjeck and Shyy 1986). The induction of heat shock proteins by a mild heat shock correlates with the tolerance of the cell to subsequent, more severe heat shock, a phenomenon often referred to as thermotolerance or, when other insults are involved, "acquired tolerance" (Landry et al. 1982; Berger and Woodward 1983; Stephanou et al. 1983; Roberts 1984; Mirkes 1987; Mosser et al. 1987; Mosser and Bols 1988). Although the molecular basis for this protective role is not known, there is evidence to suggest that both RNA processing (Yost and Lindquist 1986) and translational activity (Welch and Mizzen 1988) are protected by stress proteins. Mammalian cell lines selected for survival at high temperatures constitutively synthesize hsp70 at high levels while temperature-sensitive mutants are unable to elicit the stress protein response (Lindquist 1986). Also, temperature resistance is conferred by expression of the mammalian hsp27 gene in rodent cells (Landry et al. 1989). The hsp20-30 family correlates with acquired tolerance in *Drosophila* (Berger and Woodward 1983) and sorghum (Ougham and Stoddart 1986), and accumulates in desert succulents acclimatized to high temperatures (Kee and Noble 1986). Further, the synthesis of specific stress proteins is related to temperature tolerance and habitat in aquatic invertebrates (Bosch et al. 1988).

Methods for Detecting and Quantifying Stress Proteins

The techniques most frequently used in the study of stress proteins are metabolic labeling and assays using protein-specific antibodies or cDNA probes. Each of these techniques measures a different aspect of the response; usefulness as a biomarker assay will depend upon the context in which the technique is applied.

It appears likely that it will be the actual stress protein concentration that will be most reflective of the cell's physiological state in organisms exposed to moderate contamination over long periods. Currently, there is no accurate commercially available assay for quantifying any of the stress proteins. Such techniques must be developed before extensive research can be undertaken to determine the potential of the stress response as a biomarker of biological impact in field sampled organisms. Immunological techniques offer perhaps the most promise in developing accurate assays, and are considered last in the following discussion.

Metabolic Labeling and Electrophoresis

In metabolic labeling studies tissues are incubated with an amino acid tagged with a radioisotope (i.e., ^{35}S, ^{14}C, ^{3}H). The tissue is then homogenized and the proteins are separated by one- or two-dimensional electrophoresis, and autoradiographed to examine incorporation of the radioisotope into specific proteins. Because the technique is time consuming, expensive, and requires a high level of laboratory expertise its potential for routine monitoring is limited. The strength of the technique is that it provides information on the entire translational profile of the cell in response to a stressor, making it particularly useful for identifying new contaminant-specific inducible proteins.

It is also important to consider that changes in stress protein translational activity take place over a short timeframe (minutes to hours); even under continuous exposure to environmental stressors translational activity can revert to patterns similar to those found in controls within 24 to 48 h. As a consequence, the technique will probably be more useful for short-term, toxicological and effluent screening purposes than for long-term transplant or (in situ) studies.

cDNA Probes

Another technique often used to study the stress protein response involves using cDNA probes to measure the mRNAs that code for individual stress proteins. Stress protein mRNAs are short-lived and their increase in tissues has been found to be transient. As a consequence, they would be more useful as biomarkers in toxicological and effluent screening than for (in situ) studies. They also may be useful in conjunction with direct stress protein measurements to provide information on the time course of exposure in field sampled organisms. A major advantage in using cDNA probes is that the technique is faster and less expensive than metabolic labeling and electrophoretic techniques.

Immunological Techniques

Assays based on antibodies to the stress proteins have the best potential for environmental monitoring because they allow direct quantification of stress proteins accumulation. Of all the stress proteins, the hsp70 family, particularly the highly inducible hsp72, provides an ideal candidate for an assay of general stress because little of it is present under normal conditions. Further, since it is highly conserved, potential exists for developing antibodies against this stress protein that cross-react with hsp70 of a broad range of organisms, including invertebrate, vertebrate, and plant species. A number of antibodies have been raised against the two hsp70 gene products isolated in mammals, *Drosophila*, yeast, and bacteria. However, the extent to which they might cross-react with other organisms is not clear. We are encouraged by the fact that a monoclonal antibody against mammalian hsp70 cross-reacts with a diverse group of organisms including molluscs, fish, algae, and crustaceans (Sanders 1990).

Another promising finding is that the highly inducible hsp72 can be detected by Western blotting with this antibody in organisms exposed to heat shock, Cu, and tributyl tin (TBT), but not in the controls (Sanders 1990).

Another good candidate for a biomarker is the mitochondrial hsp60. A polyclonal antibody made against hsp60 from a moth has been shown to cross-react with hsp60 from a particularly wide variety of species ranging from the groE gene product in *Escherichia coli* to the hsp60 in rotifers, algae, *Mytilus*, and fathead minnow (Miller 1987; Sanders 1990). Currently, quantitative Western blotting, dot blots, and slot blots with these antibodies are the sole methods of semiquantifying these proteins. These techniques are cumbersome, less sensitive, and less precise than other immunological techniques such as ELISA or radioimmunoassay. Major emphasis should be placed on developing quantitative assays for these proteins to explore the full potential of stress proteins as biomarkers in environmental monitoring.

Sources of Variability

Proper interpretation of data acquired from the techniques just described must consider other factors that can modify the stress response, such as pH, temperature, nutrition, salinity, anoxia, handling, and disease. Care should be taken not to inadvertently subject organisms to heat shock during sample collection. Because water quality (pH, temperature, salinity, and dissolved oxygen) can induce the response, these factors should be monitored during laboratory experiments and field collections. Since data from mammalian systems suggest that nutritional state and disease also alter stress protein concentration, these factors should also be taken into account when interpreting data.

Reproductive state, molt cycle, and factors that change seasonally could also result in apparent seasonal variation in the stress response in field sampled organisms. Fortunately, handling stress does not appear to alter stress protein synthesis perhaps because the cellular stress response is not induced by glucocorticoids. Preliminary experiments in fathead minnows, molluscs, sea urchin embryos, and algae suggest this to be the case.

Potential of Stress Proteins as Biomarkers

Stress protein assays for some of the heat shock proteins may be useful for determining the extent to which the cell is attempting to protect itself from environmental damage. They could be well-suited for both transplant and (in situ) studies to evaluate biological impact of contaminants. Other stressor-specific stress proteins may serve as biomarkers of contaminant exposure (Table 5). Major factors must be considered in evaluating the feasibility of using increased accumulation of stress proteins as a biomarker for chemical contamination, including the environmental relevance of the response, its relationship to biological effects at higher levels of organization, and its utility in field studies.

Environmental Relevance

Two factors to consider in evaluating stress protein-based assays as biomarkers are (1) the response should be sensitive enough to be induced by contaminant concentrations found in the environment, and (2) the response being measured must be sustained over time. To date we know little about these aspects of the stress response, and much research needs to be focused along these lines.

Most of the research to date has involved exposure of cells in culture or exposure to stressors that are often extreme and unlikely to occur naturally in the environment (Krause et al. 1986; Heuss-La Rosa et al. 1987; Welch and Mizzen 1988). A few studies have been carried out at contaminant concentrations that are environmentally realistic. Metabolic labeling studies have demonstrated induction of the stress response at concentrations as low as 7.5 ng/l for TBT and 10^{-12} M free cupric ion activity following a 4-h (in vitro) exposure of *Mytilus edulis* hemolymph (Sanders 1990). These concentrations are well within the ranges measured in the environment (Sanders et al. 1983; Beaumont and Budd 1984; Cleary and Stebbing 1985; Bryan et al. 1986).

It also must be demonstrated that the aspect of the stress response measured is persistent over time. An area of confusion in terms of the persistence of the stress proteins involves the distinction between the time course associated with the increased synthesis of stress proteins and the persistence of elevated concentrations of stress proteins in the cell. There is evidence to suggest that although the induced synthesis of stress proteins is transient, the proteins themselves accumulate at high concentrations in tissues for a much longer time (Sanders 1990). Further, data on the kinetics of induction and recovery of the stress response suggest differences specific to the type of environmental insult and severity of exposure (Heikkila et al. 1982). In *Mytilus edulis*, for example, a 29°C heat shock induces the response within a few minutes, while recovery (as determined by a reversion back to translational patterns similar to controls) occurs 12 h after removal of the heat shock (Sanders 1990). The kinetics of induction are longer in response to trace metals, and recovery is not apparent for several days after removal of the stressor, presumably because the metal is still present in the cell. Similar stressor-specific kinetics have been observed in fish and other marine invertebrates.

Research will be needed to examine these relationships in a wide variety of organisms, and a broad range of contaminants, before we will be able to determine how these parameters will affect the response under specific conditions.

Relationship to Higher Levels of Biological Organization

A major advantage of using stress proteins as biomarkers is that because they are involved in protecting the cell from environmentally induced damage, they reflect the cellular physiological state. This attribute provides the potential to be more sensitive than existing organismal indices, and yet be related to adverse physiological conditions in the organism. To evaluate this aspect of the stress response, more research needs to examine the relationships between environmentally induced damage at the cellular and tissue level and the physiological condition of the organism across a broad range of organisms under many different environmental conditions.

A few studies have attempted to begin to examine these relationships. Cochrane et al. (1989) have detected hsp60 accumulation in rotifers upon exposure to TBT and Cu at about 10% of the LC_{50} for each compound. Increased accumulation of hsp60 has also been detected in *Mytilus edulis* exposed to a range of Cu concentrations for 7 d (Sanders et al. 1991). This increase was correlated with scope for growth (SFG), a common organismal index based on bioenergetic changes that measures clearance rate, respiration rate, and assimilation efficiency. The SFG index showed no significant differences among organisms exposed to 0, 1, 3.2, and 10 $\mu g/l$ Cu for one week. However those exposed to 32 and 100 $\mu g/l$ were significantly lower than controls, a condition indicative of growth inhibition. The hsp60 concentration in mantle tissue of replicate organisms did not differ between controls and organisms exposed to 1 $\mu g/l$ Cu. However, organisms exposed to 3.2 $\mu g/l$ of Cu exhibited a significant increase in hsp60 accumulation; a linear relationship existed between the log of hsp60 concentrations and the log of Cu concentration from 3.2 to 100 μg. Thus, the significant increase in the accumulation of hsp60, detected at a Cu concentration one order of magnitude lower than that which caused inhibition of organismal growth, demonstrated the sensitivity of the response as well as its relationship to higher level biological effects. Further, since hsp60 concentration correlated quantitatively with exposure level, it can provide information on the magnitude of impact.

Field Studies

Numerous laboratory and field experiments are needed before we will be able to evaluate the use of a stress protein based assay in environmental monitoring. One of the few studies that addresses in situ expression of the stress response was conducted

Table 5.   Stress Proteins as Biomarkers.

Stress Protein	Inducers	Exposure	Biological Effect
hsp90	heat, metals	++	+
hsp70	heat, metals, UV, xenobiotics, oxidative stress	+++	++
hsp60	heat, metals, xenobiotics, oxidative stress	++	+++
hsp20-30	metals, xenobiotics, heat, steroid hormones	++	++
ubiquitin	heat	+	+
grp group	anoxia, lead, glucose deprivation	+	−

Note. The +, ++, +++ indicate relative usefulness for determining exposure to contaminants and biological effects.

by Kee and Noble (1986), who demonstrated that low molecular-weight stress proteins remain abundant in desert succulents even after completion of high temperature acclimation. Clearly, long-term experiments with diverse species and under a broad range of environmental conditions need to be carried out to determine if elevated accumulation of stress proteins is a chronic condition upon exposure to sublethal concentrations of contaminants.

Conclusions

Stress proteins represent changes in gene expression in response to environmental variables. They are part of the cell's strategy to protect itself from potential damage. From a mechanistic viewpoint, the strategy for selecting potential biomarkers based on the molecular mechanisms underlying protection from environmentally induced damage is sound and offers promise for identifying environmentally relevant biomarker assays.

Much is known about regulation of gene expression of these stress responses, and we are beginning to understand their role in the physiology of normal and perturbed cells. Since very little is known about the environmental relevance of these responses in whole organisms exposed to environmental contaminants, it is currently difficult to evaluate their usefulness as biomarkers.

Preliminary data suggests that the accumulation of stress proteins has potential in environmental monitoring and toxicological screening. Much more research will be required before their usefulness can be accurately evaluated. Research on the stress response is particularly needed in fish, aquatic invertebrates, and plants, in both the laboratory and the field. The mechanisms linking tissue level stress responses and impairment of function at the organismic level will also be important for this evaluation. The persistence of elevated stress protein concentrations in native and transplanted organisms exposed to contaminants in the environment also needs to be examined in depth under different environmental conditions.

PHASE II (CONJUGATING) ENZYMES

A sequence of reactions is often involved in the biotransformation of foreign compounds. The first step (phase I) is frequently an oxidative process whereby a polar moiety, such as a hydroxyl group, is introduced. This reaction is usually catalyzed by the cytochrome P450 monooxygenase system. Subsequent reactions involve a second class of enzyme-catalyzed reactions known as phase II or conjugation reactions (Armstrong 1987; Foureman 1989; James 1987; Buhler and Williams 1988). Phase II enzymes serve to link metabolites to various water-soluble endogenous compounds present in the cell at high concentrations. These reactions generally result in further increases in water solubility and elimination rates, and reduced toxicity of the foreign compound (Buhler and Williams 1988). The most widely studied and perhaps most important of the phase II enzymes are glutathione transferases, UDP-glucuronosyltransferases, and sulfotransferases, which link metabolites to glutathione, glucuronic acid, and sulfate, respectively (Armstrong 1987; Foureman 1989; James 1987; Buhler and Williams 1988). Although not always considered to be a phase II enzyme, epoxide hydrolase catalyzes the addition of a molecule of water to an epoxide formed in phase I (Armstrong 1987; Foureman 1989) and is included in this section.

Because phase II enzymes appear, in some cases, to be influenced by exposure to various foreign compounds (Buhler and Williams 1988; Hammock and Ota 1983; Andersson et al. 1985), there has been some recent interest in their potential usefulness as indicators or biomarkers of pollution exposure and/or effects (Van Veld and Lee 1988; Suteau et al. 1988; Collier and Varanasi 1984; Andersson et al. 1988; Lindström-Seppä 1988). This section provides a brief description of the function, occurrence, and characteristics of the conjugating enzymes just listed, with emphasis placed on current knowledge of the effects of foreign compounds on the activity of these enzymes. Although biomarker programs have focused primarily on aquatic systems (Bayne et al. 1988; Giam 1977; McIntyre and Pierce 1980), much of our understanding of the characteristics and function of drug metabolizing enzymes in aquatic species has evolved from the vast mammalian literature and from our knowledge of known similarities in these enzymes among aquatic species and mammals. Therefore, information on phase II enzymes in mammals is included where appropriate. The section evaluates how well the characteristics of phase II enzymes match the features of an ideal biomarker. Suggestions for future work in this area are also given.

Glutathione Transferases

The glutathione transferases (GST) represent an important family of enzymes named for their role as catalysts for the conjugation of various electrophilic compounds (e.g., epoxides of PAH) with the tripeptide glutathione (Armstrong 1987; Foureman 1989; James 1987; Buhler and Williams 1988; Mannervik and Danielson 1988; Nimmo 1987). These proteins play additional roles in detoxification as well. The soluble GSTs increase the availability of lipophilic toxicants to phase I enzymes (e.g., monooxygenases) by serving as carrier proteins (Tipping and Ketterer 1981; Hanson-Painton et al. 1983; Van Veld et al. 1987). Also, by covalently binding to electrophilic compounds themselves, GSTs reduce the likelihood of these compounds binding to other cellular macromolecules such as DNA (Van Veld et al. 1987; Schelin et al. 1983). In both mammals (Coles and Ketterer 1990) and fish (Varanasi et al. 1987) the susceptibility of different species to chemical carcinogenesis may be modulated by the activity of GST.

GSTs appear to be present in all animal species studied thus far. Considerable species differences exist in activity towards various substrates (Stenerson et al. 1987; James et al. 1979). Although microsomal GSTs have been characterized in mammals, most GSTs are soluble (Armstrong 1987; Mannervik and Danielson 1988). Liver is a major source of GST in vertebrates. In rats, hepatic GST represents approximately 10% of the soluble hepatic protein (Armstrong 1987). In fish liver, it appears that GST accounts for a large fraction of the soluble protein as well (Nimmo 1987). GST activities in several extrahepatic tissues of mammals, fish, and invertebrates have also been described. Activities in these organs are generally lower than those reported for liver and hepatopancreas of these species (James et al. 1979; Mannervik 1985). Sex differences (Mulder 1986) as well as differences during the transition from the fetal to the adult state (Mannervik 1985) have been reported in the mammalian literature (Mulder 1986). There is a limited amount of information regarding sexual, seasonal, or developmental differences in fish. Activities of microsomal and cytosolic GST were not significantly different in sexually mature and immature female plaice, *Pleuronectes platessa* (George and Buchanan 1990). In the same study, cytosolic GST displayed a marked seasonal variation in both male and female fish. During the period of maximal activity (March to April) activity in male fish was approximately double that of females. During the remainder of the year, sexual differences were not apparent.

GSTs have been purified from several species of mammals, elasmobrachs, fishes, and invertebrates (Buhler and Williams 1988; Mannervik and Danielson 1988; Nimmo 1987; Clark 1989). In rats, the multiplicity of GST isozymes arises from dimeric combinations of at least six subunits (Mannervik and Danielson 1988). As in mammals, piscine GSTs exist as a family of dimeric enzymes with subunit molecular weights of

25 to 30 kDa (Nimmo 1987). In fish and mammals, the various isozymes exhibit broad and overlapping ranges of substrate specificities. All forms identified thus far appear to be active towards 1-chloro-2,4 dinitrobenzene (CDNB). Thus, CDNB is often the substrate of choice when total GST activity is being measured. Examples of substrates having environmental relevance include benzo[a]pyrene 4,5-oxide, styrene 7,8-oxide, methyl parathion, and halogenated hydrocarbons.

Induction of GST by various classical inducers of drug metabolism (e.g., PAH, PB, and PCB) has been demonstrated in mammals. The mechanism of regulation of various forms of GST is not clearly understood but appears to be under transcriptional control (Ding and Pickett 1985). In hepatic cytosol from PAH- or PB-treated mammals, GST activity towards various substrates is 1.5 to 2.0 times that of untreated animals (Foureman 1989; Hammock and Ota 1983; Clifton and Kaplowitz 1978). Other investigators have shown that the antioxidants BHA and BHT are more effective inducers of GST. Treatment of rodents with these compounds can result in up to a tenfold increase in total soluble GST activity (Cha and Heine 1982). Because of the role that GSTs play in conjugating reactive epoxide species, induction of these enzymes must be considered to be beneficial, although metabolic activation of some xenobiotics such as ethylene dibromide by GST is also well recognized (Armstrong 1987).

Studies involved with the inducibility of GST activity in various aquatic species have, for the most part, dealt with the PAH-type inducing agents. The studies to date have yielded inconsistent results. For example, intraperitoneal injections of rainbow trout with either β-naphthoflavone (BNF) or Clophen A50 (CL A50) resulted in approximately a twofold induction of hepatic GST activity (Andersson et al. 1985). In another study with the same species, hepatic GST activity increased by a factor of 2.5 as a result of BNF treatment (Goksøyr et al. 1987). However, in that study, treatment of cod with an equivalent dose of BNF did not induce a response. GST activity in the liver of freshwater climbing perch was increased over twofold by a variety of toxicants including phenols (Chatterjee and Bhattacharya 1984). PCB-treated channel catfish exhibited a very slight increase in hepatic GST activity (Ankley et al. 1986), while sheepshead GST appeared unaffected by treatment with either PCB or PBB (James and Little 1981). Several studies indicate that hepatic GST activity of many other fish species are not significantly elevated following intraperitoneal injections of PAH (Collier and Varanasi 1984; James et al. 1988; Balk et al. 1980). Treatment of mummichog with dietary BNF resulted in a threefold increase in intestinal GST but had no effect on hepatic GST (Van Veld et al. 1991). Studies with blue crab (*Callinectes sapidus*) indicate that GST in hepatopancreas responds to BHA and BHT; GST in this organ approximately doubled after repeated dietary exposure to these compounds (Lee et al. 1988). There is no information on the sensitivity of the response of GST to various doses of inducing agents.

Attempts to detect chemically induced levels of GST in fish collected from the field have yielded conflicting results. Levels and activity of both hepatic and intestinal GST were three- to fourfold higher in mummichog collected from a creosote-contaminated site than in those collected from a relatively clean reference site (Van Veld et al. 1991).

At a moderately contaminated site used in the same study, hepatic GST activity in English sole (Collier and Varanasi 1984) and intestinal GST activity in flounder (Van Veld and Lee 1988) collected from PAH-contaminated sites exhibited a trend towards higher activity than those in tissues of fish collected from relatively clean areas. A similar observation was reported in studies with whole mussel (*Mytilus edulis*) (Suteau et al. 1988). Using mussels from the same sites, another researcher found no significant differences in levels of GST in digestive glands between sites (Lee 1988). In that study, a trend towards higher GST activity in hepatopancreas of crab (*Cancer maenus*) collected in contaminated sites was reported. In a 1.5-year study on the effect of an oil spill in a boreal archipelago, hepatic GST activity in perch was apparently unaffected (Lindström-Seppä 1988). Rainbow trout caged at various distances from bleached kraft mill effluent did not exhibit significant changes in hepatic GST activity (Lindström-Seppä and Oikari 1988).

Epoxide Hydrolase

Within the cell, epoxide hydrolase (EH) competes with GST for epoxide substrates produced by the monooxygenase system (Armstrong 1987; Foureman 1989; Thomas and Oesch 1988). Addition of a water molecule to an epoxide group results in the formation of a diol. Although diol formation is generally a detoxification step, this reaction is also involved in the formation of carcinogenic diol epoxides of PAH (Guengerich and Liebler 1985).

EH exists in all animal species studied to date (Thomas and Oesch 1988). The liver of vertebrates and hepatopancreas of invertebrates are particularly rich sources of EH, although the enzyme is present in extrahepatic organs as well (Armstrong 1987; Buhler and Williams 1988; James et al. 1979). The competition between EH and GST for epoxide substrates is important from a toxicological standpoint and is reflected in the substrates (e.g., benzo[a]pyrene 4,5-oxide, styrene oxide, and *trans*-stilbene oxide) commonly used for measuring activity of both of these enzymes. EH is located primarily in microsomal fractions, although cytosolic forms have been described in mammals as well (Armstrong 1987; James and Little 1981; Seidegård and Depierre 1983; Bulleid et al. 1986). In mammals, microsomal EH primarily catalyzes hydration of *cis* epoxides, while the preferred substrates for cytosolic EH are *trans* epoxides (Hammock and Ota 1983).

Species and sex differences in EH activity are apparent in mammalian and aquatic species (James et al. 1979) as well as large variations between individuals of the same species (Stegeman and James 1985). Because of the role of EH in the activation of some chemicals, these differences may be expected to have significant effects on the relative susceptibility of various organisms to carcinogen exposure. For example, in fish exhibiting relatively low hepatic EH activity, the formation of promutagenic diols can be reduced to exceedingly low levels (Stegeman and James 1985).

In mammals, microsomal EH is inducible (less than twofold) by Pb and *trans*-

stilbene oxide but only slightly by PAH-type inducers if at all (Seidegård and Depierre 1983). Cytosolic EH does not appear to be inducible and in some cases has been shown to decrease following exposure to Pb, 3-MC, and PCB (Hammock and Ota 1983).

There is apparently very little evidence for chemical induction of EH activity in aquatic species. Slightly elevated levels of EH activity in mussels exposed to diesel oil were reported in one study (Suteau et al. 1988). Intraperitoneal treatment of several species of estuarine and freshwater fish with 3-MC did not induce EH activity (James et al. 1988). Similarly EH activity in northern pike was unaffected by treatment with this polyaromatic hydrocarbon (Balk et al. 1980). Sheepshead EH activity was unaffected by injections of either PCB or PBB (James and Little 1981). Treatment of stingray and sheepshead with 3-MC resulted in reduced EH activity (James and Bend 1980). EH activity in rainbow trout was unaffected by treatment with either BNF or CL A50 (Andersson et al. 1985).

UDP-glucuronosyltransferases

UDP-glucuronosyltransferase (UDPGT) enzymes catalyze the transfer of the glucuronyl groups from uridine 5'-diphosphoglucuronate to many acceptors including hydroxylated PAH and various endogenous compounds (Armstrong 1987; Foureman 1989; James 1987; Buhler and Williams 1988; Kasper and Henton 1980). Commonly used substrates for measuring UDPGT activity include p-nitrophenol, 1-naphthol, and testosterone. UDPGT enzymes are membrane-bound proteins residing primarily on the luminal side of the endoplasmic reticulum (Armstrong 1987; Foureman 1989). Because of their location, UDPGT activity is often enhanced by treatment of microsomes with surfactants (Armstrong 1987; Foureman 1989). In mammals, several forms of UDPGT have been purified, some of which are inducible by xenobiotics.

UDPGT activity is present in hepatic and extrahepatic tissues of fish as well (James 1987; Lindström-Seppä et al. 1981; Koivusaari et al. 1981). Although primarily studied in liver, UDPGT activity in other fish tissues (e.g., gill and intestine) can exceed that of liver in some cases (Lindström-Seppä et al. 1981; Koivusaari et al. 1981). UDPGT activity in fish is reported to be influenced by sex, season, pH, and temperature differences (Koivusaari and Andersson 1984; Koivusaari et al. 1984).

There are several reports of induction of UDPGT activity resulting from exposure of aquatic species to environmental toxicants. Although only a few chemicals have been tested, the compounds producing effects generally fall into the category of PAH-type inducers. Hepatic UDPGT activity towards p-nitrophenol, 1-naphthol, and testosterone were increased by a factor ranging from 1.4 to 3 following intraperitoneal exposure of rainbow trout to BNF or CL A50 (Andersson et al. 1985). Maximum induction towards p-nitrophenol and 1-naphthol increased for one to two weeks and remained maximal for approximately six weeks. In contrast UDPGT activity towards testosterone peaked in approximately one week and returned to basal levels by the end of the second week. Based upon the different patterns of induction of UDPGT activity towards these substrates, the authors suggest that multiple forms of UDPGT exist in fish. In other studies of the effect of xenobiotics on UDPGT activity, BNF treatment resulted in a twofold induction of hepatic UDPGT activity in rainbow trout but had no effect on activity in cod (Gøksøyr et al. 1987). Basal levels of UDPGT activity in cod were approximately threefold higher than that of trout, indicating large species variation in the activity of this enzyme. In another study, UDPGT activity in liver of rainbow trout approximately doubled three days after treatment with BNF (Pesonen et al. 1987). Activity in kidney of these fish was elevated approximately by a factor of 2.5 two weeks after treatment.

We should mention that while conjugates with glucuronic acid are commonly found in fish, conjugates with glucose are more frequently found in invertebrates (Dutton 1980). In the mollusc, *Arion ater*, UDP glycosyl transferase activity in the enteric gland was reported to be induced by PB (Leakey and Dutton 1975).

Although induction of UDPGT normally results in detoxification of the inducing agent and related compounds, undesirable effects may also result. For example, biliary excretion of estradiol glucuronides was found to approximately double following treatment of rainbow trout with BNF (Förlin and Haux 1985). Reduced levels of plasma steroids following treatment of fish with PCB were reported to occur concomitant with elevated P450 and UDPGT activities (Silvarajah et al. 1985). It is possible that reduced steroid levels may have profound effects on the reproductive success of these and other species.

In field studies, exposure of rainbow trout to bleached kraft mill effluent in one study resulted in approximately a twofold induction of hepatic UDPGT (Andersson et al. 1988). These results are in contrast to another study in which a similar effluent had no effect on UDPGT activity in the same species (Lindström-Seppä and Oikari 1988). In a third study of this type, hepatic UDPGT activity in vendace decreased at the beginning (14–70 d) of exposure then increased after about 120 d (Lindström-Seppä et al. 1989). Perch collected in the vicinity of an oil spill did not exhibit any changes in UDPGT activity over a 1.5-year period (Lindström-Seppä 1988).

While exposure to PAH-type inducers appears in some cases to elevate UDPGT activity, other environmental contaminants have been found to have an inhibitory effect on UDPGT activity. For example, hepatic UDPGT activity was decreased by up to 85% by treatment of rainbow trout with trichlorophenol, pentachlorophenol, and dehydroabietic acid, all of which are common chemicals in pulp mill effluents (Andersson et al. 1988). This type of inhibition could not only reduce the ability of UDPGT to metabolize xenobiotics but may affect its role in metabolizing endogenous compounds as well. For example, elevated bilirubin (a substrate for UDPGT) in blood was reported to be related to inhibition of UDPGT activity (Oikari and Nakori 1982).

Sulfotransferases

Sulfotransferases (ST) are cytosolic enzymes that catalyze the transfer of the sulfuryl group of 3′-phosphoadenosine 5′-phosphosulfate to nucleophiles such as alcohols, phenols, and amines. ST competes with UDPGT for hydroxylated metabolites of PAH and other foreign compounds (Armstrong 1987; Foureman 1989; James 1987; Buhler and Williams 1988). Several forms of ST isolated from mammalian tissues are active towards xenobiotics (Armstrong 1987). There is no evidence, however, that mammalian sulfotransferases are inducible by xenobiotics. Evidence for the presence of ST in aquatic species has been amply demonstrated, yet there is no evidence for induction in these species either (Foureman 1989).

Role as Biomarkers of Preneoplastic and Neoplastic Lesions

In the mammalian field, there has been a great deal of interest in recent reports of altered levels and activity of drug-metabolizing enzymes in various chemically-induced preneoplastic and neoplastic lesions. A common pattern emerging from these reports is that of a reduction in phase I enzymes (e.g., P450 monooxygenase) and a contrasting increase in many GST (Buchmann et al. 1985; Kitahara et al. 1984; Tatematsu et al. 1985), EH (Buchmann et al. 1985; Enomoto et al. 1981), and UDPGT (Yin et al. 1982) activities. The decrease in the primary enzymes (P450s) involved in electrophile production from chemical carcinogens, combined with increases in enzymes involved in detoxification and elimination of electrophiles, suggests an adaptive response to a toxic environment. Further, several investigators have indicated that expression of some phase II enzymes may prove to be useful markers in the clinical diagnosis of neoplastic and preneoplastic tissues (Buchmann et al. 1985; Satoh et al. 1985; Soma et al. 1986).

Studies of phase II enzymes in tumors of nonmammalian species are limited to a few recent reports of fish GST. In contrast to the trend towards elevated GST in mammalian tumors, GST activity was depressed in liver neoplasms of white sucker collected from industrially polluted areas (Hayes et al. 1990). GST in the majority of hepatic neoplasms induced by aflatoxin B, or 1,2-dimethylbenzanthracene was also depressed relative to levels observed in adjacent normal hepatic tissue of rainbow trout (Kirby et al. 1990). In tumors of mummichog collected from a creosote-contaminated site, GST activity was not significantly different from adjacent normal hepatic tissue (Van Veld et al. 1991).

Summary and Conclusions

It is well recognized that phase II enzymes play an important role in the metabolism and clearance of many exogenous and endogenous compounds. At the present time, information on the characteristics, functions, and responses of these enzymes in natural populations of organisms may be insufficient to evaluate their full potential as biomarkers of exposure and/or effects. We know that phase II enzymes are present in most species and that assays for their activity are fast, reliable, and reproducible. Some laboratory and field studies indicate that activities of GST, EH, and UDPGT are influenced by exposure to various environmental contaminants. However, compared to some of the other enzymes and proteins discussed in this chapter, the phase II enzymes have a more restricted use as indicators of either pollution exposure or effects.

In order to evaluate the contribution of phase II enzymes to a biomarker program, several factors need to be considered: sensitivity, variability, baseline levels, and methodology.

Sensitivity. It is obvious that in those cases where a response is reported we are dealing with a two- or at most threefold difference in control versus exposed organisms. This response may seem trivial compared to that of monooxygenase activity, yet a two- or threefold increase in the activity of one or more of these enzymes may have profound effects on the susceptibility of an organism to the harmful effects of environmental contaminants. As previously discussed, increased UDPGT activity may seriously alter steroid metabolism, while decreases could result in bilirubin accumulation. Further, small changes in the activity of EH can have very large effects on the production of promutagenic diols of PAH.

Variability. The subtle response of phase II enzymes to environmental contaminants may be partially or completely "masked" by a variety of factors such as sex, season, temperature, hormonal status, and starvation. We have very little information on the effects of these environmental and physiological variables and how they affect our measurements of phase II activity.

Baseline levels. True baseline levels of activity for these enzymes are as yet unknown. Fish used as "controls" may be already "induced". Fish used in monooxygenase experiments are often kept for as long as 18 months before being treated as controls (Stegman, J., unpublished). There may be unknown inducers of phase II enzymes in the aquatic environment that affect these activities.

Methodology. In aquatic species, studies dealing with the response of conjugating enzymes to toxicants have used total activity measurements. Yet we know that these are families of enzymes. There may be specific forms of phase II enzymes that are very sensitive to exposure to specific contaminants and whose induction may be detected only through the use of antibodies or cDNA probes.

Recommendations

Due to the important role of phase II enzymes in detoxification and excretion of compounds, they should be considered in future biomonitoring programs on the toxic effects of environmental contaminants on biota. Further studies in the following areas of research will undoubtedly improve the usefulness of these enzymes in biomarker programs.

Perform basic chemistry on purified forms. Specific forms of these enzymes need to be purified so basic properties of individual forms can be studied. Antibodies against these purified forms can be produced and used for sensitive detection. Future studies dealing with the induction of specific isozymes of phase II enzymes may provide a more sensitive indicator of exposure compared to measurement of total activity.

Study influencing factors and baseline activity. Basic information needs to be collected on environmental and biological factors influencing the levels and activities of specific phase II enzymes. Determinations need to be made on the influence of these factors on the response of phase II enzymes to environmental contaminants. More information is needed on natural baseline or control activities for these enzymes as well.

Relate enzyme activities to higher levels of organization. More studies are needed on the linkage between effects of altered activity of these enzymes and effects at higher levels of organization (e.g., organ or whole animal).

OXIDANT-MEDIATED RESPONSES

The areas of free radical biology and oxidative stress have received intensive investigation by the biomedical community in recent years. These studies have elucidated endogenous and xenobiotic-mediated mechanisms of oxyradical production, antioxidant defense mechanisms, and deleterious consequences of oxyradical fluxes that outstrip detoxification pathways. In recent years, basic comparative studies of these phenomena in a variety of organisms (particularly aquatic animals and plants) and oxidative stress-related responses in organisms exposed to environmental contaminants suggest a significant potential for this area to yield useful methodologies for biomonitoring. Oxidant-mediated effects with potential utility as biomarkers include adaptive responses (such as increased activities of antioxidant enzymes) and oxidant-mediated toxicities (such as oxidations of proteins, lipids, and nucleic acids; and perturbed tissue redox status). Taken together, results from studies to date strongly motivate continued research in this area. Before practical biomarkers based on free radical biology will be generally available, additional research is required concerning (1) normal physiological and environmental influences on the relevant systems, and (2) effects arising from exposure to a variety of xenobiotics. It is worth noting that oxidative stress comprises a complex set of phenomena, and it is highly unlikely that a single response will provide a general marker for it. Oxidative stress in the context of xenobiotic metabolism is briefly reviewed here, and specific biochemical responses with apparent potential as biomarkers are discussed.

Overview of Oxygen Toxicity, Redox Cycling, and Antioxidant Defenses

Oxygen toxicity is defined as injurious effects due to activated oxygen species, also referred to as oxygen free-radicals or oxyradicals (see reviews by Halliwell and Gutteridge 1985; Kappus 1987; and Di Giulio et al. 1989). Of particular interest are the one, two, and three electron reduction products of molecular oxygen (O_2); these are the superoxide anion radical (O_2^-, hydrogen peroxide (H_2O_2), and the hydroxyl radical ($\cdot OH$). While not a true free radical (i.e., one possessing an unshared electron), H_2O_2 is an important reactive oxygen species in its own right, and can also play a key role in hydroxyl radical production. Under physiological conditions, an important source of $\cdot OH$ is considered to be the iron-catalyzed Haber-Weiss reaction. In this reaction, O_2^- serves as a reductant for a transition-metal oxidation-reduction catalyst such as chelated iron (Equation 1). The reduced metal then reacts with H_2O_2 to yield $\cdot OH$ (Equation 2).

$$O_2^- + Fe^{3+} - chelate \rightarrow O_2 + Fe^{2+} - chelate \qquad (1)$$

$$Fe^{2+} - chelate + H_2O_2 \rightarrow \cdot OH + OH^- + Fe^{3+} - chelate \qquad (2)$$

Numerous endogenous sources of oxyradical production exist. These sources include the activities of various enzymes (such as xanthine oxidase, diamine oxidase, prostaglandin synthase, and glucose oxidase); the multienzyme electron transport chains of mitochondria, microsomes, and chloroplasts; and active phagocytosis by leukocytes (such as neutrophils and macrophages). Of more immediate interest from the standpoint of biomarkers for environmental contaminants, however, is the ability of a number of structurally diverse compounds to enhance intracellular oxyradical production through the process of redox cycling (Figure 2). Redox-active compounds include aromatic diols and quinones, nitroaromatics, aromatic hydroxylamines, bipyridyls (such as paraquat), and certain transition metal chelates. In the redox cycle, the parent compound typically is first enzymatically reduced by an NAD(P)H-dependent reductase (such as NADPH-cytochrome P450 reductase) to yield the xenobiotic radical. This radical donates its unshared electron of O_2, yielding O_2^- and the parent compound; the latter can undergo another cycle. At each turn of the cycle, therefore, two potentially deleterious events have occurred: a reductant has been oxidized and an oxyradical has been produced.

The cytochrome P450-dependent oxidative metabolism of aromatic hydrocarbons (previously described in this chapter) can also be an important source of oxyradicals. Many substrates of this system are not tightly coupled and a significant proportion of the O_2 consumed in their metabolism appear as oxygen free radicals. For example, microsomal benzene metabolism results in the generation of oxyradicals (Johansson and Ingelman-Sundberg 1983). Additionally, the form of cytochrome P450 induced by hydrocarbon and PCB compounds in fish shows inefficient coupling, with 80 to 90% of the reducing equivalent (NADPH) consumed unaccounted for in products (Klotz et al. 1984). The "lost" electrons might plausibly be expected to participate in the production of oxyradicals through expression of the terminal oxidase of microsomal electron transport as H_2O_2 production.

A number of biochemical perturbations have been described as consequences of oxyradicals. These include oxidations of nucleic acids, membrane lipids, and proteins (including metal centers of proteins, as in the case of methemoglobin, described below), as well as perturbed redox status. The "clinical" consequences of these effects are often not clear and vary widely depending, of course, on variables such as target tissue and species. Of great current basic research interest are the roles that oxyradicals may play in initiation and promotion stages of chemical carcinogenesis. Additionally, oxyradicals in acute and chronic toxicities are receiving considerable attention with regard to complex industrial effluents in aquatic systems and oxidizing air pollutants in terrestrial systems (particularly plants). The significance of these findings motivates the development of oxidant-mediated responses as biomarkers.

As already described, oxyradicals arise during various biochemical activities, and indeed can be viewed as a "cost" of aerobic metabolism. As a consequence, all aerobic organisms examined possess antioxidants that serve to detoxify oxyradicals. Oxidative damage is most apparent when the flux of oxyradicals overwhelm antioxidant system capacity. The antioxidant systems of animals and plants are comprised of enzymatic and nonenzymatic components. Key antioxidant enzymes include superoxide dismutase (SOD), catalase (CAT), peroxidase (Px), glutathione reductase (GR), and DT diaphorase. Numerous low molecular-weight antioxidants have been described. Widely distributed water-soluble examples include reduced glutathione (GSH) and ascorbate; important lipid soluble antioxidants include *alpha*-tocopherol and carotenoids such as *beta*-carotene and xanthophylls (Halliwell and Gutteridge 1985).

Both adaptive responses and toxic effects associated with xenobiotic-mediated enhancements of oxyradical production warrant consideration from the standpoint of biomarkers. Adaptive responses include increased activities of antioxidant enzyme activities and/or concentrations of nonenzymatic components. These responses can occur prior to the onset of observable toxicities and may, therefore, provide early warning systems for exposures to prooxidants, including redox-active xenobiotics and photochemical oxidants. Biochemical manifestations of oxidative stress (such as oxidations of lipids, proteins, and nucleic acids, as well as perturbed redox status) may also provide useful biomarkers.

Antioxidants

In the following discussion, we describe selected antioxidants and biochemical toxicities worthy of consideration as biomarkers, including: superoxide dismutases,

catalases, peroxidases, glutathione reductase, glutathione, ascorbate (Vitamin C), and
alpha-tocopherol (Vitamin E).

Superoxide Dismutases

The superoxide dismutase (SOD) enzymes are a group of metalloenzymes that
catalyze the reaction by which O_2^- is disproportioned to yield H_2O_2 (Equation 3):

$$2O_2^- \xrightarrow[2H^+]{SOD} H_2O_2 + O_2 \qquad (3)$$

Fridovich (1986) provides an excellent review of this enzyme, including isozymes,
mechanisms, and assay methods; Rabinowitch and Fridovich (1983) provide additional
information concerning plant SODs and related phenomena. Three distinct types of
SODs, with different metal centers, have been identified:

- CuZnSODs are typically associated with the cytosol of eukaryotes and chloroplasts
  of higher plants.
- MnSODs are found in bacteria and certain organelles (e.g., mitochondria and
  chloroplasts) of higher organisms.
- FeSODs are found in bacteria and a few higher plants.

SODs are considered to play a pivotal antioxidant role; their importance is indicated
by their presence in all aerobic organisms examined. Additionally, the rate of SOD-
catalyzed O_2^- dismutation approximates the diffusion limit, making it among the most
active enzymes described (Fridovich 1986).

The analysis of SOD activity is complicated by the ephemeral nature of the substrate
(O_2^-), which even in the absence of SOD will rapidly react with itself and dismutate.
Most techniques for the measurement of SOD, including those of use in biomonitoring,
are indirect assays. In these assays, a source of O_2^- and an indicating scavenger of O_2^-
are added to the medium (e.g., tissue homogenate) that is being assayed for SOD
activity. Thus, the indicating scavenger competes with endogenous SOD for O_2^-; a unit
of SOD activity is defined as the amount that causes 50% inhibition of reduction of the
scavenger under specified conditions. A number of O_2^- generating systems and
indicating scavengers have been employed. Perhaps the most frequently employed
method in environmental studies is the method of McCord and Fridovich (1969). This
assay employs a xanthine/xanthine oxidase system for O_2^- generation and cytochrome
C as the indicating scavenger. The reduction of cytochrome C can be readily followed
spectrophotometrically. Fridovich (1986) describes improvements to the original
method.

A number of modifications can be employed to distinguish particular components
contributing to the total SOD activity obtained by the previous approach. For example,
cyanide selectively inhibits CuZnSODs, and H_2O_2 inhibits both CuZnSODs and
FeSODs. With the careful use of these inhibitors, the investigator can distinguish
among the activities of SODs with different metal centers. This can be very useful, for

example, if one desires to distinguish mitochondrial SOD activity (MnSOD) from cytosolic activity (CuZnSOD). Additionally, standard methods for SOD analysis, with or without inhibitors, can be readily coupled with electrophoresis to distinguish among specific isozyme activities (Beauchamp and Fridovich 1971).

Considerable care is required to properly calibrate SOD assays for a given material, first with pure SOD, and to check for possible interfering compounds that can inhibit the assay system or give artifactual SOD activity. Dialysis of samples is often recommended. Tandy et al. (1989) provide a recent example of an approach successfully employed to measure total and isozyme-specific SOD activities in a very difficult sample type (conifer needles).

SODs are highly inducible enzymes, and this feature is the basis for their potential as biomarkers. Examples of increased activities of SODs in tissues of organisms exposed to prooxidant contaminants include hepatic tissues of paraquat-exposed mice (Matkovics et al. 1980) and ribbed mussels (*Guekensia demissa*) (Wenning et al. 1988), leaves of SO_2-exposed poplars (*Populus euramericana*) (Tanaka and Sugahara 1980), and leaves of ozone (O_3)-exposed Norway spruce (*Picea abies*) (Castillo et al. 1987) and loblolly pine (*Pinus taeda*) (Richardson et al. 1990). The increased SOD activities observed in the O_3-exposed pine were accompanied by reduced photosynthesis. Matters and Scandalios (1986) demonstrated the ability of paraquat to enhance expression of particular SOD genes in maize. In the only example of SOD measurement in a field study of which we are aware, Roberts et al. (1987) found more consistent inductions of SOD versus AHH activities in feral spot (*Leiostomus xanthurus*) inhabiting the highly PAH-polluted Elizabeth River in Virginia.

As with the bulk of oxidative-stress associated phenomena, the utility of SOD as a biomarker remains largely speculative and in need of field testing. However, available information motivates continued research. Of particular interest is the study of responses by different isozymes, especially those associated with particular organelles. This approach may yield a powerful outlet for monitoring oxidative responses at the subcellular level in organisms exposed in vivo, including field studies.

As indicated in Equation 3, the product of SOD activity is H_2O_2, which is itself an important active oxygen species. H_2O_2 is detoxified by two enzymes: catalases and peroxidases.

Catalases

Catalases (CAT) are hematin-containing enzymes that facilitate the removal of H_2O_2 by the following reaction (Equation 4):

$$2H_2O_2 \xrightarrow{\text{CAT}} 2H_2O + O_2 \tag{4}$$

Diesseroth and Dounce (1970) and Frew and Jones (1984) provide useful overviews of CAT. Compared to some peroxidases that can reduce various lipid peroxides as well as H_2O_2, CAT can decompose only H_2O_2. The bulk of CAT activity is associated with the peroxisomes or microbodies that primarily function in various unresolved aspects of fatty acid metabolism (see Fahimi and Sies 1987). During peroxisomal fatty acid oxidation, H_2O_2 is produced as a by-product; CAT apparently acts to scavenge this H_2O_2. Of great current interest are a class of nonmutagenic carcinogens referred to as peroxisome proliferating compounds that include certain hypolipidemic drugs and phthalate plasticizers (Reddy and Lalwani 1983). The mechanisms by which these compounds foster cancer in rodents are unclear; however, a role for oxyradicals has been proposed (Reddy and Lalwani 1983). These compounds typically elicit very marked increases (10 to 15 times greater) in activities of various fatty acid oxidases that generate H_2O_2 and smaller increases (2 to 3 times greater) in CAT activity. Thus a net increase in the flux of H_2O_2 may result that could initiate or promote carcinogenesis; again, this mechanism remains speculative. However, these phenomena provide additional motivation for examining CAT as a biomarker.

A commonly employed assay for the measurement of CAT is that of Bergmeyer (1955). This method follows the disappearance of exogenous H_2O_2 spectrophotometrically.

Increased CAT activities have been observed in paraquat-exposed mice (Matkovics et al. 1980) and ribbed mussels (Wenning et al. 1988). Additionally, significant increases (by a factor of 1.5 to 2.5) in channel catfish (*Ictalurus punctatus*) exposed to two complex mixtures have been observed. In one case, a bleached kraft mill effluent, the increased CAT activities were accompanied by greater increases (by a factor of 2 to 7) in activities of two fatty oxidases, results consistent with peroxisomal proliferation (Mather-Mihaich and Di Giulio 1991). In another study involving a sediment highly contaminated with PAH, no consistent differences in fatty acid oxidase activities accompanied increased catalase activities (Di Giulio and Habig, in review).

Because CAT is localized in the peroxisomes of most cells and involved in fatty acid metabolism, changes in its activities may often be difficult to interpret. However, CAT also occurs in erythrocytes independently of peroxisomes in most vertebrates where it appears to act in concert with glutathione peroxidase and methemoglobin reductase to counter the oxidative stress to which these cells are prone (Diesseroth and Dounce 1970). Therefore, CAT activities in erythrocytes may provide a more appropriate marker for oxidant exposures to vertebrates, when these cells are reasonable targets, than tissues in which CAT is principally involved in peroxisomal fatty acid metabolism. Gabryelak and Klekot (1985) observed a transitory increase erythrocyte CAT activity in crucian carp (*Carassius carrasius*) exposed to paraquat.

Another aspect of CAT activities meriting considerable attention is the topic of peroxisome proliferation in nontraditional models, such as fish. A vital question concerns whether synchronous increases in CAT and fatty acid oxidases, which serve as reliable markers for peroxisome proliferation in rodent models, can serve as biomarkers for an important group of nongenotoxic carcinogens in the aquatic environment. Recently, Yang et al. (1990) reported peroxisomal proliferation in rainbow trout exposed to a model proliferator in rodents, ciprofibrate. Considerable basic research is needed in this area.

Peroxidases

Peroxidases comprise a large and diverse array of enzymes that, collectively, reduce a variety of peroxides to their corresponding alcohols (Frew and Jones 1984). While catalase employs one molecule of H_2O_2 as the donor in the reduction of another H_2O_2 molecule, peroxidases employ other reductants. In animals, the principal peroxidase is a selenium-dependent tetrameric cytosolic enzyme that employs GSH as a cofactor; Se occurs (as selenocysteine) at the functional site of each tetramer (Flohe 1982). The net reaction catalyzed by glutathione peroxidase (GPx), with H_2O_2 as the substrate, is as follows (Equation 5):

$$2GSH + H_2O_2 \xrightarrow{\text{GPx}} GSSG + 2H_2O \tag{5}$$

Similarly, GPx can also catalyze the reduction of organic hydroperoxides to their corresponding alcohols (i.e., ROOH ——> ROH); this is considered an important mechanism for halting lipid peroxidizing chain reactions (described in the "Lipid Peroxidation" section of this chapter). Glutathione-S-transferase (GST) enzymes can also employ GSH in the reduction of a broad range of organic hydroperoxides, but cannot reduce H_2O_2. This peroxidatic activity by GST is sometimes referred to as "selenium-independent peroxidase", although GST is not a true peroxidase. However, GST can apparently serve a significant peroxidatic function, particularly in Se-restricted animals (Flohe 1982).

In plants, an ascorbate-dependent peroxidase (AsPx) appears to play a key antioxidant function, particularly in H_2O_2-scavenging in chloroplasts (Hossain et al. 1984; Alscher and Amthor 1988). In this reaction, AsPx catalyzes the reduction of H_2O_2 to H_2O using ascorbate as the reductant. Ascorbate is concomitantly oxidized in either a one or two electron transfer to monodehydroascorbate or dehydroascorbate, respectively; depending on the oxidized form, ascorbate is regenerated by either monodehydroascorbate reductase or dehydroascorbate reductase. In the case of monodehydroascorbate, the enzyme is also referred to as ascorbate free-radical reductase, AFR. GSH also plays a role here, by virtue of its role as a cofactor in dehydroascorbate reductase activity, which yields GSSG in addition to ascorbate. NADPH serves as the cofactor for AFR activity (see Gaspar et al. (1982) for a more general review of plant peroxidases.)

Currently employed techniques for the quantitation of peroxidase activities include those of Nakano and Asada (1981) for plant AsPx, and Paglia and Valentine (1967) for animal GPx. In animals, it is often of interest to distinguish between Se-dependent and Se-independent activities, since the latter comprises certain GSTs that may be of interest in toxicological and biomarker studies. This distinction can be accomplished by the use of different substrates; i.e., H_2O_2 and an organic hydroperoxide such as t-butylhydroperoxide (Lawrence and Burk 1978).

Many studies have demonstrated enhanced peroxidase activities in plants exposed to oxidizing air pollutants. Examples include O_3-exposed soybeans (Curtis et al. 1976) and Norway spruce (*Picea abies*) (Castillo et al. 1987), SO_2-exposed Scots pine (*Pinus sylvestris*) (Schulz 1986); SO_2- and NO_2-exposed peas (Horsman and Wellburn 1975), and SO_2-, NO_2-, and O_3-exposed peas (Mehlhorn et al. 1987). Additionally, plant peroxidases have been employed in field studies to "map" air pollutants (Keller 1974; Petit 1984). Gaspar et al. (1982) provide a useful discussion of plant peroxidases as sensitive biomarkers for pollutants. Among the antioxidants discussed, Px (particularly AsPx) is probably the closest to a currently usable biomarker, particularly in the context of oxidizing air pollutants. However, care must be exercised since peroxidases have also been shown to respond to a variety of factors unrelated to pollutants (see Gaspar et al. 1982). For the early detection of oxidant response in plants, AsPx may be most reliably utilized in combination with other oxidant-related indices, but further research is needed in this area.

The utility of GPx as a biomarker in animals has apparently received relatively little attention. However, marked increases in GPx activities from various tissues of different carp species exposed to paraquat have been reported (Matkovics et al. 1984; Gabryelak and Klekot 1985). Additionally, O_3-exposed rats displayed marked elevations in lung GPx (Chow and Tappel 1972). However, available information suggests that GPx in animals may be less responsive to prooxidants than plant AsPx. More research is required to determine the utility of GPx as a biomarker.

Glutathione Reductase

Although perhaps not as primary an antioxidant defense as the enzymes previously described, glutathione reductase (GR) merits attention because of its importance in maintaining proper GSH/GSSG ratios in the face of oxidative stress and reports of its inducibility. Additionally, changes in the activity of this enzyme may be very important in interpreting GSH/GSSG data, which may provide a very useful biomarker. GR catalyzes the following reaction (Equation 6):

$$GSSG + NADPH + H^+ \xrightarrow{\ GR\ } 2GSH + NADP^+ \qquad (6)$$

Staal et al. (1969) and Staal and Veeger (1969) provide descriptions of the properties and mechanisms of GR from human erythrocytes; GR from pea chloroplasts is described by Kalt-Torres et al. (1984).

Goldberg and Spooner (1983) provide a commonly employed technique for measuring GR, and Esterbauer and Grill (1978) describe a method for plants, including conifers.

Responses of GR to pollutants have apparently received little attention. However, marked increases in GR have been observed in O_3-exposed rats (Chow and Tappel 1972) and peas (Mehlhorn et al. 1987). Tanaka et al. (1988) demonstrated an increased production of GR with Western blots in O_3-exposed spinach. Additional research in this area is clearly warranted, particularly in conjunction with other aspects of glutathione metabolism, such as the maintenance of adequate GSH concentrations for Px and GST detoxification pathways.

Glutathione

Reduced glutathione (GSH), a tripeptide composed of glutamic acid, cysteine, and glycine, has received considerable attention in terms of its biosynthesis, regulation, and various intracellular functions (see reviews by Kosower and Kosower 1978; Reed and Beatty 1980; Sies et al. 1980; Rennenberg 1982; Larsson et al. 1983; and Meister 1984). Among these functions are two contrasting roles in detoxifications, as a key conjugate of electrophilic intermediates, principally via GST activities in phase II metabolism previously described, and as an important antioxidant. The antioxidant functions of GSH in the activities of GPx and AsPx have already been discussed. In addition, GSH can apparently serve as a nonenzymatic scavenger of oxyradicals (Halliwell and Gutteridge 1985).

Increased fluxes of oxyradicals might reasonably be expected to alter GSH status and/or metabolism in several ways, and experimental results demonstrate the occurrence of these effects. Perhaps the most obvious direct effect is a decrease in the ratio of GSH to GSSG, due to either direct radical scavenging or increased peroxidase activity. This effect could also occur indirectly due to reduced availability of NADPH (necessary for GR activity) resulting from oxidations in the first step of the redox cycle (Figure 2). (These changes also fall under the subject of perturbed redox status and are discussed later in the subsection "Biochemical Indices of Oxidative Damage".) Examples of enhanced GSH oxidations include O_3-exposed beans (Guri 1983) and paraquat-exposed rat liver (Brigelius et al. 1981). Additionally, Adams et al. (1983) found plasma GSSG concentrations to provide a sensitive index of whole body oxidative stress in the rat.

Alternatively, normal ratios of GSH to GSSG can be maintained due to increased activities of glutathione reductase. Also, GSH concentrations can actually be increased due to increased synthesis. This effect appears to have received little attention among medical toxicologists. In fact, GSH synthesis is considered to be tightly regulated in mammals via feedback inhibition by GSH on a rate-limiting synthetic enzyme, γ-glutamyl cysteine synthetase (Richman and Meister 1975). However, reports of the effects of prolonged exposures (more than a few days) to xenobiotics on GSH metabolism in mammals are difficult to find. Several studies in other animals and plants have demonstrated increased GSH concentrations that were maintained beyond the time likely to be due to an "overshoot" (i.e., a transient increase in GSH synthesis following depletion). Examples of studies indicating contaminant-induced enhancements of GSH synthesis include striped mullet (*Mugil cephalis*) exposed to cadmium or fuel oil (Thomas et al. 1982; Thomas and Wofford 1984), sediment-exposed channel catfish (Di Giulio and Habig, in review), and several species of SO_2-exposed plants (Grill et al. 1979; Chiment et al. 1986; Mehlhorn et al. 1986; Alscher et al. 1987).

Several methods have been employed for the measurement of glutathione. Care must be taken to select glutathione-specific assays if this specificity is required, as in determining ratios of GSH to GSSG. Many protocols employing Ellman's reagent (which reacts with -SH groups) actually measure all nonprotein sulfhydryls. In most tissues of mammals and plants examined, glutathione comprises the great majority (greater than 90%) of nonprotein sulfhydryls, but this relationship is unknown for most organisms (Kosower and Kosower 1978; Chiment et al. 1986). A widely employed technique for the spectrophotometric analysis of GSH and GSSG is described by Griffith (1980). This technique achieves specificity by employing NADPH and GR to drive the assay. This assay has been slightly modified by Smith (1985) for use in plants. Highly sensitive and accurate HPLC techniques, such as described by Fariss and Reed (1987), have also been developed for GSH and GSSG measurements. When feasible, these techniques are often preferred. The HPLC techniques, however, are far more time-consuming than the spectrophotometric assays, and consequently may not be as appropriate for many biomonitoring applications.

The key role played by glutathione in detoxifications and the responsiveness of this system to xenobiotics motivates continued research in this area. Research aims include elucidating basic mechanisms of xenobiotic metabolism as well as developing useful biomarkers.

Ascorbate (Vitamin C)

Ascorbate is an important water-soluble antioxidant (see the review by Seib and Tolbert 1982; Halliwell and Gutteridge 1985). In addition to its antioxidant role, ascorbate is a cofactor for the enzymes proline hydroxylase and lysine hydroxylase (which are involved in collagen biosynthesis) and dopamine beta-hydroxylase (which converts dopamine into noradrenaline). Plants and many animals can synthesize ascorbate, but a number of animals (including humans) lack a necessary synthetic enzyme and require vitamin C in their diet. Besides the key role of ascorbate in plant peroxidase activity, (previously described), ascorbate can directly scavenge some oxyradicals, including ·OH, to yield the monodehydroascorbate or semidehydroascorbate radical. This radical can react with itself to produce ascorbate and dehyroascorbate. Both semidehydroascorbate and dehydroascorbate can also be enzymatically reduced to ascorbate in animals and plants by their respective reductases.

While considered an antioxidant, ascorbate is also a prooxidant. Its activity as a cofactor for the biosynthetic enzymes mentioned is apparently due to its ability to act as a reducing agent and maintain the metal centers (iron or copper) in their reduced states, as required for catalytic activity. Similarly, ascorbate can replace O_2^- in Equation 1 and reduce Fe^{3+} to Fe^{2+}, which can then produce ·OH via Fenton chemistry (Equation 2). The importance of this activity in vivo is unclear; however, under certain conditions, such as iron-overloading, the prooxidant role of ascorbate appears significant (Halliwell and Gutteridge 1985).

Spectrophotometric and HPLC methods for measuring ascorbate are described by Carr et al. (1983). A spectrophotometric method developed specifically for plant tissues is provided by Foyer et al. (1983).

Studies addressing the usefulness of ascorbate as a biomarker are very limited. Increased ascorbate concentrations have been observed in conifers exposed to oxidizing air pollutants (Barnes 1972; Mehlhorn et al. 1986). The utility of ascorbate as a biomarker is limited to plants and to animals that can synthesize it. Additionally, it may often be very important to know the ascorbate content of animal diets employed experimentally in studies of oxidative stress, and perhaps the ascorbate status of field-collected organisms as well. In a large-scale field study in Sweden, perch (*Perca fluviatilis*) inhabiting waters receiving bleached kraft mill effluents consistently displayed higher ascorbate concentrations than fish from a reference site (Andersson et al. 1988).

Alpha-tocopherol (Vitamin E)

Vitamin E is a lipid-soluble antioxidant that is synthesized by plants but required in the diets of animals (see the review by De Duve and Hayaishi 1978; also Halliwell and Gutteridge 1985). This compound appears to play a major role in protecting membranes from lipid peroxidation (described below), particularly by reacting with lipid peroxy radicals to produce the relatively stable vitamin E radical, thus serving as a terminator of lipid peroxidizing chain reactions. The vitamin E radical can be reduced back to vitamin E by ascorbate; however, the significance of this reaction in vivo apparently has not been determined.

Vitamin E is most reliably measured by HPLC. Burton et al. (1985) provide a useful protocol for animal tissues, while Wise and Naylor (1987) describe an HPLC method for plant tissues. A fluorometric method is provided by Taylor et al. (1976).

Again, the utility of vitamin E as a biomarker has received little attention; Mehlhorn et al. (1986) observed increased concentrations of *alpha*-tocopherol in air pollutant-exposed conifers. Its direct application as a biomarker is probably restricted to plants; however, as in the case of ascorbate, measures of animal tissues may sometimes be useful.

Biochemical Indices of Oxidative Damage

A large number of biochemical and physiological effects have been associated with increased fluxes of oxyradicals. Some biochemical perturbations that seem particularly promising from the standpoint of biomarkers are lipid peroxidation, DNA oxidation, methemoglobinemia, and redox status.

Lipid Peroxidation

The oxidation of polyunsaturated fatty acids is a very important consequence of oxidative stress and has been investigated extensively (see reviews by Chan 1987;

Horton and Fairhurst 1987; Kappus 1985; Halliwell and Gutteridge 1985). The process of lipid peroxidation proceeds by a chain reaction and, as in the case of redox cycling, demonstrates the ability of a single radical species to propagate a number of deleterious biochemical reactions. Lipid peroxidation is initiated by the abstraction of a hydrogen atom from a methylene group ($-CH_2-$) of a polyunsaturated fatty acid (represented as "LH"); oxyradicals, particularly ·OH, can readily perform this abstraction, yielding the lipid radical, L·. The carbon-based L· radical tends to be stabilized by molecular rearrangement to a conjugated diene radical. The L· radical reacts readily with O_2 to produce the peroxy radical, LOO·. This radical can readily abstract a hydrogen from another LH, yielding a lipid hydroperoxide, LOOH, and a new L·, which can then continue the chain reaction, propagating additional LOOH and L·. LOOH is relatively stable in isolation, but can react with transition metal complexes (including cytochrome P450) to yield alkoxyl radicals (LO·). The lipid peroxidation chain reaction can be terminated by two lipid radicals reacting to form a nonradical product, or by quenching by a radical scavenger such as vitamin E. Note that the actual chemistry of lipid peroxidation and associated productions of various free-radical species is extremely complex and beyond the realm of this discussion. The interested reader is referred to Kappus (1985), Horton and Fairhurst (1987), and references therein.

Numerous studies have demonstrated enhancements of lipid peroxidation in various tissues from several different species exposed in vivo to a diverse array of chemicals. Examples include paraquat-exposed carp (Gabryelak and Klekot 1985) and ribbed mussels (Wenning et al. 1988); mullet exposed to acetaminophen, CCl_4, cadmium, and PCB (Wofford and Thomas 1988); catfish exposed to a PAH-contaminated sediment (Di Giulio and Habig, in review); and ozone-exposed loblolly pine (Richardson et al. 1990). Lipid peroxidation appears to have considerable potential as a biomarker. However, lipid peroxidation can occur as a consequence of cellular damage due to a variety of insults, and its enhancement alone is insufficient to indicate that free-radical-mediated mechanisms are the basis for the toxicity of a compound producing this effect (Kappus 1985).

Various techniques are available for measuring or indexing lipid peroxidation; each has its good and bad points. Halliwell and Gutteridge (1985) and Kappus (1985) provide lucid discussions of the most common techniques, including their advantages and disadvantages. Related discussions, together with a number of specific techniques are contained in Packer (1984) and Packer and Glazer (1990). As these authors remark, the investigator is urged to consider carefully the nature and design of a study before selecting a particular assay, or preferably, suite of assays, for lipid peroxidation.

The most widely employed assay for lipid peroxidation is the thiobarbituric acid (TBA) test for malonaldehyde (MDA, a byproduct of lipid peroxidation). Tanizawa et al. (1981) describe a relatively simple and sensitive assay for MDA. MDA measurements as an index of lipid peroxidation have produced some concerns, such as interfering compounds, sensitivity of the assay to minor changes in assay conditions, and in vivo metabolism of MDA. However, when carefully performed with proper consideration given to its limitations, this method remains a useful index for lipid peroxidation. Its simplicity relative to most alternatives contributes considerably to its utility as a biomarker. When possible, however, results from TBA analyses should be corroborated by a more specific assay, such as the HPLC method described by Esterbauer et al. (1984) or Draper and Hadley (1990). Another approach that merits attention is the measurement of fluorescent damage products arising from lipid peroxidation (Dillard and Tappel 1984). These products are Schiff bases that result from conjugation of MDA with endogenous amines. They are more stable than MDA itself and, therefore, perhaps more reliable as markers for lipid peroxidation in vivo than MDA.

DNA Oxidation

The role that free-radicals, including oxyradicals, play in chemical carcinogenesis is currently a topic of intense research interest. The phase I metabolism of some hydrocarbon procarcinogens (e.g., benzene, benzo[a]pyrene, and anthracene) produce organic free radical intermediates. These may play important roles in DNA alterations, including adduct formation, associated with some hydrocarbons (Nagata et al. 1982; Irons and Sawahata 1985; Cavalier and Rogan 1990). Furthermore, oxyradicals have been implicated in the metabolism of PAH procarcinogens into adduct-forming products (Morreal et al. 1968; Georgellis et al. 1987). As previously described, both cytochrome P450 dependent metabolism of hydrocarbons and redox cycling of a variety of compounds (including quinone derivatives of PAH) can generate oxyradicals (see Chesis et al. 1984). These oxyradicals can also be genotoxic. For example, oxyradicals can produce strand scissions and chromosomal breakage (Imlay and Linn 1988). However, these radical mediated genotoxic effects, adduct formation and strand scission, are not specific for free radicals. Other mechanisms for these effects, and methods for detecting them in environmental samples, are described in the chapter on DNA alterations.

Recently, methods for detecting DNA modifications resulting from oxyradicals have been described. These products appear generally to result from ·OH attack at various sites of DNA bases that result in the production of hydroxylated bases (Dizdaroglu and Bergtold 1986; Pryor 1988). The methods of most promise for monitoring these products in environmental samples employ HPLC separation and electrochemical detection of hydroxylated bases, such as thymine glycol (Cathcart et al. 1984) or 8-hydroxy deoxyguanosine (8-OHdG) (Floyd et al. 1986). These methods are extremely sensitive, but relatively involved. Refinements to this methodology are likely to enhance its feasibility as a biomarker in the near future. This is potentially a very powerful technique and merits scrutiny.

Cundy et al. (1988) employed measures of 8-OHdG in human urine as an index of oxidative DNA damage. Also, Floyd et al. (1989) observed elevated concentrations of 8-OHdG in chloroplasts from O_3-exposed beans. Malins et al. (1990) detected 2,6-diamino-4-hydroxy-5-formamidopyrimidine in DNA from hepatic neoplasms in English sole collected from a polluted site in the Puget Sound. This base, apparently derived from ·OH attack on the C-8 position of guanine, was not observed in adjacent, non-neoplastic tissue nor in nonneoplastic livers of sole from a reference site.

Methemoglobinemia

Under normal conditions, a very small proportion (less than 1 to 2%) of the hemoglobin in vertebrate erythrocytes exists as methemoglobin (MetHb). In MetHb, the iron centers of the heme moieties exist in the oxidized (Fe^{3+}) state; this prevents the molecules from serving their normal function of O_2 binding and transport. The enzyme methemoglobin reductase catalyzes the reduction of Fe^{3+} to Fe^{2+}, thereby regenerating functional oxyhemoglobin (Stern 1985).

Increased concentrations of MetHb (methemoglobinemia) provide for a relatively sensitive early indication of oxidative damage in red blood cells. Prolonged or more severe oxidative damage appears as Heinz bodies (denatured hemoglobin precipitates), lipid peroxidation, and finally lysis. Several compounds have been shown to enhance MetHb formation, including a number of aromatic hydrazines, quinones, nitrite, and some transition metals, particularly copper. The mechanisms underlying MetHb formation by these compounds has been shown to proceed by the production of oxyradicals that facilitate the oxidation of heme-bound Fe^{2+}.

The spectrophotometric analysis of MetHb concentrations in whole blood samples is straightforward; Hegesh et al. (1970) describe this technique. It is important to note that reliable measurements of MetHb require freshly drawn, unfrozen blood samples.

Very few studies have examined contaminant-induced changes in MetHb concentrations in vertebrates other than mammals. However, available information support careful consideration of this variable as a useful, readily measured biomarker in fish. For example, two naphthoquinones elicited marked increases in MetHb concentrations in channel catfish at toxicant concentrations far below the LC_{50} of each of these compounds (Andaya and Di Giulio 1987). Additionally, in a Swedish field study of paper mill effluents, perch inhabiting effluent-contaminated waters exhibited elevated MetHb levels compared to fish from a reference site (Andersson et al. 1988).

Redox Status

The effects just described (lipid peroxidation, DNA oxidations, and methemoglobinemia) involve oxidized products resulting from direct reactions with oxyradicals. Perhaps a more fundamental effect of oxyradical-generating compounds, however, is their impact on the redox status of cells or tissues. Healthy cells regulate and maintain a critical pool of reducing equivalents, particularly as NADH and NADPH, which ultimately drive all energy-requiring metabolic processes. GSH (previously described) also comprises a key reductant that is employed in regulating protein synthesis and transport, as well as in detoxifications. In the healthy cell, therefore, ratios of reduced to oxidized glutathione (GSH:GSSG) is typically very high, greater than 10:1 (Kosower and Kosower 1978).

Oxyradical-generating compounds, however, can impose a drain on intracellular reducing equivalents with potentially profound consequences on a variety of metabolic processes (Halliwell and Gutteridge 1985). This drain can arise directly, for example, as a consequence of electron abstractions from pyridine nucleotides by redox-active compounds; this phenomenon is facilitated by various cytochrome reductases (Figure 2). The consumption of GSH due to the direct scavenging of oxyradicals or as a cofactor for Px activities represents another important drain; NADPH must be oxidized to maintain reduced glutathione (GSH) levels. More indirectly, oxidative stress can impose a drain on the reductant pool as a consequence of the energetic costs of mounting a defense against an increased flux of oxyradicals (i.e., biosynthesis of antioxidants).

Few studies have directly addressed the issue of effects of prooxidants on redox status. We previously discussed oxidant-mediated effects on the ratios of GSH to GSSG. Results from studies of among the most intensively investigated redox-active compounds, paraquat, are instructive. Oxyradical-mediated lipid peroxidation has been proposed as the key mechanism underlying paraquat toxicity (Bus et al. 1976). However, other studies have failed to support the linkage between lipid peroxidation and cellular toxicity (Stacey and Klaassen 1981). It appears that the drain paraquat imposes on reducing equivalents such as NADPH is more fundamental to the mechanism of the toxicity of this compound (Keeling and Smith 1982).

Methods for measuring GSH and GSSG concentrations were previously described. From the standpoint of assessing redox status, however, measures of pyridine nucleotides are more fundamental. A number of methods for measuring ratios of NAD(P) to NAD(P)$^+$ are available. The spectrophotometric method of Lowry et al. (1961) has been employed widely. More recently, HPLC techniques, such as that described by Stocchi et al. (1985), have been developed. Such HPLC techniques appear to be more reliable and are generally preferred.

The effect of prooxidants on redox status is an important area of research, particularly considering the central role played by reductants in cellular metabolism.

Variables Influencing Oxidative Processes

Relatively little attention has been given to various biological (e.g., sex, age, development, reproduction, and nutrition) and environmental (e.g., temperature, salinity, dissolved oxygen content, season, and photoperiod) variables on oxidant-mediated exposures in species of interest to biomonitoring. However, the medical science literature illustrates a number of differences in the antioxidant defense system and oxidative responses due to age and nutritional status in mammalian models (Yam et al. 1978a; Frank et al. 1978; Wills 1985). Additionally, elevated atmospheric O_2 concentrations have been employed as probably the standard method of imposing oxidative stress in mammalian models. These studies have generally indicated marked increases in antioxidants due to the exposures (Stevens and Autor 1977; Yam et al. 1978b).

A few studies have examined nutritional influences on antioxidant enzyme activities in fish. Radi et al. (1985) attempted to determine for several species the influence of feeding strategy on enzyme activation, but yielded inconclusive results. Knox et al. (1981) observed altered activities of CuZnSOD and MnSOD in rainbow trout fed Mn-deficient diets, and Heisinger and Dawson (1983) reported reduced GPx activities in Se-deficient brown bullheads (*Ictalurus nebulosus*). Additionally, several studies have investigated effects of varying dietary vitamin C and E concentrations on oxidative processes such as lipid peroxidation and methemoglobin in fish (e.g., Hung and Slinger 1982; Wilson et al. 1984; Wise et al. 1988). These studies have generally indicated a greater propensity for tissues from antioxidant vitamin-restricted fish to undergo oxidations compared to tissues from fish adequately provided with these vitamins in their diets.

Several studies have also examined effects of environmental variables on antioxidants. Kobayashi (1955) examined the effects of salinity in CAT activities in *Oncorhynchus masou*; CAT activities displayed a transitory rise when fish were transferred from fresh to salt water. In a rare study demonstrating antioxidant adaptation to hyperoxia in aquatic animals, Dykens and Shick (1982) reported SOD activities in sea anemones (*Anthopleura elegantissima*) with and without symbiotic dinoflagellates (*Symbiodinium microadriacticum*). Anemones containing symbionts displayed SOD activities that were two orders of magnitude greater than those without symbionts, apparently due to the hyperoxic intracellular conditions produced in the anemones by the algae.

A marked seasonal trend in antioxidants has been observed in several conifer tree species. Increased activities of SOD and AsPx and increased concentrations of GSH and ascorbate have been observed during the transition from late summer to early winter (Madamachi et al. 1991; Richardson et al. 1990; Tandy et al. 1989). These changes are thought to be adaptations associated with winter hardening and appear to be controlled by photoperiod rather than temperature.

The influence of significant biological and environmental variables on antioxidants and oxidative damage are important research needs in this area. In aquatic animals, dissolved oxygen and temperature are variables that are particularly likely to influence oxidative processes. The importance of seasonality on plant antioxidants has been described, and in the case of conifers, needle age. Drought, soil nutrient status, and pathogenic organisms also are key variables that may influence oxidative processes in plants.

Summary and Conclusions

Oxidant-mediated responses provide a potentially powerful, mechanistically-based methodology for assessing organismal responses to a diverse set of prooxidant chemical pollutants. This set of chemicals includes oxidizing air pollutants (such as O_3, NO_x, and SO_2), other direct acting oxidants (such as H_2O_2, organic peroxides, and water-borne nitrite and chlorine); and redox-active compounds (including aromatic diols and quinones, nitroaromatics, azo dyes and hydrazines, bipyridyl herbicides, and transition metals). The methodology provided by oxidant-mediated responses comprises both adaptive responses (i.e., antioxidant inductions) and clearly deleterious effects (such as oxidations of various biochemical and cellular structures).

The diversity of compounds that elicit oxidant-mediated responses entails both advantages and disadvantages for this approach to biomonitoring. Without analytical chemistry data, little can be surmised concerning the chemical agents responsible for observed responses. However, when chemical data are available, one can make reasonable predictions concerning causative compounds, and with further testing, may critically test these predictions.

At least in the near future, the greatest utility of this methodology as a useful marker for monitoring environmental health relates to complex mixtures of industrial effluents entering aquatic systems and photochemical oxidants emitted into the atmosphere. To date, the most promising results from this approach are those that have emerged from studies of biochemical effects of air pollutants on higher plants. This may reflect in part the relative absence of analogous studies in aquatic systems. Available results, as well as the underlying mechanistic basis, strongly motivate research in this field.

Recommendations

In order for this approach to reach its potential as a biomarker for environmental contamination, considerable research is needed. First, much basic research is needed to elucidate fundamental processes such as glutathione metabolism; tissue, cellular, and subcellular distributions of antioxidants; and inherent inducibilities of antioxidants in a selected array of organisms. Dose-response studies should be performed with a carefully selected set of model prooxidants in selected organisms. Research in methodologies is needed to discern or develop the most appropriate assays for certain responses in the context of biomonitoring. The greatest needs here appear to be in the areas of SOD, lipid peroxidation, and DNA oxidations. Considerable research is needed to determine natural fluctuations in oxidant-mediated responses due to organismal variables such as species, genetics, sex, reproductive status, age, and nutritional status; and environmental variables such as temperature, dissolved oxygen, salinity, and photoperiod. Finally, field validation studies in various aquatic and terrestrial settings are required to test the utility of this approach for biomonitoring. As with most biomarkers described in this book, it is likely that this methodology will achieve its greatest utility when performed in conjunction with other biomarkers.

HEME AND PORPHYRINS

Utilization of chemically induced alterations in the heme pathway as biomarkers of both exposure and effect has been the subject of intense study over the past 15 years in mammalian toxicology. These studies have delineated a number of chemical-specific responses that have proven extremely useful in early detection of low-dose chemical effects in mammals. Such alterations in this essential cellular pathway have demonstrated clear-cut effects for both metals/metalloids and organics.

The heme/porphyrin pathway is essential for synthesis of hemoproteins (such as hemoglobin) and various cytochromes (such as cytochrome P450). A diagram of this pathway, marked with sites of lead (Pb) inhibition, is shown in Figure 3.

A review of chemical-induced disturbances in the heme pathway has been published by the New York Academy of Sciences (Silbergeld and Fowler 1987). It summarizes studies showing that a number of agents, such as lead, arsenic, mercury, PCBs, dioxin, hexachlorobenzene, and alcohol, each produce highly specific alterations in only certain heme pathway enzymes with concomitant increases in selected heme precursors and porphyrins in urine. The relationships between these disturbances and cell injury have also been extensively studied and ultrastructural, morphometric correlations of cell injury made for several agents in target tissues (Fowler and Woods 1987).

Metal-metal interaction studies (Fowler and Mahaffey 1978; Mahaffey et al. 1981) have shown mixture-specific porphyrinuria patterns in rats exposed to a combination of lead, cadmium, and arsenic (Figure 4). These data provide further evidence for the applicability of this approach to chemical mixtures. Since the heme pathway is essential to life and is highly conserved in mammals (Chang et al. 1984), it is reasonable to expect that it should react in a similar manner to chemical-induced disturbances in aquatic species.

In addition, studies by a number of investigators have also shown similar specific changes in heme pathway enzyme activities of aquatic species following waterborne exposures to sublethal concentrations of lead (Jackim 1973; Hodson 1976) and cadmium (Dalwani et al. 1985; Arriyoshi et al. 1990) in the blood and liver, respectively. The effects of lead and other experimental variables on the activity of fish δ-aminolevulinic acid dehydratase have recently been studied in rainbow trout (Johansson-Sjobeck and Larson 1979; Sordyl and Osterland, 1990; Addison et al. 1990; Gonzalez et al. 1987) and channel catfish (Connor and Fowler 1991) and other vertebrate fish species (Rodriques et al. 1989) (Figure 5). Similar findings have been reported with birds (Fox et al. 1988; Rattner et al. 1989a; 1989b; Roscoe et al. 1979). These data have recently been extended by in vitro studies (Conner et al. 1991; Conner and Fowler 1991) that have demonstrated at least some enzyme activities such δ-aminolevulinic acid dehydratase (ALAD) show inhibition patterns to lead exposure similar to those observed in mammals. Recent studies in plants (Jacobs et al. 1990; Matringe et al. 1989a; 1989b; Varsano et al. 1990) and bacteria (Jacobs et al. 1990) have demonstrated the sensitivity of protoporphyrinogen oxidase to peroxidizing herbicides. These multiphyla data clearly point to the potential utility of chemical-induced disturbances in this essential pathway in aquatic species, provided a firm scientific data base is established.

The expanded application of these indicators to aquatic species would speed the development of sensitive tests, since virtually all assay procedures are already well-developed and would necessitate only a mechanical adaptation to these species. Since a number of relationships between these indicators and cell injury are already known in mammals, interpretation of the alterations in a given heme pathway enzyme activity/ porphyrinuria and an injurious process will also be more readily interpretable. This information should hence greatly expedite the interpretation of changes in heme pathway indicators in aquatic species.

Heme Enzyme Porphyrin Assays

This section describes current knowledge on heme enzyme/porphyrin assays in fish. To perform these assays, we recommend the following method of preparing vertebrate fish tissues. Invertebrate species may require other modifications.

1.   After the stipulated exposure, vertebrate fish from control and experimental tanks should be removed and sacrificed by cervical severing of the spinal cord in accordance with AALAC and NIH guidelines.
2.   Tissues should be excised and washed with $0.05M$ Tris-HCL buffer (pH 7.5) and then blotted dry using filter paper placed on ice.
3.   The tissues (liver, kidney, blood) should then be weighed, minced and homogenized in nine volumes of $0.25\ M$ sucrose in $0.05\ M$ Tris-HC1 buffer (pH 7.5) using a homogenizer fitted with a Teflon pestle held in ice. Invertebrate tissues may require other modifications.
4.   The preparation of subcellular fractions should be performed as described by Fowler et al (1989).

The heme pathway enzyme assays are listed below in sequence of action within the pathway:

1.   Delta-aminolevulinic Acid Synthase: this enzyme, which is the first step in the heme/porphyrin pathway and catalyzes the condensation reactions of glycine with succinyl/COA, should be assayed according to the method of Woods (1974).

2.   Delta-aminolevulinic Acid Dehydratase (porphrobilinogen synthase): tissue (liver, kidney, blood) postmitochondrial supernatants should be prepared using the methods of Gibson et al. (1955) and Baron and Tephly (1969) as modified by Conner and Fowler (1991).

3.   Porphobilinogen Deaminase: Porphobilinogen deaminase activity should be measured in 9000-g tissue supernatant fractions (Fowler et al. 1989) after being heated at $65°C$ for 15 minutes, as previously described by Woods and Fowler (1978).

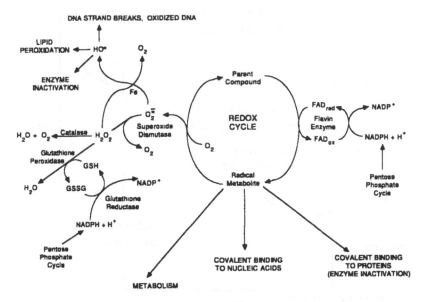

Figure 3.   Redox cycle. (Adapted from Kappas 1985.)

4.   Uroporphyrinogen Decarboxylase: assay of this cytosolic enzyme should be by the method of Elder and Matlin (1976). This method measures the activity in tissue supernatant fractions utilizing the 5-carboxyl porphyrinogen III as a substrate.

5.   Coproporphyrinogen Oxidase: the HPLC method of Li et al. (1986) should be used to measure coproporphyrinogen oxidase activity.

6.   Protoporphyrinogen Oxidase: the spectrofluorometric assay procedure of Matringe et al. (1989) should be employed for assay of this enzyme activity.

7.   Ferrochelatase: Ferrochelatase (Protoheme Ferrolyase) should be measured by a modification of the method of Porra and Jones (1963), as modified by Woods and Fowler (1978).

Urinary Porphyrin Assay

To perform a urinary porphyrin assay, we recommend that fish be cannulated (Hunn et al. 1966) and urine collected in lightproof containers wrapped in aluminum foil. Specimens may be stored at $-70°C$ for a week to several months with minimal loss of porphyrins. Analysis of urine samples for various porphyrins should be conducted by HPLC using the method of Schreiber et al. (1983).

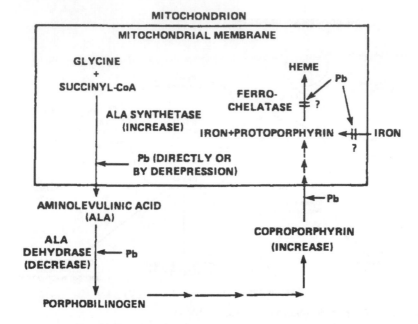

Figure 4.   Effects of lead (Pb) on heme biosynthesis. (From U.S. EPA Air Quality Criterion for Lead 1986.)

In addition to the heme biosynthetic pathway previously noted, many metals/metalloids (such as cobalt, mercury, lead, cadmium, zinc, tin, indium, nickel, arsenic, and platinum) have been demonstrated to induce microsomal heme oxygenase (Silbergeld and Fowler 1987), the rate-limiting enzyme in the heme degradative pathway. A number of investigators have shown that induction of this enzyme is closely correlated with concomitant decreases in total microsomal cytochrome P450 content and differential inhibition of a number of microsomal monooxygenase activities. Ultrastructural morphometric studies conducted on hepatocytes of rats injected with indium (Fowler et al. 1983) have shown concomitant dose-related destruction of the endoplasmic reticulum and an increase in the relative volume density of iron containing phagolysosomes. These data provide clear evidence that induction of heme oxygenase and associated reductions in cytochrome P450-mediated microsomal enzyme activities is correlated with morphological evidence of cell injury. Such data strongly suggest that induction of this enzyme may be interpreted as a biomarker of both exposure and deleterious effect.

Further evidence for this idea is provided by recent studies which have demonstrated that heme oxygenase is the 32-kDa member of the stress protein family induced during cell injury by a variety of agents, including metals, sodium arsenite, and thiol-reactive agents (Caltabiano et al. 1986a; Sargan et al. 1986). This protein has recently been isolated and identified as heme oxygenase, an enzyme essential for heme catabolism; this enzyme cleaves heme to form biliverdin, which is subsequently reduced to bilirubin (Caltabiano et al. 1988; Keyse and Tyrrell 1989). It is considered by some to

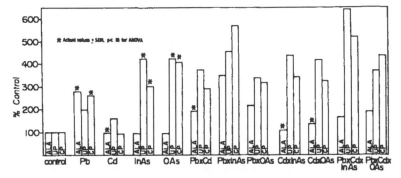

Figure 5.   Effects of dietary metal exposure to Pb (200 ppm), Cd (50 ppm), inorganic arsenic (InAs), or organic arsenic (OAs, arsanillic acid - 50 ppm), for 10 weeks on urinary excretion of aminolevulinic acid (ALA), total uroporphyrin (UP), and coproporphyrin (CP), showing distinctive porphyrin "fingerprints" for each combination.

be a minor heat shock protein because of limited evidence that it can be induced to some extent by heat in rats and humans (Keyse and Tyrrell 1989; Shibahara et al. 1987; Shuman and Przybyla 1988); however, others report that heat treatment has no effect on heme oxygenase activity or mRNA levels (Yoshida et al. 1988; Caltabiano et al 1986b). This protein is most highly inducible by a variety of stressors which cause oxidative damage, such as UVA radiation, sodium arsenite, and hydrogen peroxide (Keyse and Tyrrell 1989). It has been suggested that since breakdown products of heme can readily react with peroxyl radicals, they may play an important role in protecting cells from oxidative damage as free radical scavengers in concert with glutathione (Keyse and Tyrrell 1989). Interestingly, Cd and other metals (such as Cu, Zn, Pb, and Ag) have been shown to be particularly effective inducers of heme oxygenase, and it is the most prominent stress protein induced by these metals upon in vitro exposure (Caltabiano et al. 1986a; Sargan et al. 1986; Keyse and Tyrrell 1989).

GENERAL DISCUSSION

Interrelationships Between Biomarkers

The biochemical systems discussed in the subsections of this chapter are intricately linked. This can be illustrated by several simple examples: (1) cytochrome P450 isozymes are capable of mediating toxicity of chemicals via generation of reactive oxygen species, as well as by formation of reactive metabolites, (2) reactive metabolites can be detoxified by glutathione transferase, (3) formation of the active cytochrome P450 is absolutely dependent upon the heme biosynthetic pathway, (4) evidence suggests that glutathione transferases play a role in the transport of heme between sites of synthesis (mitochondria) to heme-containing proteins, (5) the toxic action of a given chemical species (a metal or an oxygen radical) against the enzymes in the heme pathway can influence the availability of heme for P450 synthesis, (6) conversely, chemical damage to P450 will cause the induction of the heme degradative enzyme heme oxygenase, (7) heme oxygenase has recently been identified as the 32 kDa stress protein, which is also induced by a wide variety of metals, and (8) several metals inducing heme oxygenase Cu, Cd also cause the induction of metallothionein. The description of connections can go on.

These few examples provide ample evidence for functional interrelationships between various protein systems considered as biomarkers. It should be equally clear from our discussion that a number of relationships exist between the various biomarkers discussed in this chapter, and the classes of biomarkers covered by other chapters. This is biologically reasonable in a general sense, since many, perhaps most, biomarker endpoints involve interactions with or are mediated by proteins, either directly or indirectly. For example, P450 transforms hydrocarbons into products that, if not conjugated with glutathione, sulfate, or glucuronide via conjugating enzymes, form DNA adducts. Reactive oxygen species generated by the P450 system are capable of damaging DNA or protein and causing lipid peroxidation in membranes. P450 action can initiate processes culminating in histological disease. Similarly, metal-binding proteins such as metallothionein may play important roles in the availability of Zn to DNA polymerases, or of other metals such as Hg to DNA repair enzymes, and thereby influence fidelity of DNA synthesis or repair. Activities of enzymes in the heme pathway regulate the production and persistence of porphyrins within cells.

The obvious linkages between broad classes of diverse biomarkers should greatly expand the utility and range of approaches for any given problem of environmental contamination. On a practical basis, this means that it is also probable that results derived by a single biomarker may be independently confirmed by other biomarker approaches. Such combined approaches should add enormous confidence to management or regulatory decisions based on these procedures. Combined studies should also suggest new research directions to understanding mechanisms of chemically induced cell injury.

Integration and Higher-Order Effects

Understanding the nature of linkages between various biomarkers could disclose the connections to effects at higher levels of organization. The function of these proteins can be intimately connected to other aspects of toxic injury, such as the role of P450 in activation of carcinogens, or the development of porphyrinuria following lead or alcohol exposure. It is difficult at present to demonstrate clearly that changes in the proteins discussed in this chapter are linked to overt disease or reproductive problems in aquatic species. However, such linkages can be rationalized from the functions and interactions that are known for the proteins. More importantly, some such linkages have been demonstrated explicitly in mammalian systems, including humans. For example, inducibility of cytochrome P450 in humans has been linked to a greater susceptibility to lung cancer. Most of the protein systems dealt with in this chapter—metallothioneins, stress proteins, oxidant-mediated responses, and the heme/ porphyrin pathway—are very highly conserved in phylogeny, and we can expect similar functions in nonmammalian systems. Our knowledge of the basic biochemistry of these systems in fish and invertebrates is still sparse. Yet, it is not arguable that the systems considered here are directly relevant to the individual and its response to chemical exposure. Direct linkages to higher order effects are certain to be disclosed as research on fundamental properties progresses.

Biomarker Use and Interpretation

Aspects related to these protein systems as biomarkers, and the interpretation of changes in these systems resulting from chemical exposure, are discussed in the context of two tables. Table 6 considers the attributes of the various systems as biomarkers, summarizing many of the aspects discussed in the subsections of this chapter. Table 7 addresses questions about some of the ways in which these systems can be applied as biomarkers.

Attributes of Biomarkers

The summary of information provided in Table 6 suggests that each of the systems discussed in this chapter could be employed as biomarkers. However, there are at least three important considerations which must be borne in mind regarding Table 6.

First, each of the systems in this chapter is a multicomponent system, and the attributes listed reflect the knowledge only for a limited number of these components. Cytochrome P450, for example, refers only to the cytochrome P4501A proteins; other P450 proteins that might be used as biomarkers could have a different set of qualities. Eventually, separate tables like this should be possible for each of these systems, with entries for each component.

Second, the responses for some of the systems are based in large part on knowledge obtained with "traditional" mammalian species, or other species not of concern in the environment. This is most true for the heat shock proteins and for the heme/porphyrin pathway. These two points emphasize the need and value of continued basic research on all the systems.

Third, there are important considerations regarding the methodology for using protein systems as biomarkers, and the interpretation of changes resulting from chemical exposure. These considerations include the practical aspects involved in methods of analyses, and their reliability.

Methods development. Methods for analysis of the proteins or their activity, as discussed in this chapter, have relied primarily on chromatographic, spectrophotometric, and other traditional techniques. These methods will continue to have utility. As antibodies have become available for specific proteins of interest, explicit detection of changes in protein content has become possible. Immunological methods are now being widely used in studies of P450, stress proteins, and metallothioneins. The use of cDNA probes to detect expression at the mRNA level is also now being employed in studies of P450, stress protein, and metallothionein induction in fish and/or mammals. Since the various approaches reveal different features of gene expression, they provide different information in experimental studies, and thus differ in their utility in monitoring. The application of any of these methods in a routine way at present requires a well-equipped laboratory and investigators experienced in the analyses and knowledgeable about the protein systems. Research is needed on the development of more simplified methods with increased utility and decreased cost.

Sample handling. The validity of results obtained by analysis of all the systems discussed can be strongly influenced by the conditions of animals collected and methods of tissue handling and sample preparation. Studies on P450, for example, have shown that methods of subcellular fractionation may need to be modified for different organs or developmental stages (Stegeman 1979; Binder and Stegeman 1984), and that storage conditions for archived samples can affect their function, at least for some species (Förlin and Andersson 1985). Many of the conditions for P450 systems are considered in detail in Stegeman (1987). Attention must be given to similar concerns for each of the systems treated in other chapters of this text.

Environmental and biological sources of variability. Environmental and biological variables are known to influence the levels and functions of some of these systems, as well as their response to chemicals. Environmental variables include temperature, O_2 content, and salinity. Biological variables include hormonal status, particularly as related to sex and reproduction. For example, reproductive status can strongly influence the response of P4501A to inducers in fish (Förlin et al. 1985), and copper-binding to metallothionein in oysters. Molting in crustacea affects metallothionein levels. There are also unexplained seasonal changes in content, functions, or responses for some proteins; these may be linked primarily to seasonal changes in temperature and/or hormonal status, but other seasonal factors could be involved. Developmental status is known to affect these systems in aquatic species. Genetic variation could alter the function of these proteins and their response. This area is virtually unknown in nonmammalian species, but in mammals there is clear demonstration of the importance of genetic factors affecting, for example, P450 induction in mice. Nutrition, disease states, and tissue damage can also affect these systems and their responses to chemicals. The influence of these factors, however, is still poorly understood in aquatic species. Although many factors can influence the expression or function of these proteins, understanding the mechanisms and magnitude of such influences can reduce the uncertainty.

Utility of Protein Systems as Biomarkers

Table 7 offers our opinions about the potential for application of protein biomarkers in various ways, based on current information from nonmammalian and, where necessary, mammalian studies. We emphasize that some of these systems are already being used in some of these ways; for example, P450. The following are general comments regarding each of the categories in Table 7.

1. The potential to relate the magnitude of changes in the biomarkers to the magnitude of exposure is strong (see Table 6).

2. The relationship to health may depend upon linkages being identified. Even without such linkages, responses can be interpreted as an indication that biochemically significant exposure has occurred; a clear sign that other effects could ensue.

3. The responses of these protein systems may in many cases be the most sensitive and earliest effects detectable.

4. The biochemical biomarkers we have considered are ideal for the evaluation of chemical risk in natural populations. They can accurately signal the presence of biochemically significant levels of contaminants. As such, they can be used as the first screen for effects. Conversely and equally important, a low or absent signal would indicate that effective chemicals are below active concentrations. These systems could be employed in detecting new problems, as well as monitoring for "recovery" in an ecosystem following clean-up. Other potential or actual uses of these systems include the monitoring of effluents; for example, with caged fish or fish maintained in receiving ponds. These uses of protein biomarkers could be considered in light of a modified interpretation of a "no-effect" level, that of no-effect associated with an effluent, and in an ecosystem. The early, sensitive responses of some of these systems lend themselves ideally to that use. Establishing a no-effect level in presently used "standard" toxicity tests has limited environmental relevance, as the potential for effects resulting from toxicant accumulation in receiving waters is not considered.

5. Use of these proteins might be considered in a tiered approach. The biomarker concept involves the use of biochemical, cellular, and physiological parameters as diagnostic screening tools in environmental monitoring. Biomarkers might

Table 6. Protein/Enzyme Biomarkers.

Attribute	Cytochrome P450	Metal Binding Proteins	Heat Shock Protein	Phase II Enzymes	Oxidative Damage Response	Heme and Porphyrins
Exposure marker	yes	yes	yes	yes	yes	yes
Effects marker	yes	?	?	?	yes	yes
Biological specificity	vertebrates	Eukaryotes	Prokaryotes Eukaryotes	vertebrates	broad	Prokaryotes, Eukaryotes
Chemical specificity	PAH, planar HHC	selected metals	metals, organics, inorganics	PAH, PCB	oxidants, redox-active compounds	Metals, some organics
Sensitivity	high	high	high	low	high in plants untested in animals	mod. to high
Time of response	6-24 h	hours-days	min.-hours	days	variable	hours-days
Reliability	high	high	high	high	untested	high
General or point source	both	both	both	both	both	both
Linked to further effect	yes	yes	yes	yes	yes	yes
Field trials	positive	positive	no data	–	positive (for those tested)	little data
Assay cost - present	moderate	expensive	moderate	moderate	low to moderate	moderate
- future	low	moderate	low	low	variable	low
Ease of assay - present	sophisticated	sophisticated	sophisticated	sophisticated	from easy to difficult	simple/ sophisticated
- future	may be simplified	simple	may be simplified	may be simplified		may be simplified
Research needed	yes	yes	yes	yes	yes	yes

Table 7. Potential Uses of Protein Biomarkers.

Area of Use	Cytochrome P4501A1	Metal Binding Protein	Heat Shock Protein	Phase II Enzymes	Oxidative Damage Response	Heme and Porphyrins
• Quantify impact or exposure	+	−	+	?	+	(+)
• Monitor organism and ecosystem "health"	+	?	+	+?	+	+
• Identify subtle, early effects	++	?	++	?	+	+
• Risk assessment	+	?	?	?	+	+
• Trigger regulatory action	+	−?	+?	−	?	(+)
• Identify exposure to specific compounds	++	+	(+)	?	?	−
• Toxicological screening	+	−	+	+	−	(+)
• Research on toxic mechanisms	++	++	++	++	++	++

Note: + indicates demonstration of highly likely usefulness in the indicated area; ++ indicates very strongly useful; ? indicates uncertainty; − indicates not useful. The parentheses indicate that information is not available in species of concern, but may be available in model species.

be applied to detect sublethal stress in an organism (designated as a Tier I biomarker) and to detect exposure to specific contaminants (designated as a Tier II biomarker). The utility of a Tier I biomarker for detecting sublethal stress would depend on its ability to be used in a broad range of organisms when exposed to a wide variety of stress conditions in their environment; in practicality, it should also be easily measured, in a cost-efficient manner. Tier II biomarkers, used to identify exposure to specific types of contaminants, should be detected in organisms exposed to a particular class of contaminants in their environment, and be easily measured. By integrating both kinds of biomarkers into a multitiered approach to environmental monitoring, one could develop a series of assays in which organisms are initially screened with biomarkers to detect general stress; if the results were positive, they could be assayed with an array of Tier II biomarkers, and of markers that may explicitly reveal health effects (e.g., histopathology). Such a strategy would provide a comprehensive overview of both the extent of biological damage and the contaminants responsible. Research is needed on the relative sensitivity of various systems. Depending on circumstances, effects on these protein systems could trigger either management action or analysis of additional biomarkers.

6. Some of the responses, e.g., P4501A and metallothionen induction, can be highly specific, others are not. Research is needed on the relative specificity and sensitivity of P450 or MT induction and Hsp induction.

7. Each of these systems has potential in screening new or untested compounds. Screening for induction of P450 or HSP, for example, would indicate the potential for a compound to cause these changes and any effects that would result from that induction. Such screening might be done in whole animals or isolated cell cultures.

8. One of the most significant uses for these systems is in research on toxic mechanisms of chemicals. As discussed throughout this chapter, these systems play vital dual roles in protection against, and causation of, chemical injury. Understanding the mechanism of toxic effect involving these systems is essential to understanding effects at higher levels of organization.

RESEARCH NEEDS AND RECOMMENDATIONS

The development of basic knowledge and technical advances involving analytical methods will improve the ability to use these proteins as biomarkers and to interpret results of that use. Specific recommendations regarding research needs can be found in each of the individual subsections. At this point, we give general recommendations for improving the utility of these proteins as biomarkers. These are grouped according to their significance to basic research needs and applied aspects.

Basic Research Needs

1. The systems considered are multicomponent systems, each representing protein families or groups; however, detailed information is available for only a few components in each system. There is an urgent need for studies on the basic biochemistry of the better known components, such as P4501A, as well as additional components, such as the P4504A and P4502B proteins now known to occur in fish.

2. The molecular mechanisms by which chemicals elicit the responses in the various systems need to be established. Structure-activity relationships must be determined where relevant, in light of these mechanisms.

3. The mechanistic bases by which biological and environmental variables influence these systems, as well as their responses to chemicals, requires research.

4. Linkages and integration between these various systems, as well as linkage between the protein systems and other biochemical, physiological, and cellular changes, need to be understood.

5. Physiological and/or temporal relationships between changes in these biochemical systems and toxic effects at higher levels of organization, long a research goal, continue to need vigorous research.

Applied Aspects

1. The responses of these systems need to be linked experimentally and in-field studies need to measure internal and external doses of chemicals.

2. Field studies are needed to validate the utility of these systems, and to compare responses of these biochemical systems and other biomarkers. Several combined field studies have been done (e.g., Bayne et al, 1988; National Oceanographic and Atmospheric Administration 1989), but such additional studies are required. Experimental studies to compare the sensitivity of different biomarkers are also required.

3.   Convenient, inexpensive methods for measurement of changes in these systems need to be developed.

4.   The effects of combinations of chemicals need to be established. In fact, the biochemical systems discussed in this chapter could finally offer an approach to understanding the mechanisms of chemical interactions.

REFERENCES

Adams, J.D., Jr., B.H. Lauterburg and J.R. Mitchell. "Plasma Glutathione and Glutathione Disulfide in the Rat: Regulation and Response to Oxidative Stress," *J. Pharmacol. Exp. Ther.* 227:749–754 (1983).

Addison, R.F., and A.J. Edwards. "Hepatic Microsomal Monooxygenase Activity in Flounder *Platichthys flesus* from Polluted Sites in Langesundfjord and from Mesocosms Experimentally Dosed with Diesel Oil and Copper," *Mar. Ecol. Prog. Ser.* 46:52–54 (1988).

Addison, R.F., D. Fitzpatrick and K.W. Renton. "Distribution of δ-Aminolevulinic Acid Synthetase and δ-Aminolevulinic Acid Dehydratase in Liver and Kidney of Rainbow Trout (*Salmo gairdnerii*)," *Comp. Biochem. Physiol.* 95B:317–319 (1990).

Alscher, R.G., and J.S. Amthor. "The Physiology of Free-radical Scavenging: Maintenance and Repair Processes," in *Air Pollution and Plant Metabolism*, S. Schulte-Hostede, N.M. Darrall, L.W. Blank and A.R. Wellburn, Eds. (New York: Elsevier Applied Science, 1988), pp. 95–115.

Alscher, R., J. Bower and W. Zippel. "The Basis for Different Sensitivities of Photosynthesis to SO2 in Two Cultivars of Pea," *J. Exp. Bot.* 38:99–108 (1987).

Andaya, A.A., and R.T. Di Giulio. "Acute Toxicities and Hematological Effects of Two Substituted Naphthoquinones in Channel Catfish," *Arch. Environ. Contam. Toxicol.* 16:233–238 (1987).

Andersson, T., B. Bengtsson, L. Förlin, J. Härdig and Å. Larsson. "Long-term Effects of Bleached Kraft Mill Effluents on Carbohydrate Metabolism and Hepatic Xenobiotic Biotransformation Enzymes in Fish," *Ecotoxicol. Environ. Saf.* 13:53–60 (1987).

Andersson, T., L. Förlin, J. Härdig and Å. Larsson. "Physiological Disturbances in Fish Living in Coastal Water Polluted with Bleached Kraft Pulp Mill Effluents," *Can. J. Fish. Aquat. Sci.* 45:1525–1536 (1988).

Andersson, T., M. Pesonen and C. Johansson. "Differential Induction of Cytochrome P-450 Monooxygenase, Epoxide Hydrolase, Glutathione Transferase and UDP Glucuronosyl Transferase Activities in the Liver of Rainbow Trout by β-Naphthoflavone or Clophen A50," *Biochem. Pharmacol.* 34:3309–3314 (1985).

Ankley, G.T., V.S. Blazer, R.E. Reinert and M. Agosin. "Effects of Arochlor 1254 on Cytochrome P-450 Dependant Monooxygenase Glutathione S-Transferase and UDP-Glucuronosyltransferase Activities in Channel Catfish Liver," *Aquat. Toxicol.* 9:91–103 (1986).

Armstrong, R.N. "Enzyme-catalyzed Detoxication Reactions: Mechanisms and Stereochemistry," *CRC Crit. Rev. Biochem.* 22:39–88 (1987).

Arrigo, A.P., J.P. Suhan and W.J. Welch. "Dynamic Changes in the Structure and Intracellular Locale of the Mammalian Low-molecular-weight Heat Shock Protein," *Mol. Cell. Biol.* 8:5059–5071 (1988).

Arriyoshi, T., S. Hiba, H. Hasegawa and K. Arizono. "Profiles of Metal-binding Proteins and Heme Oxygenase in Red Carp," *Bull. Environ. Contam. Toxicol.* 44:643–649 (1990).

Atkinson, B.G., and D.B. Walden, Eds. *Changes in Eukaryotic Gene Expression in Response to Environmental Stress* (New York: Academic Press, Inc., 1985), p. 313.

Bakka, A., A.S. Johnson, L. Endresen and H.E. Rugstad. "Radioresistance in Cells with High Content of Metallothionein," *Experientia* 38:381–383 (1982).

Balk, L., J. Meijer, J. Seidegård, R. Morgenstein and J.W. Depierre. "Initial Characterization of Drug-metabolizing Systems in the Liver of the Northern Pike (*Esox lucius*)," *Drug Metab. Dispos.* 8:98–103 (1980).

Barnes, R.L. "Effects of Chronic Exposure to Ozone on Soluble Sugar and Ascorbic Acid Contents of Pine Seedlings," *Can. J. Bot.* 50:215–219 (1972).

Baron, J., and T.R. Tephly. "Effect of 9-Amino-1,2,4-triazole on the Stimulation of Hepatic Microsomal Heme Synthesis and Induction of Hepatic Microsomal Oxidases Produced by Phenobarbital," *Mol. Pharmacol.* 5:10–20 (1969).

Bayne, B.L., K.R. Clarke and J.S. Gray, Eds. "Biological Effects of Pollutants: Results of a Practical Workshop, Oslo, Norway, 1986," *Mar. Ecol. Prog. Ser.* 46 (1988).

Beauchamp, C.O., and I. Fridovich. "Superoxide Dismutase: Improved Assays and an Assay Applicable to Acrylamide Gels," *Anal. Biochem.* 44:276–287 (1971).

Beaumont, A.R., and M.D. Budd. "High Mortality of the Larvae of the Common Mussel at Low Concentrations of Tributylin," *Mar. Poll. Bull.* 15:402–405 (1984).

Beckmann, R.B., L.A. Mizzen and W.J. Welch. "Interactions of Hsp 70 with Newly Synthesized Proteins: Implications for Protein Folding and Assembly," *Science* 248:850–853 (1990).

Bend, J.R., and M.O. James. "Xenobiotic Metabolism in Marine and Freshwater Species," in *Biochemical and Biophysical Perspectives in Marine Biology*, D.C. Malins, and J.R. Sargent, Eds. (New York: Academic Press, 1978), pp. 128–172.

Berger, H.M., and M.P. Woodward. "Small Heat Shock Proteins in Drosophila May Confer Thermal Tolerance," *Exp. Cell Res.* 147:437–442 (1983).

Bergmeyer, H.U. "Zur Messung von Katalase-aktivitaten," *Biochem. Z.* 327:255–258 (1955).

Binder, R.L., and J.L. Lech. "Xenobiotics in Gametes of Lake Michigan Lake Trout (*Salvelinus namaycush*) Induce Hepatic Monooxygenase Activity in their Offspring," *Fund. Appl. Toxicol.* 4:1042–1054 (1984).

Binder, R.L., and J.J. Stegeman. "Induction of Aryl Hydrocarbon Hydroxylase Activity in Embryos of an Estuarine Fish," *Biochem. Pharmacol.* 29:949–951 (1980).

Binder, R.L., and J.J. Stegeman. "Microsomal Electron Transport and Xenobiotic Monooxygenase Activities During the Embryonic Period of Development in the Killifish, *Fundulus heteroclitus*," *Toxicol. Appl. Pharmacol.* 73:432–443 (1984).

Bosch, T.C., S.M. Krylow, H.R. Bode and R.E. Steele. "Thermotolerance and Synthesis of Heat Shock Proteins: These Responses Are Present in *Hydra attenuata* but Absent in *Hydra oligactis*," *Proc. Natl. Acad. Sci. U.S.A.* 85:2727–2731 (1988).

Bouquegneau, J.M. "Evidence for the Protective Effect of Metallothioneins Against Inorganic Mercury Injuries to Fish," *Bull. Environ. Contam. Toxicol.* 23:218–219 (1979).

Bournias-Vardiabasis, N., and C.H. Buzin. "Developmental Effects of Chemicals and the Heat Shock Response in Drosophila Cells," *Teratogen. Carcinogen. Mutagen.* 6:523–536 (1986).

Bournias-Vardiabasis, N., R.L. Teplitz, G.F. Chernoff and R.L. Seecof. "Detection of Teratogens in the Drosophila Embryonic Cell Culture Test: Assay of 100 Chemicals," *Teratology* 28:109–122 (1983).

Bracken, W.M., and C.D. Klaassen. "Induction of Metallothionein in Rat Primary Hepatocytes Culture: Evidence for Direct and Indirect Induction," *J. Toxicol. Environ. Health* 22:163–174 (1987).

Brady, F.O., and B.S. Helvig. "Effect of Epinephrine and Norepinephrine on Zinc Thionein Levels and Induction in Rat Liver," *Am. J. Physiol.* 247:E318–322 (1984).

Brady, F.O., and M. Webb. "Metabolism of Zinc and Copper in the Neonate," *J. Biol. Chem.* 256:3931–3935 (1981).

Bremner, I., W.G. Hoekstra, N.T. Davies and N.T. Young. "Effect of Zinc Status of Rats on the Synthesis and Degradation of Copper Induced Metallothionein," *Biochem. J.* 174:883–892 (1978).

Bremner, I., Williams, R.B. and B.W. Young. "Effects of Age, Sex, and Zinc Status on the Accumulation of (Copper-zinc) Metallothionein in Rat Kidneys," *J. Inorg. Biochem.* 14:135–146 (1981).

Brigelius, R., A. Hashem and E. Lengfelder. "Paraquat-induced Alterations of Phospholipids and GSSG-Release in the Isolated Perfused Rat Liver, and the Effect of SOD-Active Copper Complexes," *Biochem. Pharmacol.* 30:349–354 (1981).

Brouwer, M., T.M. Brouwer-Hoexum and D.W. Engel. "Cadmium Accumulation by the Blue Crab, *Callinectes sapidus*: Involvement of Hemocyanin and Characterization of Cadmium-binding Proteins," *Mar. Environ. Res.* 14:71–88 (1984).

Brouwer, M., P. Whaling and D.W. Engel. "Copper-metallothioneins in the American Lobster, *Homarus americanus*: Potential Role as Cu(I) Donors to Apohemocyanin," *Environ. Health Persp.* 65:93–100 (1986).

Brouwer, M., D.R. Winge and W.R. Gray. "Structural and Functional Diversity of Copper-metallothionein from the American Lobster, *Homarus americanus*," *J. Inorg. Biochem.* 35:289–303 (1989).

Brouwer, M., and Brouwer-Hoexum. "Interaction of Copper-Metallthionein from the American Lobster, *Homarus americanus*, with Glutathione," *Arch. Biochem. Biophys.* 290:207–213 (1991)

Brown, D.A., S.M. Bay, D.J. Greenstein, P. Szalay, G.P. Hershelman, C.F. Ward, A.M. Westcott and J.N. Cross. "Municipal Wastwater Contamination in the Southern California Bight: Part II. Cystolic Distribution of Contaminants and Biochemical Effects in Fish Livers," *Mar. Environ. Res.* 21:135–161 (1987).

Bryan, G.W., P.E. Gibbs, L.G. Hummerstone and G.R. Burt. "The Decline of the Gastropod *Nucella lapillus* Around South-west England: Evidence for the Effect of Tributyltin from Antifouling Paints," *J. Mar. Biol. Assoc. U.K.* 66:611–640 (1986).

Buchmann, A., W. Kuhlmann, M. Schwarz, W. Kunz, C.R. Wolf, E. Moll, T. Friedberg and F. Oesch. "Regulation and Expression of Four Cytochrome P-450 Isoenzymes. NADPH-cytochrome P-450 Reductase, the Glutathione Transferases B and C and Microsomal Epoxide Hydrolase in Preneoplastic and Neoplastic Lesions in Rat Liver," *Carcinogenesis* 6:513–521 (1985).

Buhler, D.R., and D.E. Williams. "The Role of Biotransformation in the Toxicity of Chemicals," *Aquat. Toxicol.* 11:19–28 (1988).

Bulleid, N.J., A.B. Graham and J.A. Craft. "Microsomal Epoxide Hydrolase of Rat Liver. Purification and Characterization of Enzyme Fractions with Different Chromatographic Characteristics," *Biochem. J.* 233:607–611 (1986).

Burton, G.W., A. Webb and K.U. Ingold. "A Mild, Rapid, and Efficient Method of Lipid

Extraction for Determining Vitamin E/Lipid Ratios," *Lipids* 20:29–39 (1985).

Bus, J.S., S.D. Aust and J.E. Gibson. "Paraquat Toxicity: Proposed Mechanism of Action Involving Lipid Peroxidation," *Environ. Health Perspect.* 16:139–146 (1976).

Cairns, V.W., P.V. Hodson and J.O Nriagu. *Contaminant Effects on Fisheries* (Toronto: John Wiley & Sons, 1984).

Caltabiano, M.M., T.P. Koestler, G. Poste and R.G. Greig. "Induction of 32 & 34–kDa Stress Proteins by Sodium Arsenite, Heavy Metals, and Thiol-reactive Agents," *J. Biol. Chem.* 261:13381–13386 (1986a).

Caltabiano, M.M., T.P. Koestler, G. Poste and R.G. Greig. "Induction of Mammalian Stress Proteins by a Triethylphosphine Gold Compound Used in the Therapy of Rheumatoid Arthritis," *Biochem. Biophys. Res. Comm.* 138:1074–1080 (1986b).

Caltabiano, M.M., G. Poste and R.G. Greig. "Induction of the 32-kD Human Stress Protein by Auranofin and Related Triethylphosphine Gold Analogs," *Biochem. Pharmacol.* 37:4089–4093 (1988).

Carmichael, N.G., and B.A. Fowler. "Cadmium Accumulation and Toxicity in the Kidney of the Bay Scallop *Argopecten irradians*," *Mar. Biol.* 65:35–43 (1981).

Carmichael, N.G., K.S. Squib, D.W. Engel and B.A. Fowler. "Metals in the Molluscan Kidney: Uptake and Subcellular Distributions of ^{109}Cd, ^{54}Mn and ^{65}Zn by the Clam *Mercenaria mercenaria*," *Comp. Biochem. Physiol.* 65A:203– 206 (1980).

Carmichael, N.G., K.S. Squib and B.A. Fowler. "Metals in the Molluscan Kidney: A Comparison of the Closely Related Bivalve Species (*Argopecten*) Using X-ray Microanalysis and Atomic Absorption Spectroscopy," *J. Fish. Res. Bd. Can.* 39:1149–1155 (1979).

Carr, R.S., M.B. Bally, P. Thomas and J.M. Neff. "Comparison of Methods for Determination of Ascorbic Acid in Animal Tissues," *Anal. Chem.* 55:1229–1232 (1983).

Carr, B.I., T.H. Huang, C.H. Buzin and K. Itakura. "Induction of Heat Shock Gene Expression Without Heat Shock by Hepatocarcinogens and During Hepatic Regeneration in Rat Liver," *Cancer Res.* 46:5106–5111 (1986).

Castillo, F.J., P.R. Miller and H. Greppin. "Extracellular Biochemical Markers of Photochemical Oxidant Air Pollution Damage to Norway Spruce," *Experientia* 43:111–120 (1987).

Catelli, M.G., N. Binart, I. Jung-Testas, J.M. Renoir, E.E. Baulieu, J.R. Feramisco and W.J. Welch. "The Common 90KD Protein Component of Nontransformed '8S' Steroid Receptors Is a Heat Shock Protein," *EMBO J.* 4:3131–3137 (1985).

Cathcart, R., E. Schwiers, R.L. Saul and B.N. Ames. "Thymine Glycol and Thymidine Glycol in Human and Rat Urine: A Possible Assay for Oxidative DNA Damage," *Proc. Natl. Acad. Sci. U.S.A.* 81:5633–5637 (1984).

Cavalieri, E.L., and E.G. Rogan. "Radical Cation in Aromatic Hydrocarbon Carcinogenesis," *Free Rad. Res. Commun.* 11:77–87 (1990).

Cha, Y.N., and H.S. Heine. "Comparative Effects of Dietary Administration of 2(3)-Tert-4-Hydroxyanisole and 3,5-Di-Tert-Butyl-4-Hydroxytoluene on Several Hepatic Enzyme Activities in Mice and Rats," *Cancer Res.* 42:2609–2615 (1982).

Chan, H.W.S., Ed. *Autoxidation of Unsaturated Lipids* (New York: Academic Press, 1987) p.296.

Chang, S.C., S. Sassa and D. Doyle. "An Immunological Study of δ-Aminolevulinic Acid Dehydratase Specificity with the Phylogeny of Species," *Biochem. Biophys. Acta* 797:291–301 (1984).

Chatterjee, S., and S. Bhattacharya. "Detoxication of Industrial Pollutants by the Glutathione S-Transferase System in the Liver of *Anabas testrudineus* (Bloch)," *Toxicol. Lett.* 22:187–198 (1984).

Cheng, M.Y., F-U. Hartl, J. Martin, R.A. Pollack, F. Kalousek, W. Neupert, E.M. Hallberg, R.L.

Hallberg and A.L. Norwich. "Mitochondrial Heat Shock Protein HSP 60 Is Essential for Assembly of Proteins Imported Into Yeast Mitochondria," *Nature (London)* 337:620–624 (1989).

Chesis, P.L., D.E. Levin, M.T. Smith, L. Ernster and B.N. Ames. "Mutagenicity of Quinones: Pathways of Metabolic Activation and Detoxification," *Proc. Natl. Acad. Sci. U.S.A.* 81:1696–1700 (1984).

Chiment, J.J., R. Alscher and P.R. Hughes. "Glutathione as an Indicator of SO_2-Induced Stress in Soybeans," *Environ. Exp. Bot.* 26:147–152 (1986).

Chirico, W.J., M.G. Waters and G. Blobel. "70K Heat Shock Related Proteins Stimulate Protein Translocation into Microsomes," *Nature (London)* 333:805–810 (1988).

Chow, C.K., and A.L. Tappel. "An Enzymatic Protective Mechanism Against Lipid Peroxidation Damage to Lungs of Ozone-exposed Rats," *Lipids* 7:518–524 (1972).

Clark, A.G. "The Comparative Enzymology of the Glutathione S-Transferases from Non-vertebrate Organisms," *Comp. Biochem. Physiol.* 92B:419–446 (1989).

Cleary, J.J., and A.R.D. Stebbing. "Organotin and Total Tin in Coastal Waters of Southwest England," *Mar. Poll. Bull.* 16:350–355 (1985).

Clifton, G., and N. Kaplowitz. "Effect of Dietary Phenobarbital, 3,4 Benzo[a]pyrene and 3-Methylcholanthrene on Hepatic, Intestinal and Renal Glutathione S-Transferase Activities in the Rat," *Biochem. Pharmacol.* 27:1284–1287 (1978).

Cochrane, B.J., R.B. Irby and T.W. Snell. "Stress Protein Synthesis in Response to Toxicant Stress in Two Species of Rotifer," ASTM Symposium on Aquatic Toxicology, Atlanta, 1989.

Coles, B., and B. Ketterer. "The Role of Glutathione and Glutathione Transferases in Chemical Carcinogenesis," *Crit. Rev. Biochem. Mol. Biol.* 25:47–70 (1990).

Collier, T.K., and U. Varanasi. "Field and Laboratory Studies of Xenobiotic Metabolizing Enzymes in English Sole," *Proc. Pac. Northwest Assoc. Toxicol.* 1:16 (1984).

Conner, E.A., and B.A. Fowler. "Biochemical and Immunological Properties of Fish Heptic Delta-aminolevulinic Acid Dehydratase (Porphobilinogen Synthetase)," *Biochem. J.*, submitted (1991).

Conner, E.A., B.A. Fowler and S. Sassa. "Properties of f-Aminolevulinate from Fish Liver," *The Toxicologist* 11:200 (1991).

Conner, E.A., and B.A. Fowler. "Biochemical and Imnunological Properties of Fish Hepatic δ-Aminolevulinic Acid Dehydratase," Proceedings of the Sixth International Symposium on Responses of Marine Organisms to Pollutants, Woods Hole, Massachusetts, 1991.

Cousins, R.J. "Absorption, Transport, and Hepatic Metabolism of Copper and Zinc: Special Reference to Metallothionein and Ceruloplasmin," *Physiol. Rev.* 65:238–309 (1985).

Cousins, R.J., and A.S. Leinart. "Tissue-specific Regulation of Zinc Metabolism and Metallothionein Genes by Interleukin 1," *FASEB J.* 2:2884–2890 (1988).

Craig, E.A. "Essential Roles of 70KDal Heat Inducible Proteins," Bioessays 11:48–52 (1989).

Craig, E.A., J. Kramer and J. Kosic-Smithers. "SSC1, a Member of the 70kDa Heat Shock Protein Multigene Family of *Saccharomyces cerevisae* Is Essential for Growth," *Proc. Natl. Acad. Sci. U.S.A.* 84:4156–4160 (1987).

Cundy, K.C., R. Kohen and B.N. Ames. "Determination of 8-Hydroxyde-Oxyguanosine in Human Urine: A Possible Assay for In Vivo Oxidative Damage," in *Oxygen Radicals in Biology and Medicine*, M.G. Simic, K.A. Taylor, J.E. Ward and C. von Sonntag, Eds. (New York: Plenum Press, 1988), pp. 479–482.

Curtis, C.R., R.K. Howell and D.F. Kremer. "Soybean Peroxidases from Soybean Injury," *Environ. Pollut.* 11:189–194 (1976).

Dalwani, R., J.M. Dave and K. Datta. "Alterations in Hepatic Heme Metabolism in Fish Exposed to Sublethal Cadmium Levels," *Biochem. Int.* 10(1):33–42 (1985).

Danielson, K.G., S. Ohi and P.C. Huang. "Immunochemical Detection of Metallothionein in Specific Epithelial Cells of Rat Organs," *Proc. Natl. Acad. Sci. U.S.A.* 79:2301–2304 (1982).

De Duve, C., and O. Hayaishi, Eds. *Tocopherol, Oxygen and Biomembranes* (Amsterdam: Elsevier/North Holland, 1978).

Deshaies, R.J., B.D. Koch, M. Weiner-Washiburne, E. Craig and R. Schekman. "A Subfamily of Stress Proteins Facilitates Translocation of Secretory and Mitochondrial Precursor Polypeptides," *Nature (London)* 332:800–805 (1988).

Diesseroth, A., and A.L. Dounce. "Catalase: Physical and Chemical Properties, Mechanisms of Catalysis, and Physiological Role," *Physiol. Rev.* 50:319–375 (1970).

Di Giulio, R.T., P.C. Washburn, R.J. Wenning, G.W. Winston and C.S. Jewell. "Biochemical Responses in Aquatic Animals: A Review of Determinants of Oxidative Stress," *Environ. Toxicol. Chem.* 8:1103–1123 (1989).

Dillard, C.J., and A.L. Tappel. "Fluorescent Damage Products of Lipid Peroxidation," in *Oxygen Radicals in Biological Systems*, L. Packer, Ed. (New York: Academic Press, 1984), pp. 337–341.

Ding, V.D.-H., and C.B. Pickett. "Transcriptional Regulation of Rat Liver Glutathione S-Transferase Genes by Phenobarbital and 3-Methylcholanthrene," *Arch. Biochem. Biophys.* 240:553–559 (1985).

DiSilvestro, R.A., and R.J. Cousins. "Mediation of Endotoxin-Induced Changes in Zinc Metabolism in Rats," *Am. J. Physiol.* 247:E436–441 (1984).

Dizdaroglu, M., and D.S. Bergtold. "Characterization of Free Radical-Induced Base Damage in DNA at Biologically Relevant Levels," *Anal. Biochem.* 156:182–188 (1986).

Draper, H.H., and M. Hadley. "Malondialdehyde Determination as Index of Lipid Peroxidation," in *Oxygen Radicals in Biological Systems, Part B, Oxygen Radicals and Antioxidants*, L. Packer, and A.N. Glazer, Eds. (New York: Academic Press, 1990), pp. 421–431.

Dutton, G.J. *Glucuronidaton of Drugs and Other Foreign Compounds*, (Boca Raton, FL: CRC Press, 1980).

Dykens, J.A., and M. Shick. "Oxygen Production by Endosymbiotic Algae Controls Superoxide Dismutase Activity in Their Animal Host," *Nature (London)* 297:579–580 (1982).

Eaton, D.L., and B.F. Toal. "Evaluation of the Cd/Hemoglobin Affinity Assay for the Rapid Determination of Metallothionein in Biological Tissues," *Toxicol. Appl. Pharmacol.* 66:134–142 (1982).

Edwards, D.P., D.J. Adams, N. Savage and W.L. McGuire. "Estrogen Induced Synthesis of Specific Proteins in Human Breast Cancer Cells," *Biochem. Biophys. Res. Comm.* 93:804–812 (1980).

Elcombe, C.R., and J.J. Lech. "Induction of Microsomal Hemoprotein(s) P450 in the Rat and Rainbow Trout by Polyhalogenated Biphenyls: A Comparative Study," *Environ. Health Perspect.* 23:309 (1979).

Elder, G.H., and S.A. Matlin. "The Effect of the Porphyrogenic Compound, Hexachlorobenzene, on the Activity of Hepatic Uroporphyrinogen Decarboxylase in the Rat," *Biol. Sci. Mol. Med.* 51:71–80 (1976).

Elskus, A.A., and J.J. Stegeman. "Induced Cytochrome P-450 in *Fundulus heteroclitus* Associated with Environmental Contamination by Polychlorinated Biphenyls and Polynuclear Aromatic Hydrocarbons," *Mar. Environ. Res.* 27:31–50 (1989).

Engel, D.W. "The Effect of Biological Variability on Monitoring Strategies: Metallothioneins as an Example," *Water Resour. Bull.* 24:981–987 (1988).

Engel, D.W., and M. Brouwer. "Trace Metal-Binding Proteins in Marine Molluscs and Crustaceans," *Mar. Environ. Res.* 13:177–194 (1984a).

Engel, D.W., and M. Brouwer. "Cadmium-Binding Proteins in the Blue Crab, *Callinectes sapidus*: Laboratory-Field Comparison," *Mar. Environ. Res.* 14:139–151 (1984b).

Engel, D.W., and M. Brouwer. "Metal Regulation and Molting in the Blue Crab, Callinectes sapidus: Metallothionein Function in Metal Metabolism," *Biol. Bull.* 172:239–251 (1987).

Engel, D.W., and M. Brouwer. "Metallothionein and Metallothionein-Like Proteins: Physiological Importance," *Adv. Comp. Environ. Physiol.* 4:53–75 (1989).

Engel, D.W., and B.A. Fowler. "Factors Influencing the Accumulation and Toxicity of Cadmium to Marine Organisms," *Environ. Health Perspect.* 28:81–88 (1979).

Engel, D.W., and G. Roesijadi. "Metallothioneins: A Monitoring Tool," in *Pollution and Physiology of Estuarine Organisms*, F.J. Vernberg, F.P. Thurberg, A. Calabrese and W.B. Vernberg, Eds. (Columbia: University of South Carolina Press, 1987), pp. 421–438.

Enomoto, K., T.S. Ying, M.J. Griffin and E. Farber. "Immunohistochemical Study of Epoxide Hydrolase During Experimental Liver Carcinogenesis," *Cancer Res.* 41: 3281–3287 (1981).

Esterbauer, H., and D. Grill. "Seasonal Variation of Glutathione and Glutathione Reductase in Needles of *Picea abies*," *Plant Physiol.* 61:119–121 (1978).

Esterbauer, H., J. Lang, S. Zadravec and T.F. Slater. "Detection of Malonaldehyde by High-Performance Liquid Chromatography," in *Oxygen Radicals in Biological Systems*, L. Packer, Ed. (New York: Academic Press, 1984), pp. 319–328.

Fahimi, H. D., and H. Sies, Eds. *Peroxisomes in Biology and Medicine* (New York: Springer-Verlag, 1987).

Fariss, M.W., and D.J. Reed. "High-Performance Liquid Chromatography of Thiols and Disulfides: Dinitrophenol Derivatives," *Methods Enzymol.* 143:101–109 (1987).

Feldman, S.L., and R.J. Cousins. "Degradation of Hepatic Zinc-Thionein Following Parenteral Zinc Administration," *Biochem. J.* 160:583–590 (1976).

Finley, D., E. Ozkaynak and A. Varshavsky. "The Yeast Polyubiquitin Gene Is Essential for Resistance to High Temperatures, Starvation, and Other Stresses," *Cell* 48:1035–1046 (1987).

Flohe, L. "Glutathione Peroxidase Brought Into Focus," in *Free Radicals in Biology*, W. A. Pryor, Ed. (New York: Academic Press, 1982), pp. 223–254.

Floyd, R.A., J.J. Watson, P.K. Wong, D.H. Altmiller and R.C. Rickard. "Hydroxyl Free Radical Adduct of Deoxyguanosine: Sensitive Detection and Mechanisms of Formation," *Free Rad. Res. Comm.* 1:163–172 (1986).

Floyd, R.A., M.S. West, W.E. Hogsett and D.T. Tingey. "Increased 8-Hydroxyguanosine Content of Chloroplast DNA from Ozone-Treated Plants," *Plant Physiol.* 91:644–647 (1989).

Förlin, L. and T. Andersson. "Influence of Biological and Environmental Factors on Hepatic

Steroid and Xenobiotic Metabolism in Fish: Interaction with PCB and β-Naphthoflavone,"
Mar. Environ. Res. 14:47–58 (1984).

Förlin, L. and T. Andersson. "Storage Conditions of Rainbow Trout Liver Cytochrome P-450 and Conjugating Enzymes," *Comp. Biochem. Physiol.* 80B:569–572 (1985).

Förlin, L., T. Andersson, U. Koivusaari and T. Hansson. "Influence of Biological and Enviornmental Factors on Hepatic Steroid and Xenobiotic Metabolism in Fish: Interaction with PCB and β-Naphthoflavone," *Mar. Environ. Res.* 14:47-58 (1984).

Förlin, L., and C. Haux. "Increased Excretion in the Bile of 17β-^{3}H-Estradiol-Derived Radioactivity in Rainbow Trout Treated with β-Naphthoflavone," *Aquat. Toxicol.* 6:197–208 (1985).

Förlin, L., C. Haux, T. Andersson, P.-E. Olsson and A. Larsson. "Physiological Methods in Fish Toxicology: Laboratory and Field Studies," in *Fish Physiology: Recent Advances*, S. Nilsson, and S. Holmgren, Eds. (London: Croom Helm, 1986), pp. 158–169.

Foulkes, E.C., Ed. *The Biological Roles of Metallothionein*, (Amsterdam: Elsevier/North-Holland Biomedical Press, 1982).

Foureman, G.L. "Enzymes Involved in Metabolism of PAH by Fishes and Other Aquatic Animals: Hydrolysis and Conjugation Enzymes (or Phase II Enzymes)," in *Metabolism of Polycyclic Aromatic Hydrocarbons in the Aquatic Environment*, U. Varanasi, Ed. (Boca Raton, FL: CRC Press Inc., 1989), pp. 185–202.

Foureman, G.L., N.B. White and J.R. Bend. "Biochemical Evidence that Winter Flounder (*Pseudopleuronectes americanus*) Have Induced Hepatic Cytochrome P-450-Dependent Monooxygenase Activities," *Can. J. Fish. Aquat. Sci.* 40:854–865 (1983).

Fowler, B.A. "Intracellular Compartmentation of Metals in Aquatic Organisms: Relationships to Mechanisms of Cell Injury," *Environ. Health Perspect.* 71:121–128 (1987).

Fowler, B.A., and E. Gould. "Ultrastructural and Biochemical Studies of Intracellular Metal-Binding Patterns in Kidney Tubule Cells of the Scallop *Platopecten magellanius* Following Prolonged Exposure to Cadmium," *Mar. Biol.* 97:207–212 (1988).

Fowler, B.A., C.E. Hildebrand, Y. Kojima and M. Webb. "Nomenclature of Metallothionein," *Experientia Suppl.* 52:19–22 (1987).

Fowler, B.A., G.W. Lucier and A.W. Hayes. "Organelles as Tools in Toxicology." in *Principles and Methods of Toxicology*, A.W. Hayes, Ed. (New York: Raven Press, 1989), pp. 815–833.

Fowler, B.A., R. Kardish and J.S. Woods. "Alteration of Hepatic Microsomal Structure and Function by Acute Indium Administration: Ultrastructural Morphometric and Biochemical Studies," *Lab. Invest.* 48:471–478 (1983).

Fowler, B.A., and K.R. Mahaffey. "Interactions Among Lead, Cadmium, Arsenic in Relation to Porphyrin Excretion Patterns," *Environ. Health Perspect.* 25:87–90 (1978).

Fowler, B.A., A. Oskarsson and J.S. Woods. "Metal and Metalloid-Induced Porplyrincerias: Relationship to Cell Drying," *Ann. N.Y. Acad. Sci.* 514:172–182 (1987).

Fox, G.A., S.W. Kennedy, R.J. Norstrom and D.C. Wigfield. "Porphyria in Herring Gulls: A Biochemical Response to Chemical Contamination of Great Lakes Food Chains," *Environ. Toxicol. Chem.* 7:831–839 (1988).

Foyer, C.H., J. Rowell and D.A. Walker. "Measurement of the Ascorbate Content of Spinach Leaf Protoplasts and Chloroplasts During Illumination," *Planta* 157:239–244 (1983).

Frank, L., J.R. Bucher and R.J. Roberts. "Oxygen Toxicity in Neonatal and Adult Animals of

Various Species," *J. Appl. Physiol. Respir. Environ. Exercise Physiol.* 45:699–704 (1978).

Freedman, J.H., L. Powers and J. Peisach. "Structure of the Copper Cluster in Canine Hepatic Metallothionein Using X-ray Absorption Spectroscopy," *Biochemistry* 25:2342–2349 (1986).

Frew, J.E., and P. Jones. "Structure and Functional Properties of Peroxidases and Catalases," in *Advances in Inorganic and Bioinorganic Mechanisms, Vol. 3.*, A.G. Sykes, Ed. (New York: Academic Press, 1984), pp. 319–328.

Fridovich, I. "Superoxide Dismutases," *Adv. Enzymol.* 58:61–97 (1986).

Friedman, R.L., and K.T. Stark. "Alpha-Interferon-Induced Transcription of HLA and Metallothionein Genes Containing Homologous Upstream Sequences," *Nature (London)* 314:637–639 (1985).

Furey, W.F., A.H. Robbins, L.L. Clancy, D.R. Winge, B.C. Wang and C.D. Stout. "Crystal Structure of Cd, Zn Metallothionein," *Science* 231:704–710 (1986).

Gabryelak, T., and J. Klekot. "The Effect of Paraquat on the Peroxide Metabolism Enzymes in Erythrocytes of Freshwater Fish Species," *Comp. Biochem. Physiol.* 81C:415–418 (1985).

Gaitanaris, G.A., A.G. Papavassiliou, P. Ruback, S.J. Silverstein and M.E. Gottesman. "Renaturation of Denatured Repressor Requires Heat Shock Proteins," *Cell* 61: 1013–1020 (1990).

Garvey, J. "Metallothionein: Structure/Antigenicity and Detection/Quantitation in Normal Physiological Fluids," *Environ. Health Perspect.* 54:117–127 (1984).

Gaspar, T., C. Penel, T. Thorpe and H. Greppin. *Peroxidases, 1970–1980. A Survey of Their Biochemical and Physiological Roles in Higher Plants,* (Geneva: Universite de Geneve, Centre de Botanique, 1982).

George, S.G., and G. Buchanan. "Isolation, Properties and Induction of Plaice Liver Cytosolic Glutathione-S-Transferases," *Fish Physiol. Biochem.* 8:437–449 (1990).

George, G.N., J. Byrd and D.R. Winge. "X-ray Absorption Studies of Yeast Copper Metallothionein," *J. Biol. Chem.* 263:8199–8203 (1988).

George, S.G., E. Carpene, T.L. Coombs, J. Overnell and A. Youngson. "Characterisation of Cadmium-Binding Proteins from Mussels, *Mytilus edulis* (L), Exposed to Cadmium," *Biochim. Biophys. Acta* 580:225–233 (1979).

George, S.G., B.J.S. Pirie and T.L. Coombs. "Isolation and Elemental Analysis of Metal-Rich Granules from the Kidney of the Scallop, *Pecten maximus* (L)," *J. Exp. Mar. Biol. Ecol.* 42:143–156 (1980).

Georgellis, A., J. Montelius and J. Rydstrom. "Evidence for a Free-Radical-Dependent Metabolism of 7,12-Dimethylbenzo(a)anthracene in Rat Testis," *Toxicol. Appl. Pharmacol.* 87:141–154 (1987).

Giam, C.S., Ed. *Pollutant Effects on Marine Organisms,* (Lexington, MA: Lexington Books, 1977), p. 213.

Giam, C.S., and L.E. Ray, Eds. *Pollutant Studies in Marine Animals,* (Boca Raton, FL: CRC Press, 1987).

Gibson, K.D., A. Neuberger and J.J. Scott. "The Purification and Properties of Delta-Aminolevulinic Acid Dehydratase," *Biochem. J.* 61:618–621 (1955).

Goksøyr, A. "Purification of Hepatic Microsomal Cytochromes P-450 from β-Naphthoflavone-Treated Atlantic Cod (*Gadus morhua*), a Marine Teleost Fish," *Biochim. Biophys. Acta* 840:409–417 (1985).

Goksøyr, A., T. Andersson, D.R. Buhler, J.J. Stegeman, D.E. Williams and L. Förlin. "Immunochemical Cross-Reactivity of β-Naphthoflavone-Inducible Cytochrome P450

(P4501A) in Liver Microsomes from Different Fish Species and Rat," *Fish Physiol. Biochem.* 9 (1):1–13 (1991).

Goksøyr, A., T. Andersson, L. Förlin, E.A. Snowberger, B.R. Woodin and J.J. Stegeman. "Cytochrome P-450 Monooxygenase Activity and Immunochemical Properties of Adult and Foetal Piked (Minke) Whales, *Balaenoptera acutorostrata*," in *Cytochrome P-450, Biochemistry and Biophysics*, I. Schuster, Ed. (Bassingstoke, UK: Taylor and Francis, 1989b), pp. 698–701.

Goksøyr, A., T. Andersson, T. Hansson, Klungsøyr, Y. Zhang and L. Förlin. "Species Characteristics of the Hepatic Xenobiotic and Steroid Biotransformation Systems of Two Teleost Fish, Atlantic Cod (*Gadus morhua*) and Rainbow Trout (*Salmo gairdneri*)," *Toxicol. Appl. Pharmacol.* 89:347–360 (1987).

Goksøyr, A., B. Serigstad, T.S. Solberg and J.J. Stegeman. "Response of Cod (*Gadus morhua*) Larvae and Juveniles to Oil Exposure Detected with Anti-Cod Cytochrome P-450c and Anti-Scup Cytochrome P-450E MAb 1-12-3," *Mar. Environ. Res.* 24:31–35 (1988).

Goldberg, D.M., and R.J. Spooner. "Glutathione Reductase," in *Methods of Enzymatic Analysis*, H.U. Bergmeyer, Ed. (Deerfield Beach, FL: Verlag Chemie, 1983), pp. 258–265.

Goldfarb P., J.A. Spry, D. Dunn, D. Livingstone, A. Wiseman and G.G. Gibson. "Detection of mRNA Sequences Homologous to the Human Glutathione Peroxidase and Rat Cytochrome P-450IVA1 genes in *Mytilus edulis*," *Mar. Environ. Res.* 28:57–60 (1989).

González, O., J. Fernández and M. Martín. "Inhibition of Trout (*Salmo gairdneri* R.) PBG-Synthase by Some Metal Ions (Mg^{2+}, Pb^{2+}, Zn^{2+})," *Comp. Biochem. Physiol.* 86C:163–167 (1987).

Gooch, J.W., A.A. Elskus, P.J. Kloepper-Sams, M.E. Hahn and J.J. Stegeman. "Effects of Ortho and Non-Ortho Substituted Polychlorinated Biphenyl Congeners on the Hepatic Monooxygenase System in Scup (*Stenotomus chrysops*)," *Toxicol. Appl. Pharmacol.* 98:422–433 (1989).

Griffith, O.W. "Determination of Glutathione and Glutathione Disulfide Using Glutathione Reductase and 2-Vinylpyridine," *Anal. Biochem.* 106:207–212 (1980).

Grill, D., H. Esterbauer and U. Klosch. "Effect of Sulphur Dioxide on Glutathione in Leaves of Plants," *Environ. Pollut.* 18:187–194 (1979).

Grill, E., E.L. Winnacker and M.H. Zenk. "Phytochelatins: The Principal Heavy-Metal Complexing Peptides of Higher Plants," *Science* 230:674–676 (1985).

Guengerich, F.P., and D.C. Liebler. "Enzymatic Activation of Chemicals to Toxic Metabolites," *CRC Crit. Rev. Toxicol.* 14:259–307 (1985).

Guri, A. "Variation in Glutathione and Ascorbic Acid Content Among Selected Cultivars of *Phaseolus vulgaris* Prior to and after Exposure to Ozone," *Can. J. Plant Sci.* 63:733–737 (1983).

Haasch, M.L., P.J. Wejksnora, J.J. Stegeman and J.J. Lech. "Cloned Rainbow Trout Liver P_1450 Complementary DNA as a Potential Environmental Monitor," *Toxicol. Appl. Pharmacol.* 98:362–368 (1989).

Hahn, M.E., B.R. Woodin and J.J. Stegeman. "Induction of P450E (P450IA1) by 2,3,7,8-Tetrachlorodibenzofuran (2,3,7,8-TCDF) in the Marine Fish Scup (*Stenotomus chrysops*)," *Mar. Environ. Res.* 28:61–65 (1989).

Hahn, M.E., A. Poland, E. Glover and J.J. Stegeman. "The Ah Receptor in Marine Animals: Phylogenetic Distribution and Relationship to Cytochrome P4501A Inducibility," *Mar. Environ. Res.*, in press (1991).

Halliwell, B., and J.M.C. Gutteridge. *Free Radicals in Biology and Medicine*, (Oxford: Clarendon Press, 1985).

Hamer, D.H. "Metallothionein," *Ann. Rev. Biochem.* 55:913–951 (1986).

Hamilton, S.J., and P.M. Mehrle. "Metallothionein in Fish: Review of Its Importance in Assessing Stress from Metal Contaminants," *Trans. Am. Fish. Soc.* 115:596–609 (1986).

Hammock, B.D., and K. Ota. "Differential Induction of Cytosolic Epoxide Hydrolase, Microsomal Epoxide Hydrolase and Glutathione S-Transferase Activities," *Tox. Appl. Phamacol.* 71:254–265 (1983).

Hammond, G.L., Y.K. Lai and C.L. Market. "Diverse Forms of Stress Lead to New Patterns of Gene Expression Through a Common and Essential Metabolic Pathway," *Proc. Natl. Acad. Sci. U.S.A.* 79:3485–3488 (1982).

Hanson-Painton, O., M.J. Griffin and J. Tang. " Involvement of a Cytosolic Carrier Protein in the Microsomal Metabolism of Benzo[a]pyrene in Rat Liver," *Cancer Res.* 43:4198–4206 (1983).

Hansson, T., J. Rafter and J.-A. Gustafsson. "Effects of Some Common Inducers on the Hepatic Microsomal Metabolism of Androstenedione in Rainbow Trout with Special Reference to Cytochrome P-450-Dependent Enzymes," *Biochem. Pharmacol.* 29:583–587 (1980).

Haux, C., and L. Förlin. "Biochemical Methods for Detecting Effects of Contaminants on Fish," *Ambio* 17:376–380 (1988).

Hayes, M.A., I.R. Smith, T.H. Rushmore, T.L. Crane, C. Thorn, T.E. Kocal and H.W. Ferguson. "Pathogenesis of Skin and Liver Neoplasms in White Suckers from Industrially Polluted Areas in Lake Ontario," *Sci. Total Environ.* 94:105–123 (1990).

Hegesh, E., N. Gruener, S. Cohen, R. Bochkovsky and H. Shuval. "A Sensitive Micromethod for Determination of Methemoglobin in Blood," *Clin. Chim. Acta* 30:679–682 (1970).

Heikkila, J.J., G.A. Schultz, K. Iatrou and L. Gedamu. "Expression a Set of Fish Genes Following Heat or Metal Ion Exposure," *J. Biol. Chem.* 257:12000–12005 (1982).

Heilmann, L.J., Y.-Y. Sheen, S.W. Bigelow and D.W. Nebert. "Trout P4501A1: cDNA and Deduced Protein Sequence Expression in Liver, and Evolutionary Significance," *DNA* 7:379–387 (1988).

Heisinger, J.F., and S.M. Dawson. "Effect of Selenium Deficiency on Liver and Blood Glutathione Peroxidase Activity in the Black Bullhead," *J. Expt. Zool.* 225:325–328 (1983).

Heuss-La Rosa, K., R.R. Mayer and J.H. Cherry. "Synthesis of Only Two Heat Shock Proteins Is Required for Thermoadaptation in Cultured Cowpea Cells," *Plant Physiol.* 85:4–7 (1987).

Higham, D.P., J.K. Nicholson, J. Overnell and P.J. Sadler. "NMR Studies of Crab and Plaice Metallothionein," *Environ. Health Perspect.* 65:157–165 (1986).

Hodson, P.V. "δ-Aminolevulinic Acid Dehydratase Activity of Fish Blood as an Indicator of a Harmful Exposure to Lead," *J. Fish. Res. Bd. Can.* 33:268–271 (1976).

Hogstrand, C., and C. Haux. "Radioimmunoassay for Perch (*Perca fluviatilis*) Metallothionein," *Toxicol. Appl. Pharmacol.* 103:56–65 (1990).

Horsman, D.C., and A.R. Wellburn. "Synergistic Effect of SO_2 and NO_2 Polluted Air Upon Enzyme Activity in Pea Seedlings," *Environ. Pollut.* 8:123–133 (1975).

Horton, A.A., and S. Fairhurst. "Lipid Peroxidation and Mechanisms of Toxicity," *CRC Crit. Rev. Toxicol.* 18:27–79 (1987).

Hossain, M.A., Y. Nakano and K. Asada. "Monodehydroascorbate Reductase in Spinach Chloroplast and Its Participation in Regeneration of Ascorbate for Scavenging Hydrogen Peroxide," *Plant Cell Physiol.* 25:385–395 (1984).

Hung, S.S.O., and S.J. Slinger. "Effect of Dietary Vitamin E on Rainbow Trout (*Salmo gairdneri*)," *Int. J. Vit. Nutr. Res.* 52:119–124 (1982).

Hunn, J.B., R.A. Schoettger and W.A. Willford. "Turnover and Urinary Excretion of Free M.S. 222 by Rainbow Trout, *Salmo gairdneri*," *J. Fish. Res. Bd. Can.* 25:25–31 (1966).

Imlay, J.A., and S. Linn. "DNA Damage and Oxygen Radical Toxicity," *Science* 240:1302–1309 (1988).

Ireland, R., and E. Burger. "Synthesis of Low Molecular Weight Heat Shock Proteins Stimulated by Ecdysterone in a Cultured Drosophila Cell Line," *Proc. Natl. Acad. Sci. U.S.A.* 79:855–859 (1982).

Irons, R.D., and T. Sawahata. "Phenols, Catechols, and Quinones," in *Bioactivation of Foreign Compounds*, M. W. Anders, Ed. (Orlando,FL: Academic Press, 1985), pp. 259–281.

Jackim, E. "Influence of Lead and Other Metals on Fish Delta-Aminolevulinate Dehydratase Activity," *J. Fish. Res. Bd. Can.* 30:560–562 (1973).

Jacobs, J. M., N. J. Jacobs, S. E. Borotz and M. L. Guerinot. "Effects of the Photobleaching Herbicide, Aciflurofen-Methyl, on Protoporphyrinogen Oxidation in Barley Organelles, Soybean Root Mitochondria, Soybean Root Nodules and Bacteria," *Arch. Biochem. Biophys.* 280:369–375 (1990).

James, M.O. "Conjugation of Organic Pollutants in Aquatic Species," *Environ. Health Perspect.* 71:97–103 (1987).

James, M.O. "Cytochrome P450 Monooxygenase in Crustaceans," *Xenobiotica* 19:1063–1076 (1989).

James, M.O., and J.R. Bend. "Polyaromatic Hydrocarbon Induction of Cytochrome P-450-Dependent Mixed Function Oxidases in Marine Fish," *Toxicol. Appl. Pharmacol.* 54:117–133 (1980).

James, M.O., E.R. Bowen, P.M. Dansette and J.R. Bend. "Epoxide Hydrase and Glutathione S-Transferase Activities with Selected Alkene and Arene Oxides in Several Marine Species," *Chem.-Biol. Interact.* 25:321–344 (1979).

James, M.O., C.S. Heard and W.E. Hawkins. "Effect of 3-Methylcholanthrene on Monooxygenase, Epoxide Hydrolase, and Glutathione S-Transferase Activities in Small Estuarine and Freshwater Fish," *Aquat. Toxicol.* 12:1–15 (1988).

James, M.O., and P.J. Little. "Polyhalogenated Biphenyls and Phenobarbital: Evaluation as Inducers of Drug Metabolizing Enzymes in the Sheepshead (*Archosargus probatocephalus*)," *Chem.-Biol. Interact.* 36:229–248 (1981).

James, M.O., and K. Shiverick. "Cytochrome P-450-Dependent Oxidation of Progesterone, Testosterone, and Ecdysone in the Spiny Lobster, *Panulirus argus*," *Arch. Biochem. Biophys.* 233:1–9 (1984).

Johansson, I., and M. Ingelman-Sundberg. "Hydroxyl Radical-Mediated, Cytochrome P-450-Dependent Metabolic Activation of Benzene in Microsomes and Reconstituted Enzyme Systems from Rabbit Liver," *J. Biol. Chem.* 258:7311–7316 (1983).

Johansson-Sjobeck, M.L., and A. Larsson. "Effects of Inorganic Lead on Delta-Aminolevulinic Acid Dehydratase Activity and Hematological Variables in Rainbow Trout, *Salmo gairdneri*," *Arch. Environ. Contam. Toxicol.* 8:419–431 (1979).

Kagi, J.H.R., and Y. Kojima. "Chemistry and Biochemistry of Metallothionein," *Experientia Suppl.* 52:35–61 (1987).

Kagi, J.H.R., and M. Nordberg. Eds. *Metallothionein*, (Basel: Birkhauser, 1978).

Kagi, J.H.R., M. Vasak, K. Lerch, D.E.O. Gilg, P. Hunyiker, W.R. Bernhard and M. Good. "Structure of Mammalian Metallothionein," *Environ. Health Perspect.* 54:93 103 (1984)

Kalt-Torres, W., J.J. Burke and J.M. Anderson. "Chloroplast Glutathione Reductase: Purification and Properties," *Plant Physiol.* 61:271–278 (1984).

Kapoor, M., and J. Lewis. "Alteration of the Protein Synthesis Pattern in *Neurospora crassa* Cells by Hyperthermal and Oxidative Stress," *Can. J. Microbiol.* 33:162–168. (1987).

Kappus, H. "Lipid Peroxidation: Mechanisms, Analysis, Enzymology and Biological Relevance," in *Oxidative Stress*, H. Sies, Ed. (London: Academic Press, 1985), pp. 273–310.

Kappus, H. "Oxidative Stress in Chemical Toxicity," *Arch. Toxicol.* 60:144–149 (1987).

Karin, M. "Metallothionein: Proteins in Search of Function," *Cell* 41:9–10 (1985).

Kasper, C.B., and D. Henton. "Glucuronidation," in *Enzymatic Basis of Detoxification, Vol. II*, W.B. Jakoby, Ed. (New York: Academic Press, 1980), pp. 4–36.

Kee, S.C., and P.S. Noble. "Concomitant Changes in High Temperature Tolerance and Heat-Shock Proteins in Desert Succulents," *Plant Physiol.* 80:596–598 (1986).

Keeling, P.L., and L.L. Smith. "Relevance of NADPH Depletion and Mixed Disulphide Formation in Rat Lung to the Mechanism of Cell Damage Following Paraquat Administration," *Biochem. Pharmacol.* 31:3243–3249 (1982).

Keller, T. "The Use of Peroxidase for Monitoring and Mapping Air Pollution Areas," *Eur. J. Forest Pathol.* 4:11–19 (1974).

Keyse, S.M., and R.M. Tyrrell. "Heme Oxygenase Is the Major 32-kDa Stress Protein Induced in Human Skin Fibroblasts by UVA Radiation, Hydrogen Peroxide, and Sodium Arsenite," *Proc. Natl. Acad. Sci. U.S.A.* 86:99–103 (1989).

Kirby, G. M., M. Stalker, C. Metcalfe, T. Kocal, H. Ferguson and M. A. Hayes. "Expression of Immunoreactive Glutathione S-Transferases in Hepatic Neoplasms Induced by Aflatoxin B1 or 1,2-Dimethylanthracene in Rainbow Trout (*Oncorhyncus mykiss*)," *Carcinogenesis* 11:2255–2257 (1990).

Kitahara, A., K. Satoh, K. Nishimura, T. Ishikawa, K. Riuke, K. Sato, H. Tsuda and N. Ito. "Changes in Molecular Forms of Rat Hepatic Glutathione S-Transferase During Chemical Hepatocarcinogenesis," *Cancer Res.* 44:2698–2703 (1984).

Kleinow, K., M.J. Melancon and J.J. Lech. "Biotransformation and Induction: Implications for Toxicity, Bioaccumulation and Monitoring of Environmental Xenobiotics in Fish," *Environ. Health Perspect.* 71:105–119 (1987).

Klerks, P.L., and J. Levinton. "Effects of Heavy Metals in a Polluted Aquatic Environment," in *Ecotoxicology: Problems and Approaches*, S.A. Levin, G.R. Kelly and M.A. Harwell, Eds. (Berlin: Springer, 1988), pp. 57–84.

Kloepper-Sams, P.J., and J.J. Stegeman. "The Temporal Relationships Between P450E Protein Content, Catalytic Activity, and mRNA Levels in the Teleost *Fundulus heteroclitus* Following Treatment with β-Naphthoflavone," *Arch. Biochem. Biophys.* 268:525–535 (1989).

Klotz, A., J.J. Stegeman and C. Walsh. "An Aryl Hydrocarbon Hydroxylating Hepatic Cytochrome P-450 from the Marine Fish *Stenotomus chrysops*," *Arch. Biochem. Biophys.* 226:578–592 (1983).

Klotz, A.V., J.J. Stegeman and C. Walsh. "An Alternative 7-Ethoxyresorufin O-Deethylase Activity Assay: A Continuous Visible Spectrophotometric Method for Measurement of Cytochrome P-450 Monooxygenase Activity," *Anal. Biochem.* 140:138–145 (1984).

Klotz, A., J. Stegeman, B. Woodin, E. Snowberger, P. Thomas and C. Walsh. "Cytochrome P450 Isozymes from the Marine Teleost *Stenotomus chrysops*: Their Roles in Steroid Hydroxylation and the Influence of Cytochrome b_5," *Arch. Biochem. Biophys.* 249: 326–338 (1986).

Knight, G.C., and C.H. Walker. "A Study of the Hepatic Microsomal Monooxygenase of Sea Birds and Its Relationship to Liposoluble Pollutants," *Comp. Biochem. Physiol.* 73C:211–221 (1982).

Knox, D., C. B. Cowey and J. W. Adron. "The Effect of Low Dietary Manganese Intake on Rainbow Trout (*Salmo gairdneri*)," *Br. J. Nutr.* 46:495–501 (1981).

Kobayashi, S. "Changes in Catalase Activity of the Tissues and Blood of "Masu", *Oncorhynchus masou*, when Transferred from Fresh Water to Sea Water," *Bull. Fac. Fish.* 6:1–6 (1955).

Koivusaari, U., and T. Andersson. "Partial Temperature Compensation of Hepatic Biotransformation Enzymes in Juvenile Rainbow Trout (*Salmo gairdneri*) During the Warming of Water in Spring,"*Comp. Biochem. Physiol.* 78B:223–226 (1984).

Koivusaari, U., P. Lindstrom-Seppä and O. Hänninen. "Xenobiotic Metabolism in Rainbow Trout Intestine," *Adv. Physiol. Sci.* 29:433–440 (1981a).

Koivusaari, U., M. Harri and O. Hänninen. "Seasonal Variation of Hepatic Biotransformation in Female and Male Rainbow Trout (*Salmo gairdneri*)," *Comp. Biochem Physiol.* 79C: 149–157 (1981b).

Koivusaari, U., M. Pesonen and O. Hänninen. "Polysubstrate Monooxygenase Activity and Sex Hormones in Pre- and Postspawning Rainbow Trout (*Salmo gairdneri*)," *Aquat. Toxicol.* 5:67-76 (1984).

Kosower, N.S., and E.M. Kosower. "The Glutathione Status of Cells," *Int. Rev. Cytol.* 54:109-160 (1978).

Krause, K.W., E.M. Hallberg and R.L. Hallberg. "Characterization of a *Tetrahymena thermophila* Mutant Strain Unable to Develop Normal Thermotolerance," *Mol. Cell Biol.* 6:3854-3861 (1986).

Landry, J., D. Bernier, P. Cretien, L.M. Nicole, R.M. Tanguay and N. Marceau. "Synthesis and Degradation of Heat Shock Proteins During Development and Decay of Thermotolerance," *Cancer Res.* 42:2457-2461 (1982).

Landry, J., P. Chretien, H. Lambert, E. Hickey and L. Weber. "Heat Shock Resistance Conferred by Expression of the Human Hsp27 Gene in Rodent Cells," *J. Cell Biol.* 109:7–15 (1989).

Langston, W.J., and M. Zhou. "Evaluation of the Significance of Metal-Binding Proteins in the Gastropod *Littorina littorea*," *Mar. Biol.* 92:505–515 (1986).

Larsson, Å., S. Orrenius, A. Holmgren and B. Mannervik, Eds. *Functions of Glutathione*, (New York: Raven Press, 1983).

Lawrence, R.A., and R.F. Burk. "Species, Tissue and Subcellular Distribution of Non Se-Dependent Glutathione Peroxidase Activity," *J. Nutr.* 108:211–215 (1978).

Leakey, J.E.A., and G.J. Dutton. "Effect of Phenobarbital on UDP-Glucosyltransferase Activity and Phenolic Glucosidation in the Mollusc *Arion ater*," *Comp. Biochem. Physiol.* 51C:215–217 (1975).

Lech, J.J., M.J. Vodicnik and C.R. Elcombe. "Induction of Monooxygenase Activity in Fish," in *Aquatic Toxicology*, L.J. Weber, Ed. (New York: Raven Press, 1982), pp. 107–148.

Lee, R.F. "Mixed Function Oxygenase (MFO) in Marine Invertebrates," *Mar. Biol. Let.* 2:87–105 (1982).

Lee, R.F. "Glutathione S-Transferase in Marine Invertebrates from Langesundfjord," *Mar. Ecol. Prog. Ser.* 46:33–36 (1982).

Lee, R.F., W.S. Keeran and G.V. Pickwell. "Marine Invertebrate Glutathione S-Transferase: Purification, Characterization and Induction," *Mar. Environ. Res.* 24:97–100 (1988).

Lerch, K., D. Ammer and R.W. Olafson. "Crab Metallothionein. Primary Structures of Metallothioneins 1 and 2," *J. Biol. Chem.* 257:2420-2426 (1982).

Li, F., C.K. Lim and T.J. Peters. "A High-Performance Liquid Chromatographic Method for the Assay of Coproporphyrinogen Oxidase Activity in Rat Liver," *Biochem. J.* 239:481–484 (1986).

Li, T.Y., A.J. Kraker, C.F. Shaw and D.H. Petering. "Ligand Substitution Reactions of

Metallothioneins with EDTA and Apo-Carbonic Anhydrase," *Proc. Natl. Acad. Sci. U.S.A.* 77:6334–6338 (1980).

Lindquist, S. "The Heat Shock Response," *Ann. Rev. Biochem.* 55:1151–1191 (1986).

Lindström-Seppä, P. "Seasonal Variation of the Xenobiotic Metabolizing Enzyme Activities in the Liver of Male and Female Vendace (*Coregonus albula* L.)," *Aquat. Toxicol.* 6: 323–331 (1985).

Lindström-Seppä, P. "Biomonitoring of Oil Spill in Boreal Archipelago by Xenobiotic Biotransformation in Perch (*Perca fluviatilis*)," *Ecotoxicol. Environ. Saf.* 15:162–170 (1988).

Lindström-Seppä, P., U. Koivusaari and O. Hänninen. "Extrahepatic Xenobiotic Metabolism in North-European Freshwater Fish," *Comp. Biochem. Physiol.* 69C:259–263 (1981).

Lindström-Seppä, P., and A. Oikari. "Hepatic Biotransformation in Fishes Exposed to Pulp Mill Effluents," *Water Sci. Tech.* 20:167–170 (1988).

Lindström-Seppä, P., P.J. Vuorinen, M. Vuorinen and O. Hänninen. "Effect of Bleached Kraft Pulp Mill Effluent on Hepatic Biotransformation Reactions in Vendace (*Coregonus albula* L.)," *Comp. Biochem. Physiol.* 92C:51–54 (1989).

Livingstone, D.R. "Cytochrome P-450 and Oxidative Metabolism in Molluscs," *Xenobiotica* 19:1041–1062 (1989).

Livingstone, D.R. "Organic Xenobiotic Metabolism in Marine Invertebrates," in *Advances in Comparative and Environmental Physiology*, R. Gilles, Ed. (Berlin: Springer-Verlag, 1991), pp. 45–185.

Lowry, O.H., J.V. Passonneau, D.W. Schulz and M.K. Rock. "The Measurement of Pyridine Nucleotides by Enzymatic Cycling," *J. Biol. Chem.* 236:2746 (1961).

Lucier, G.W., K.G. Nelson, R.B. Everson, T.K. Wong, R.M. Philpot, T. Tiernan, M. Taylor and G.I. Sunahara. "Placental Markers of Human Exposure to Polychlorinated Biphenyls and Polychlorinated Dibenzofurans," *Environ. Health Perspect.* 76:79–87 (1987).

Madamanchi, N.R., A. Hausladen, R.G. Alscher, R. Amundson and S. Fellows. "Seasonal Changes in Antioxidant Activities in Red Spruce (*Picea rubens* Sarg.) from Three Field Sites in the Northeastern United States," *New Phytol.* 118:331–338 (1991).

Mahaffey, K.R., S.G. Capar, B.C. Gladen and B.A. Fowler. "Concurrent Exposure to Lead, Cadmium, and Arsenic: Effects on Toxicity and Tissue Metal Concentrations in the Rat," *J. Lab. Clin. Med.* 98:463–481 (1981).

Malins, D.C., G.K. Ostrander, R. Haimanot and P. Williams. "A Novel DNA Lesion in Neoplastic Livers of Feral Fish: 2,6-Diamino-4-Hydroxy-5-Formamidopyrimidine," *Carcinogensis* 11:1045–1047 (1990).

Mannervik, B. "The Isoenzymes of Glutathione Transferase," *Adv. Enzymol.* 57:357–417 (1985).

Mannervik, B., and U.H. Danielson. "Glutathione Transferases—Structure and Catalytic Activity," *CRC Crit. Rev. Biochem.* 23:283–337 (1988).

Maroni, G., J. Wise, J.E. Young and E. Otto. "Metallothionein Gene Duplications and Metal Tolerance in Natural Populations of *Drosophila melanogaster*," *Genetics* 117:739–744 (1987).

Mather-Mihaich, E., and R.T. Di Giulio. "Oxidant, Mixed-Function Oxidase, and Peroxisomal Responses in Channel Catfish Exposed to Bleached Kraft Mill Effluent," *Arch. Environ. Contam. Toxicol.* 20:391–397 (1991).

Matkovics, B., L. Szabo, Sz.I. Varga, K. Barabas and G. Berensci. "In Vivo Effects of Paraquat on Some Oxidative Enzymes in Mice and Experiments to Support the Defence Against the Poisoning," in *Biological and Clinical Aspects of Superoxide and Superoxide*

Dismutase, W. H. Bannister, and J.V. Bannister, Eds. (New York: Elsevier/North Holland Press, 1980), pp. 367–380.

Matkovics, B., L. Szabo, Sz.I. Varga, K. Barabas, G. Berensci and J. Nemcsok. "Effects of a Herbicide on the Peroxide Metabolism Enzymes and Lipid Peroxidation in Carp Fish (*Hypophthalmichthys molitrix*)," *Acta Biol. Hung.* 35:91–96 (1984).

Matringe, M., J.-M. Camadro, P. Labbe and R. Scalla. "Photoporphyrinogen Oxidase as a Molecular Target for Diphenyl Either Herbicides," *Biochem. J.* 260:231–235 (1989).

Matters, G.L., and J.G. Scandalios. "Effect of the Free Radical Generating Herbicide Paraquat on the Expression of the Superoxide Dismutase (SOD) Genes in Maize," *Biochem. Biophys. Acta* 882:29–38 (1986).

McCord, J.M., and I. Fridovich. "Superoxide Dismutase: An Enzymatic Function of Erythrocuprein," *J. Biol. Chem.* 244:6049–6055 (1969).

McIntyre, A.D., and J.B. Pierce. "Biological Effects of Marine Pollution and Monitoring," *Rapp. Proc.-Verb. Reun. Cons. Int. pour L'Explor. de la Mer.* 179:1–346 (1980).

McLemore, L., S. Adelberg, M. Czerqinski, W.C. Hubbard, S.J. Yu, R. Storeng, T.G. Wood, R.N. Hines and M.R. Boyd. "Altered Regulation of the Cytochrome P4501A1 Gene: Novel Inducer-Independent Gene Expression in Pulmonary Carcinoma Cell Lines," *J. Natl. Cancer Inst.* 81:1787–1794 (1989).

Mehlhorn, H., D.A. Cottam, P.W. Lucas and A.R. Wellburn. "Induction of Ascorbate Peroxidase and Glutathione Reductase Activities by Interactions of Mixtures of Air Pollutants," *Free Rad. Res. Comm.* 3:193–197 (1987).

Mehlhorn, H., G. Seufert, A. Schmidt and K.J. Kunert. "Effect of SO_2 and O_3 on Production of Antioxidants in Conifers," *Plant Physiol.* 82:336–338 (1986).

Meister, A. "New Aspects in Glutathione Regulation," *Fed. Proc.* 43:3031–3041 (1986).

Melancon, M.J., K.A. Turnquist and J.J. Lech. "Relation of Hepatic Microsomal Monooxygenase Activity to Tissue PCBs in Rainbow Trout (*Salmo gairdneri*) Injected with [^{14}C]PCBs," *Environ. Toxicol. Chem.* 8:777–782 (1989).

Miller, M.R., D.E. Hinton and J.J. Stegeman. "Cytochrome P-450E Induction and Localization in Gill Pillar (Endothelial) Cells of Scup and Rainbow Trout," *Aquat. Toxicol.* 14:307–322 (1989).

Miller, S.G. "Association of a Sperm-Specific Protein with the Mitochondrial F1F0-Atpase in Heliothis," *Insect Biochem.* 17:417–432 (1987).

Miranda, C. L., J.-L. Wang. M.C. Henderson and D.R. Buhler. "Purification and Characterization of Hepatic Steroid Hydroxylases from Untreated Rainbow Trout," *Arch. Biochem. Biophys.* 268: 227–238 (1989).

Mirkes, P.E. "Hyperthermia-Induced Heat Shock Response and Thermotolerance in Postimplantation Rat Embryos," *Dev. Biol.* 119:115–122 (1987).

Moromoto, R., A. Tissieres and C. Georgopoulos, Eds. *The Role of the Stress Response in Biology and Disease* (Cold Springs Harbor, New York: Cold Spring Harbor Laboratory, 1990).

Morreal, C.E., T.L. Dao, K. Eskins, C.L. King and J. Dienstag. "Peroxide Induced Binding of Hydrocarbons to DNA," *Biochim. Biophys. Acta* 169:224–229 (1968).

Mosser, D.D., and N.C. Bols. "Relationship Between Heat-Shock Protein Synthesis and Thermotolerance in Rainbow Trout Fibroblasts," *J. Comp. Physiol.* 158B:457–467 (1988).

Mosser, D.D., J. van Oostrom and N.C. Bols. "Induction and Decay of Thermotolerance in Rainbow Trout Fibroblasts," *J. Cell. Physiol.* 132:155–160 (1987).

Mulder, G.J. "Sex Difference in Drug Conjugation and Their Consequences for Drug Toxicity.

Sulfation, Glucuronidation and Glutathione Conjugation," *Chem.-Biol. Interact.* 57:1–15 (1986).

Munro, S., and H.R.B. Pelham. "An HSP-70 Like Protein in the ER: Identify with the 78kD Glucose-Regulated Protein and Immunoglobulin Heavy Chain Binding Protein," *Cell* 46:291–300 (1986).

Nagata, C., M. Kodama, Y. Ioki and T. Kimura. "Free Radicals Produced from Chemical Carcinogens and Their Significance in Carcinogenesis," in *Free Radicals and Cancer*, R.A. Floyd, Ed. (New York: Marcel Dekker, Inc., 1982), pp. 1–62.

Nakano, Y., and K. Asada. "Hydrogen Peroxide Is Scavenged by an Ascorbate-Specific Peroxidase in Spinach Chloroplasts," *Plant Cell Physiol.* 22:867–880 (1981).

"An Evaluation of Candidate Measures of Biological Effects for the National Status and Trends Program," National Oceanographic and Atmospheric Administration, NOAA Tech. Memorandum NOS OMA 45 (1989).

Nebert, D.W., M. Adesnik, M.J. Coon, R.W. Estabrook, F.J. Gonzalez, F.P. Guengerich, I.C. Gunsalus, E.F. Johnson, B. Kemper, W. Levin, I.R. Phillips, R. Sato and M.R. Waterman. "The P450 Gene Superfamily: Recommended Nomenclature," *DNA* 6:1–11 (1987).

Nebert, D.W., D. Nelson, M. Coon, R. Estabrook, R. Feyereisen, Y. Fujii-Kuriyama, F. Gonzalez, F. Guengerich, I. Gunsalus, E. Johnson, J. Loper, R. Sato, M. Waterman and D. Waxman. "The P450 Superfamily: Update on New Sequences, Gene Mapping, and Recommended Nomenclature," *DNA Cell Biol.* 10:1–14 (1991).

Nemer, M., D.G. Wilkinson, E.C. Travaglini and E.J. Sternberg. "Sea Urchin Metallothionein Sequence: Key to an Evolutionary Diversity," *Proc. Natl. Acad. Sci. U.S.A.* 82:4992–4994 (1985).

Nielson, K.B., C.L. Atkin and D.R. Winge. "Distinct Metal-Binding Configurations in Metallothionein," *J. Biol. Chem.* 260:5342–5350 (1985).

Nimmo, I.A. "The Glutathione S-Transferases of Fish," *Fish Physiol. Biochem.* 3:163–172 (1987).

Noel-Lambot, F., J.M. Bouquegneau, F. Frankenne and A. Disteche. "Cadmium, Zinc, and Copper Accumulation in Limpets (*Patella vulgata*) from the Bristol Channel with Special Reference to Metallothioneins," *Mar. Ecol. Prog. Ser.* 2:81–89 (1980).

Nolan, C.V., and E.J. Duke. "Cadmium-Binding Proteins in *Mytilus edulis*: Relation to Mode of Administration and Significance in Tissue Retention of Cadmium," *Chemosphere* 12:65–74 (1983).

Nordberg,, M., I. Nuottaniemi, A. Nordberg, P. Kjellstrom and J.S. Garvey. "Characterization Studies on the Cadmium-Binding Proteins from Two Species of New Zealand Oysters," *Environ. Health Perspect.* 65:57–62 (1986).

Oh, S.H., J.T. Deagen, P.D. Whanger and P.H. Weswig. "Biological Function of Metallothionein. V. Its Induction in Rats by Various Stresses," *Am. J. Physiol.* 3:E282–285 (1978).

Ohtake, H., K. Hasegawa and M. Koga. "Zinc-Binding Protein in the Livers of Neonatal, Normal and Partially Hepactomized Rats," *Biochem. J.* 174:999–1005 (1978).

Oikari, A.D.J., and T. Nakori. "Kraft Pulp Mill Effluent Components Cause Liver Dysfunction in Trout," *Bull. Environ. Contam. Toxicol.* 28:266–270 (1982).

Olafson, R.W. "Differential Pulse Polarographic Determination of Marine Metallothionein Induction Kinetics," *J. Biol. Chem.* 256:1263–1268 (1981).

Olafson, R.W., and R.G. Sim. "An Electrochemical Approach to Quantification and Characterization of Metallothioneins," *Anal. Biochem.* 100:343–351 (1979).

Olsson, P.E., and C. Haux. "Increased Hepatic Metallothionein Content Correlates to Cadmium Accumulation in Environmentally Exposed Perch (*Perca fluviatilis*)," *Aquat. Toxicol.* 9:231–242 (1986).

Olsson, P.E., S.J. Hyllner, M. Zafarullah, T. Andersson and L. Gedamu. "Differences in Metallothionein Gene Expression in Primary Cultures of Rainbow Trout Hepatocytes and the RTH-149 Cell Line," *Biochim. Biophys. Acta* 1049:78–82 (1990).

Ortiz de Montellano, P.R. *Cytochrome P-450. Structure, Mechanism, and Biochemistry* (New York: Plenum Press, 1986).

Ostermann, J., A.L. Horwich, W. Neupert and F.-U. Hartl. "Protein Folding in Mitochondria Requires Complex Formation with Hsp60 and ATP Hydrolysis," *Nature (London)* 341: 125–130 (1989).

Otvos, J.D., and I.M. Armitage. "Structure of the Metal Clusters in Rabbit Liver Metallothionein," *Proc. Natl. Acad. Sci. U.S.A.* 77:7094–7098 (1980).

Otvos, J.D., R.W. Olafson and I.M. Armitage. "Structure from an Invertebrate Metallothionein from *Scylla serrata*," *J. Biol. Chem.* 257:2427–2431 (1982).

Ougham, H.J., and J.L. Stoddart. "Synthesis of Heat-Shock Protein and Acquisition of Thermotolerance in High-Temperature Tolerant and High-Temperature Susceptible Lines of Sorghum," *Plant Sci.* 44:163–167 (1986).

Overnell, J., M. Good and M. Vasak. "Spectroscopic Studies on Cadmium(II) and Cobalt(II)-Substituted Metallothionein from the Crab *Cancer pagurus*: Evidence for One Additional Low-Affinity Metal-Binding Site," *Eur. J. Biochem.* 172:171–177 (1988).

Overnell, J., and E. Trewhella. "Evidence for the Natural Occurrence of (Cadmium-Copper)-Metallothionein in the Crab *Cancer pagurus*," *Comp. Biochem. Physiol.* 64C:69–76 (1979).

Packer, L., Ed. "Oxygen Radicals in Biological Systems," in *Methods in Enzymology*, Vol. 105, S.P. Colowick, and N.O. Kaplan, Eds. (New York: Academic Press, 1984).

Packer, L., and A.N. Glazer, Eds. "Oxygen Radicals in Biologial Systems, Part B, Oxygen Radicals and Antioxidants, in *Methods in Enzymology*, Vol. 186, J.N. Abelson, and M.I. Simon, Eds. (New York: Academic Press, 1990).

Paglia, D.E., and W.N. Valentine. "Studies on the Quantitative and Qualitative Characterization of Erythrocyte Glutathione Peroxidase," *J. Lab. Clin. Med.* 70:158–169 (1967).

Pajor, A.M., J.J. Stegeman, P. Thomas and B.R. Woodin. "Feminization of the Hepatic Microsomal Cytochrome P-450 System in Brook Trout by Estradiol, Testosterone, and Pituitary Factors," *J. Exp. Zool.* 253:51–60 (1990).

Payne, J.F., L.L. Fancey, A.D. Rahimtula and E.L. Porter. "Review and Perspective on the Use of Mixed-Function Oxygenase Enzymes in Biological Monitoring," *Comp. Biochem. Physiol.* 86C:233–245 (1987).

Payne, J.F., and W.R. Penrose. "Induction of Aryl Hydrocarbon Benzo[a]pyrene Hydroxylase in Fish by Petroleum," *Bull. Environ. Contam. Toxicol.* 14:112–116 (1975).

Pelham, H.R.B. "Coming in from the Cold," *Nature (London)* 332:776–777 (1988).

Pesonen, M., M. Celander, L. Förlin and T. Andersson. "Comparison of Xenobiotic Biotransformation Enzymes in Kidney and Liver of Rainbow Trout (*Salmo gairdneri*)," *Toxicol. Appl. Pharmacol.* 91:75–84 (1987).

Pesonen, M., T. Hansson, L. Förlin and T. Andersson. "Regional Distribution of Xenobiotic and Steroid Metabolism in Kidney Microsomes from Rainbow Trout," *Fish Physiol. Biochem.* 8:141–145 (1990).

Petering, D.H., and B.A. Fowler. "Discussion Summary. Roles of Metallothionein and Related Proteins in Metal Metabolism and Toxicity. Problems and Perspectives," *Environ. Health Perspect.* 65:217–224 (1986).

Petit, J.L. "Scales of Air Pollution Related to Peroxidase Activity of Radish," *Sci. Total Environ.* 39:189–208 (1984).

Pryor, W.A. "Why is the Hydroxyl Radical the Only Radical that Commonly Adds to DNA?

Hypothesis: It Has a Rare Combination of High Electrophilicity, High Thermochemical Reactivity, and a Mode of Production that Can Occur Near DNA," *Free Rad. Biol. Med.* 4:219–223 (1988).

Poland, A., and J.C. Knutson. "2,3,7,8-Tetrachlorodibenzo-p-Dioxin and Related Halogenated Aromatic Hydrocarbons: Examination of the Mechanism of Toxicity," *Ann. Rev. Pharmacol. Toxicol.* 22:517-554 (1982).

Porra, R.J., and O.T.G. Jones. "Assay and Properties of Ferrochelatase from Pig-Liver Mitochondrial Extract," *Biochem. J.* 87:181–185 (1963).

Rabinowitch, H.D., and I. Fridovich. "Superoxide Radicals, Superoxide Dismutases, and Oxygen Toxicity in Plants," *Photochem. Photobiol.* 37:679–690 (1983).

Radi, A.A.R., D.Q. Hai, B. Matkovics and T. Gabryelak. "Comparative Antioxidant Enzyme Study in Freshwater Fish with Different Types of Feeding Behaviour," *Comp. Biochem. Physiol.* 81C:395–399 (1985).

Ramagopal, S. "Salinity Stress Induced Tissue-Specific Proteins in Barley Seedlings," *Plant Physiol.* 84:324–331 (1987).

Rattner, B.A., D.J. Hoffman and C.M. Marn. "Use of Mixed-Function Oxygenases to Monitor Contaminant Exposure in Wildlife," *Environ. Toxicol. Chem.* 8:1093–1102 (1989a).

Rattner, B.A., W.J. Flemming and C.M. Bunck. "Comparative Toxicity of Lead Shot in Black Ducks (*Anas rubripes*) and Mallards (*Anas platyrhynchos*)," *J. Wildl. Dis.* 25:175–183 (1989b).

Rauser, W.E. "Phytochelatins," *Ann. Rev. Biochem.* 59:61–86 (1990).

Reddy, J.K., and N.D. Lalwani. "Carcinogenesis by Hepatic Peroxisome Proliferators: Evaluation of the Risk of Hypolipidemic Drugs and Industrial Plasticizers to Humans," *CRC Crit. Rev. Toxicol.* 12:1–68 (1983).

Reed, D.J., and P.W. Beatty. "Biosynthesis and Regulation of Glutathione," in *First Reviews in Biochemical Toxicology*, E. Hodgson, J.R. Bend and R. Phillpot, Eds. (New York: Elsevier/North Holland, 1980), pp. 213–241.

Reese, R.N., R.K. Mehra, E.B. Tarbet and D.R. Winge. "Studies on the Gamma-Glutamyl Cu-Binding Peptide from *Schizosacharomyces pombe*," *J. Biol. Chem.* 263:4186–4192 (1988).

Rennenberg, H. "Glutathione Metabolism and Possible Biological Roles in Higher Plants," *Phytochemistry* 21:2771–2781 (1982).

Richards, R.I., A. Heguy and M. Karin. "Structural and Functional Analysis of the Human Metallothionein Ia Gene: Differential Induction by Metal Ions and Glucocorticoids," *Cell* 37:263–272 (1988).

Richardson, C.J., T.W. Sasek and R.T. Di Giulio. "The Use of Physiological and Biochemical Markers for Assessing Air Pollutant Stress in Trees," in *Plants for Toxicity Assessment, ASTM STP 1091*, W. Wang, J.W. Gorsuch and W.R. Lower, Eds. (Philadelphia: American Society of Testing Materials, 1990), pp. 143–155.

Richman, P.G., and A. Meister. "Regulation of γ-Glutamyl-Cysteine Synthetase by Nonallosteric Feedback Inhibition by Glutathione," *J. Biol. Chem.* 250:1422–1426 (1975).

Ridlington, J. W., D.R. Winge and B.A. Fowler. "Long-Term Turnover of Cadmium Metallothionein in Liver and Kidney Following a Single Low Dose of Cadmium in Rats," *Biochim. Biophys. Acta* 673:177–183 (1981).

Roberts, M.H., D.W. Sveedand and S.P. Felton. "Temporal Changes in AHH and SOD Activities in Feral Spot from the Elizabeth River, a Polluted Subestuary," *Mar. Environ. Res.* 23:89–101 (1987).

Roberts, P.B. "Growth in Cadmium-Containing Medium Induces Resistance to Heat in *E. coli*," *Int. J. Radiat. Biol.* 45:27–31 (1984).

Roch, M., and J.A. McCarter. "Hepatic Metallothionein Production and Resistance to Heavy Metals by Rainbow Trout (*Salmo gairdneri*)-II. Held in a Series of Contaminated Lakes," *Comp. Biochem. Physiol.* 77C:77–82 (1984).

Roch, M., J.A. McCarter, A.T. Matheson, M.J.R. Clark and R.W. Olafson. "Hepatic Metallothionein in Rainbow Trout (*Salmo gairdneri*) as an Indicator of Metal Pollution in the Campbell River System," *Can. J. Fish. Aquat. Sci.* 39:1596–1601 (1982).

Rodrigues, A.L., M.L. Bellinaso and T. Dick. "Effect of Some Metal Ions on Blood and Liver Delta-Aminolevulinate Dehydratase of *Pimelodus maculatus* (Pisces, Pimelodidae)," *Comp. Biochem. Physiol.* 94B:65–69 (1989).

Roesijadi, G. "The Significance of Low Molecular Weight, Metallothionein-Like Proteins in Marine Invertebrates: Current Status," *Mar. Environ. Res.* 4:167–179 (1981).

Roesijadi, G., A.S. Drum, J.T. Thomas and G.W. Fellingham. "Enhanced Mercury Tolerance in Marine Mussels and Relationship to Low Molecular Weight Mercury-Binding Proteins," *Mar. Pollut. Bull.* 13:250–253 (1982).

Roesijadi, G., and G.W. Fellingham. "Influence of Cu, Cd and Zn Preexposure on Hg Toxicity in the Mussel *Mytilus edulis*," *Can. J. Fish. Aquat. Sci.* 44:680–684 (1987).

Roesijadi, G., S. Kielland and P.L. Klerks. "Purification and Properties of Novel Molluscan Metallothioneins," *Arch. Biochem. Biophys.* 273:403–413 (1989).

Roesijadi, G., and P.L. Klerks. "Kinetic Analysis of Cadmium Binding to Metallothionein and Other Intracellular Ligands in Oyster Gills," *J. Exp. Zool.* 251:1–12 (1989).

Roesijadi, G., M.E. Unger and J.E. Morris. "Immunochemical Quantification of Metallothioneins of a Marine Mollusc," *Can. J. Fish. Aquat. Sci.* 45:1257–1263 (1988).

Ronis, M.J.J., J. Borlakoglu, C.H. Walker, T. Hansson and J.J. Stegeman. "Expression of Orthologues to Rat P-450IA1 and IIB1 in Sea Birds from the Irish Sea 1978–1988. Evidence for Environmental Induction," *Mar. Environ. Res.* 28:123–130 (1989).

Roscoe, D.E., S.W. Nielsen, H.H. Lamola and D. Zuckerman. "A Simple Quantitative Test for Erythrocytic Protoporphyrin in Lead Poisoned Ducks," *J. Wildl. Dis.* 15:127–136 (1979).

Rose, D.W., W.J. Welch, G. Kramer and B. Hardesty. "Possible Involvement of the 90kDa Heat Shock Protein in the Regulation of Protein Synthesis," *J. Biol. Chem.* 264:6239–6244 (1988).

Rothman, J.E. "Polypeptide Chain Binding Proteins: Catalysts of Protein Folding and Related Processes in Cells," *Cell* 59: 591–601 (1989).

Rupp, H., and U. Weser. "Circular Dichroism of Metallothioneins. A Structural Approach," *Biochim. Biophys. Acta* 533:209–226 (1978).

Sadhu, C., and L. Gedamu. "Regulation of Human Metallothionein Genes," *J. Biol. Chem.* 263:2679–2684 (1988).

Sanders, B.M. "Stress Proteins: Potential as Multitiered Biomarkers," in *Environmental Biomarkers*, L. Shugart, and J. McCarthy, Eds. (Chelsea, MI: Lewis Publishers. Inc., 1990), pp. 165–191.

Sanders, B.M., L.S. Martin, W.G. Nelson, D.K. Phelps and W. Welch. "Relationships Between Accumulation of a 60 kDa Stress Protein and Scope-for-Growth in *Mytilus edulis* Exposed to a Range of Copper Concentrations," *Mar. Environ. Res.* 31:81–97 (1991).

Sanders, B.M., and K.D. Jenkins. "Relationships Between Free Cupric Ion Concentrations in Sea Water and Copper Metabolism and Growth in Crab Larvae," *Biol. Bull.* 167:704–712 (1984).

Sanders, B.M., K.D. Jenkins, W.G. Sunda and J.D. Costlow. "Free Cupric Ion Activity in Sea Water: Effects on Metallothionein and Growth in Crab Larvae," *Science* 222:53–55 (1983).

Sargan, D.R., M.J. Tsai and B.W. O'Malley. "HSP 108, a Novel Heat Shock Inducible Protein of Chicken," *Biochemistry* 25:625–629 (1986).

Satoh, K., A. Kitihara, Y. Soma, Y. Inaba, I. Hatayama and K. Sato. "Purification, Induction and Distribution of Placental Glutathione Transferase: A New Marker Enzyme for Preneoplastic Cells in the Rat Chemical Hepatocarcinogenesis," *Proc. Natl. Acad. Sci. U.S.A.* 82:3964–3968 (1985).

Schelin, C., A. Tunek and B. Jergil. "Covalent Binding of Benzo[a]pyrene to Rat Liver Cytosolic Proteins and Its Effect on the Binding to Microsomal Proteins," *Biochem. Pharmacol.* 32:1501–1506 (1983).

Schlenk, D., and D.R. Buhler. "Determination of Multiple Forms of Cytochrome P-450 in Microsomes from the Digestive Gland of *Cryptochiton stelleri*," *Biochem. Biophys. Res. Commun.* 163:476–480 (1989).

Schlesinger, M.J. "Function of Heat Shock Proteins," *Atlas of Sci. Biochem.*, 161–164 (1988).

Schlesinger, M.J., M. Ashburner and A. Tissieres. *Heat Shock from Bacteria to Man* (Cold Springs Harbor, N.Y.: Cold Spring Harbor Laboratory, 1982), pp.1–440.

Schoor, W.P., D.E. Williams and J.J. Lech. "Combined Use of Biochemical Indicators to Assess Sublethal Pollution Effects on the Gulf Killifish (*Fundulus grandis*)," *Arch. Environ. Contam. Toxicol.* 17:437–441 (1988).

Schreiber, W.E., V.A. Raisys and R.F. Labbe. "Liquid-Chromatographic Profiles of Urinary Porphyrins," *Clin. Chem.* 29:527–530 (1983).

Schulz, H. "Biochemische and Faktoranalytische Untersuchungen zur Interpretation von SO$_2$-Indikationen Nadeln von Pinus Sylvestris," *Biochem. Physiol. Pflanzen.* 181:241–256 (1983).

Seib, P.A., and B.M. Tolbert, Eds. "Ascorbic Acid: Chemistry, Metabolism and Uses," in *Advances in Chemistry, Series 200* (Washington, D.C.: American Chemical Society, 1982).

Seidegård, J.E., and J.W. Depierre. "Microsomal Epoxide Hydrolase Properties, Regulation and Function," *Biochim. Biophys. Acta* 695:251–270 (1983).

Shelton, K.R., J.M. Todd and P.M. Egle. "The Induction of Stress-Related Proteins by Lead," *J. Biol. Chem.* 261:1935–1940 (1986).

Shibahara, S. R. Muller and H. Taguchi. "Transcriptional Control of Rat Heme Oxygenase by Heat Shock," *J. Biol. Chem.* 262:12889–12892 (1987).

Shuman, J., and A. Przybyla. "Expression of the 31-kD Stress Protein in Rat Myoblasts and Hepatocytes," *DNA* 7:475–482 (1988).

Sies, H., A. Wahllander, C. Waydhas, S. Soboll and D. Haberle. "Functions of Intracellular Glutathione in Hepatic Hydroperoxide and Drug Metabolism and the Role of Extracellular Glutathione," *Adv. Enzyme Regul.* 18:303–320 (1980).

Silbergeld, E.K., and B.A. Fowler, Eds. "Mechanisms of Chemical-induced Porphyrinopathies," *Ann. N.Y. Acad. Sci.* 514:352 (1987).

Silvarajah, K., C.S. Franklin and W.P. Williams. "The Effects of Polychlorinated Biphenyls on

Plasma Steroid Levels and Hepatic Microsomal Enzymes in Fish," *J Fish Biol*. 13:401–409 (1985).

Sinclair, P., J. Frezza, J. Sinclair, W. Bement, S. Haugen, J. Healey and H. Bonkovsky. "Immunochemical Detection of Different Isoenzymes of Cytochrome P-450 Induced in Chick Hepatocyte Cultures," *Biochem. J.* 258:237–245 (1989).

Skowyra, D., C. Georgopoulos and W. Zylicz. "The E. Coli dnaK Gene Product, the HSP70 Homolog, Can Reactivate Heat-Inactivated RNA Polymerase in an ATP Hydrolysis-Dependent Manner," *Cell* 62:939–944 (1990).

Smolowitz, R.M., M.E. Hahn and J.J. Stegeman. "Immunohistochemical Localization of Cytochrome P4501A1 Induced by 3,3′,4,4′-Tetrachlorobiphenyl and by 2,3,7,8-Tetrachlorodibenzofuran in Liver and Extrahepatic Tissues of the Teleost *Stenotomus chrysops* (Scup)," *Drug. Metab. Dispos.* 19: 113–123 (1991).

Soma, Y., K. Satoh and K. Sato. "Purification and Subunit-Structural and Immunological Characterization of Five Glutathione S-Transferases in Human Liver, and the Acidic Form as a Hepatic Tumor Marker," *Biochim. Biophys. Acta* 869:247–258 (1986).

Sordyl, H., and A. Osterland. "Investigation of Factors Influencing Delta-Aminolevulinic Acid Dehydratase Activity (ALA-D) of Rainbow Trout," *Zool. JB. Physiol.* 94:445–460 (1990).

Sparla, A.M., and J. Overnell. "The Binding of Cadmium to Crab Cadmium Metallothionein. A Polarographic Investigation," *Biochem. J.* 267:539–540 (1990).

Spector, M.P., Z. Aliabadi, T. Gonzalez and J.W. Foster. "Global Control in *Salmonella typhimurium*: Two-Dimensional Electrophoretic Analysis of Starvation-, Anaerobiosis-, and Heat Shock-Inducible Proteins," *J. Bacteriol.* 168:420–424 (1986).

Spies, R.B., J.S. Felton and L. Dillard. "Hepatic Mixed Function Oxidases in California Flatfishes Are Increased in Contaminated Environments and by Oil and PCB Ingestion," *Mar. Environ. Res.* 14:412–413 (1984).

Squibb, K.S., J.B. Pritchard and B.A. Fowler. "Cadmium-Metallothionein Nephropathy: Relationships Between Ultrastructural/Biochemical Alterations and Intracellular Cadmium Binding," *J. Pharmacol. Exp. Therap.* 229:311–321 (1984).

Staal, G.E.J., and C. Veeger. "The Reaction Mechanism of Glutathione Reductase from Human Erythrocytes," *Biochim. Biophys. Acta* 185:49–57 (1969).

Staal, G.E.J., J. Visser and C. Veeger. "Purification and Properties of Glutathione Reductase of Human Erythrocytes," *Biochim. Biophys. Acta* 185:39–48 (1969).

Stacey, N.H., and C.D. Klaassen. "Inhibition of Lipid Peroxidation without Prevention of Cellular Injury in Isolated Rat Hepatocytes," *Toxicol. Appl. Pharmacol.* 58:8–18 (1981).

Statham, C.N., C.R. Elcombe, S.P. Szyjka and J.J. Lech. "Effect of Polycyclic Aromatic Hydrocarbons on Hepatic Microsomal Enzymes and Disposition of Methyl-Naphthalene in Rainbow Trout In Vivo," *Xenobiotica* 8:65 (1978).

Stegeman, J.J. "Temperature Influence on Basal Activity and Induction of Mixed Function Oxygenase Activity in *Fundulus heteroclitus*," *J. Fish. Res. Bd. Can.* 36:1400–1405 (1979).

Stegeman, J.J. "Mixed-Function Oxygenase Studies in Monitoring for Effects of Organic Pollution," *Rapp. Proc.-Verb. Reun. Cons. Int. L'Explor. Mer.* 179:33–38 (1980).

Stegeman, J. "Polycyclic Hydrocarbons and Their Metabolism in the Marine Environment," in *Polycyclic Aromatic Hydrocarbons and Cancer, Vol. 3*, H. Gelboin, and P. Ts'o, Eds. (New York: Academic Press, 1981), pp. 1–60.

Stegeman, J.J. "Monooxygenase Systems in Marine Fish," in *Pollutant Studies in Marine Animals*, C.S. Giam, and L. Ray, Eds. (Boca Raton, FL: CRC Press, 1989), pp. 65–92.

Stegeman, J.J. "Cytochrome P450 Forms in Fish: Catalytic, Immunological and Sequence Similarities," *Xenobiotica* 19:1093–1110 (1989).

Stegeman, J.J., R.L. Binder and A. Orren. "Hepatic and Extrahepatic Microsomal Electron Transport Components and Mixed-Functions Oxygenases in the Marine Fish *Stenotomus versicolor*," *Biochem. Pharmacol.* 28:3431–3439 (1979).

Stegeman, J.J., and M.O. James. "Individual Variation in Patterns of Benzo[a]pyrene Metabolism in the Marine Fish Scup (*Stenotomus chrysops*)," *Mar. Environ. Res.* 17:122–124 (1985).

Stegeman, J.J., and P.J. Kloepper-Sams. "Cytochrome P-450 Isozymes and Monooxygenase Activity in Aquatic Animals," *Environ. Health Perspect.* 17:87–95 (1987).

Stegeman, J.J., P.J. Kloepper-Sams and J. Farrington. "Monooxygenase Induction and Chlorobiphenyls in the Deep-Sea Fish *Coryphaenoides armatus*," *Science* 231:1287–1289 (1986).

Stegeman, J.J., M.R. Miller and D.E. Hinton. "Cytochrome P450-IA1 Induction and Localization in Endothelium of Vertebrate Heart," *Mol. Pharmacol.* 36:723–729 (1989).

Stegeman, J.J., K.W. Renton, B.R. Woodin, Y-S. Zhang and R.F. Addison. "Experimental and Environmental Induction of Cytochrome P450E in Fish from Bermuda Waters," *J. Exp. Mar. Biol. Ecol.* 138: 49–67 (1990).

Stegeman, J.J., R.M. Smolowitz and M.E. Hahn. "Immunohistochemical Localization of Environmentally Induced Cytochrome P4501A1 in Multiple Organs of the Marine Teleost *Stentomus chrysops* (Scup)," *Toxicol. Appl. Pharmacol.* 110:486–504 (1991).

Stegeman, J.J., F. Teng and E.A. Snowberger. "Induced Cytochrome P-450 in Winter Flounder (*Pseudopleuronectes americanus*) from Coastal Massachusetts Evaluated by Catalytic Assay and Monoclonal Antibody Probes," *Can. J. Fish. Aquat. Sci.* 44:1270–1277 (1987).

Stegeman, J.J., B.R. Woodin and A. Goksøyr. "Apparent Cytochrome P-450 Induction as an Indication of Exposure to Environmental Chemicals in the Flounder *Platichthys flesus*," *Mar. Ecol. Prog. Ser.* 46:55–60 (1988).

Stenerson, J., S. Kobro, M. Bjerk and U. Arenal. "Glutathione Transferases in Aquatic and Terrestrial Animals from Nine Phyla," *Comp. Biochem. Physiol.* 86C:73–82 (1987).

Stephanou, G., S.N. Alahiotis, C. Christodoulou and V.J. Marmaras. "Adaptation of Drosophila to Temperature: Heat-Shock Proteins and Survival in *Drosophila melanogaster*," *Dev. Genet.* 299–308 (1983).

Stern, A. "Red Cell Oxidative Damage," in *Oxidative Stress*, H. Sies, Ed. (London: Academic Press, 1985), pp. 331–349.

Stevens, J.B., and A.P. Autor. "Induction of Superoxide Dismutase by Oxygen in Neonatal Rat Lung," *J. Biol. Chem.* 252:3509–3514 (1977).

Stocchi, V., L. Cucchiarini, M. Magnani, L. Chiarantini, P. Palma and G. Crescentini. "Simultaneous Extraction and Reverse-Phase High-Performance Liquid Chromatographic Determination of Adenine and Pyridine Nucleotides in Human Red Blood Cells," *Anal. Biochem.* 146:118–124 (1985).

Subjeck, J.R., and T.-T. Shyy. "Stress Protein Systems of Mammalian Cells," *Cell. Physiol.* 19:C1–C17 (1986).

Suteau, P., M. Daubeze, M.L. Miguad and J.F. Narbonne. "PAH-Metabolizing Enzymes in

Whole Mussels as Biochemical Tests for Chemical Pollution Monitoring," *Mar. Ecol. Prog. Ser.* 46:45–49 (1988).

Talbot, V., and R.J. Magee "Naturally-Occurring Heavy Metal Binding Proteins in Invertebrates," *Arch. Environ. Contam. Toxicol.* 7:73–81 (1978).

Tanaka, K., H. Saji and N. Kondo. "Immunological Properties of Glutathione Reductase and Inductive Biosynthesis of the Enzyme with Ozone," *Plant Cell Physiol.* 29:637–642 (1988).

Tanaka, K., and K. Sugahara. "Role of Superoxide Dismutase in Defense Against SO_2 Toxicity and an Increase in Superoxide Dismutase Activity with SO_2 Fumigation," *Plant Cell Physiol.* 21:601–611 (1980).

Tandy, N.E., R.T. Di Giulio and C.J. Richardson. "Assay and Electrophoresis of Superoxide Dismutase from Red Spruce (*Picea rubens* Sarg.), Loblolly Pine (*Pinus taeda* L.), and Scotch Pine (*Pinus sylvestris* L.). A Method for Biomonitoring," *Plant Physiol.* 90:742–748 (1989).

Tanizawa, H., Sazuka, Y. and Y. Takino. "Micro-Determination of Lipoperoxide in the Mouse Myocardium by Thiobarbituric Acid Fluorophotometry," *Chem. Pharmacol. Bull.* 29:2910–2914 (1981).

Tatematsu, M., Y. Mera, N. Ito, K. Satoh and K. Sato. "Relative Merits of Immunohistochemical Demonstrations of Placental A, B and C Forms of Glutathione S-Transferase and Histochemical Demonstration of Gamma-Glutamyl Transferase as Markers of Altered Foci During Liver Carcinogenesis in Rats," *Carcinogenesis* 6:1621–1626 (1985).

Taylor, S.L., M.P. Lamden and A.L. Tappel. "A Sensitive Fluorometric Method for Tissue Tocopherol Analysis," *Lipids* 11:530–538 (1976).

Thomas, H., and F. Oesch. "Functions of Epoxide Hydrolase," *ISI Atlas Sci. Biochem.* 1:287–291 (1988).

Thomas, P., and H.W. Wofford. "Effects of Metals and Organic Compounds on Hepatic Glutathione, Cysteine, and Acid-Soluble Thiol Levels in Mullet (*Mugil cephalus* L.)," *Toxicol. Appl. Pharmacol.* 76:172–182 (1984).

Thomas, P., H.W. Wofford and J.M. Neff. "Effect of Cadmium on Glutathione Content of Mullet (*Mugil cephalus*) Tissues," in *Physiological Mechanisms of Marine Pollutant Toxicity,* W.B. Vernberg, A. Calabrese, F.P. Thurnberg and F.J. Vernberg, Eds. (New York: Academic Press, 1982), pp. 109–125.

Thompson, J.A.J., and R.P. Cosson. "An Improved Electrochemical Method for the Quantification of Metallothioneins in Marine Organisms," *Mar. Environ. Res.* 11:137–152 (1984).

Thornalley, P.J., and M. Vasak. "Possible Role for Metallothionein in Protection Against Radiation-Induced Oxidative Stress. Kinetics and Mechanism of Its Reaction with Superoxide and Hydroxyl Radicals," *Biochim. Biophys. Acta* 827:36–44 (1985).

Tipping, E., and B. Ketterer. "The Influence of Soluble Binding Proteins on Lipophile Transport and Metabolism in Hepatocytes," *Biochem. J.* 195:441–452 (1981).

U.S. EPA. "Air Quality Criteria for Lead Vol. 4," U.S. EPA Report 600/8–83-0286 (1986).

Vandermallie, R.J., and J. Garvey. "Production and Study of Antibody Produced Against Rat Cadmium Thionein," *J. Biol. Chem.* 254:8416–8421 (1979).

Van Veld, P.A., U. Ko., W.K. Vogelbein and D.J. Westbrook. "Glutathione Transferase in Intestine, Liver and Hepatic Lesions of Mummichog," *Fish Physiol. Biochem.* 9:369–376 (1991).

Van Veld, P.A., and R.F. Lee. "Intestinal Glutathione S-Transferase Activity in Flounder (*Platichthys flesus*) Collected from Contaminated and Reference Sites," *Mar. Ecol. Prog Ser.* 46:61–63 (1988).

Van Veld, P.A., J.J. Stegeman, B.R. Woodin, J.S. Patton and R.F. Lee. "Induction of Monooxygenase Activity in the Intestine of Spot (*Leiostomus xenthurus*), a Marine Teleost, by Dietary PAH," *Drug Metab. Dispos.* 16:659–665 (1988).

Van Veld, P.A., R.D. Vetter, R.F. Lee and J.S. Patton. "Dietary Fat Inhibits the Intestinal Metabolism of the Carcinogen Benzo[a]pyrene in Fish," *J. Lipid Res.* 28:810–817 (1987).

Van Veld, P.A., D.J. Westbrook, B.R. Woodin, R. C. Hale, C.L. Smith, R.J. Huggett and J.J. Stegeman. "Induced Cytochrome P-450 in Intestine and Liver of Spot (*Leisostomus xanthurus*) from a Polycyclic Aromatic Hydrocarbon Contaminated Environment," *Aquat. Toxicol.* 17:119–132 (1990).

Varanasi, U., T. Collier, D. William and D. Buhler. "Hepatic Cytochrome P-450 Isozymes and Aryl Hydrocarbon Hydroxylase in English Sole (*Parophrys vetulus*)," *Biochem. Pharmacol.* 35:2967–2971 (1986).

Varanasi, U., J.E. Stein, M. Nishimoto, W.L. Reichert and T.K. Collier. "Chemical Carcinogenesis in Feral Fish: Uptake, Activation and Detoxication of Organic Xenobiotics," *Environ. Health Perspect.* 71:155–170 (1987).

Varsano, R., M. Matringe, N. Magnin, R. Mornet and R. Scalla. "Competitive Interaction of Three Peroxidizing Herbicides with Binding of [^{3}H] Aciflurofen to Corn Etioplast Membranes," *FEBS Lett.* 272:106–108 (1990).

Vasak, M., and J.H.R. Kagi. "Spectroscopic Properties of Metallothionein," *Metal Ions Biol. Syst.* 15:213–273 (1983).

Waalkes, M.P., and P.L. Goering. "Metallothionein and Other Cadmium-Binding Proteins: Recent Developments," *Chem Res. Toxicol.* 3:281–288 (1990).

Weber, D.N., C.F. Shaw III and D.H. Petering. "*Euglena gracilis* Cadmium-Binding Protein-II Contains Sulfide Ion," *J. Biol. Chem.* 262:6962–6964 (1987).

Welch, W.J. "The Mammalian Stress Response: Cell Physiology and Biochemistry of Stress Proteins," in *The Role of the Stress Response in Biology and Disease*, R. Moromoto, A. Tissieres, and C. Georgopoulos, Eds. (Cold Springs Harbor, N.Y.: Cold Spring Harbor Laboratory, 1990).

Welch, W.J., and J.R. Feramisco. "Nuclear and Nucleolar Localization of the 72,000 Dalton Heat Shock Protein in Heat Shocked Mammalian Cells," *J. Biol. Chem.* 259:4501–4510 (1987).

Welch, W.J., and L.A. Mizzen. "Characterization of the Thermotolerant Cell. II. Effects on the Intracellular Distribution of Heat Shock Protein 70, Intermediate Filaments and Small Ribonucleoprotein Complexes," *J. Cell Biol.* 106:1117–1130 (1988).

Wenning, R.J., R.T. Di Giulio and E.P. Gallagher. "Oxidant-Mediated Biochemical Effects of Paraquat in the Ribbed Mussel, *Guekensia demissa*," *Aquat. Toxicol.* 12:157–170 (1988).

Williams, D., and D. Buhler. "Comparative Properties of Purified Cytochrome P-448 from β-Naphthoflavone Treated Rats and Rainbow Trout," *Comp. Biochem. Physiol.* 75C: 25–32 (1983a).

Williams, D., and D. Buhler. "Purified Forms of Cytochrome P450 from Rainbow Trout Liver with High Activity Toward Conversion of Aflatoxin B$_1$ to Aflatoxin B$_1$-2,3-Epoxide," *Cancer Res.* 43: 4752–4756 (1983b).

Williams, D., and D. Buhler. "Benzo[a]pyrene Hydroxylase Catalyzed by Purified Isozymes of Cytochrome P450 from β-Naphtho-Flavone-Fed Rainbow Trout," *Biochem. Pharmacol.* 33:3742 (1984).

Wills, E.D. "The Role of Dietary Components in Oxidative Stress in Tissues," in *Oxidative Stress*, H. Sies, Ed. (London: Academic Press, 1985), pp. 197–218.

Wilson, R.P., P.R. Bowser and W.E. Poe. "Dietary Vitamin E Requirement of Fingerling Channel Catfish," *J. Nutr.* 114:2053–2058 (1984).

Winge, D.R., K.B. Nielson, W.R. Gray and D.H. Hamer. "Yeast Metallothionein. Sequence and Metal-Binding Properties," *J. Biol. Chem.* 260:14464–14470 (1985).

Wise, D.J., J.R. Tomasso and T.M. Brandt. "Ascorbic Acid Inhibition of Nitrite-Induced Methemoglobinemia in Channel Catfish," *Prog. Fish-Cult.* 50:77–80 (1988).

Wise, R.R., and A.W. Naylor. "Chilling-Enhanced Photooxidation. Evidence for the Role of Singlet Oxygen and Superoxide in the Breakdown of Pigments and Endogenous Antioxidants," *Plant Physiol.* 83:278–282 (1987).

Wofford, H.W., and P. Thomas. "Effect of Xenobiotics on Peroxidation of Hepatic Microsomal Lipids from Striped Mullet (*Mugil cephalus*) and Atlantic Croaker (*Micropogonias undulatus*)," *Mar. Environ. Res.* 24:285–289 (1988).

Woods, J.S. "Studies on the Role of Heme in the Regulation of δ-Aminolevulinic Acid Synthetase During Fetal Hepatic Development," *Mol. Pharm.* 10:389–397 (1974).

Woods, J.S., and B.A. Fowler. "Altered Regulation of Mammalian Hepatic Heme Biosynthesis and Urinary Porphyrin Excretion During Prolonged Exposure to Sodium Arsenate," *Toxicol. Appl. Pharmacol.* 43:361–371 (1978).

Worgotter, E., G. Wagner, M. Vasak, J.H.R. Kagi and K. Wutrich. "Heteronuclear Filters for Two-Dimensional H NMR. Identification of the Metal-Bound Amino Acids in Metallothionein and Observation of Small Heteronuclear Long-Range Couplings," *J. Am. Chem. Soc.* 110:2388–2393 (1988).

Yam, J., L. Frank and R.J. Roberts. "Age-Related Development of Pulmonary Antioxidant Enzymes in the Rat," *Proc. Soc. Exp. Biol. Med.* 157:293–296 (1978a).

Yam, J., L. Frank and R.J. Roberts. "Oxygen Toxicity: Comparison of Lung Biochemical Responses in Neonatal and Adult Rats," *Pediat. Res.* 12:115–119 (1978b).

Yang, J.H., P.T. Kostecki, E.J. Calabrese and L.A. Baldwin. "Induction of Peroxisome Proliferation in Rainbow Trout Exposed to Ciprofibrate," *Toxicol. Appl. Pharmacol.* 104:476–482 (1990).

Yin, A., K. Sato, H. Tsuda and N. Ito. "Changes in Activities of Uridine Diphosphate Glucuronyltransferase During Chemical Hepatocarcinogenesis," *Gann.* 73:239–248 (1982).

Yoshida, T., P. Biro, T. Cohen, R.M. Muller and S. Shibahara. "Human Heme Oxygenase cDNA and Induction of Its mRNA by Hemin," *Eur. J. Biochem.* 171:457–461 (1988).

Yost, H.J., and S. Lindquist. "RNA Splicing Is Interrupted by Heat Shock and Is Rescued by Heat Shock Protein Synthesis," *Cell* 45:185–193 (1986).

Young, J.S., and G. Roesijadi. "Reparatory Adaptation to Copper-Induced Injury and Ooccurrence of a Copper-Binding Protein in the Polychaete, *Eudystilia vancouveri*," *Mar. Pollut. Bull.* 14:30-32 (1983).

Zafarullah, M., P.E. Olsson and L. Gedamu. "Differential Regulation of Metallothionein Genes in Rainbow Trout Fibroblast, RTG-2," *Biochim. Biophys. Acta* 1046:318–323 (1990).

Zhang, Y.S., T. Andersson and L. Förlin. "Induction of Hepatic Xenobiotic Biotransformation Enzymes in Rainbow Trout by β-Naphthoflavone. Time Course Studies," *Comp. Biochem. Physiol.* 95B:247–253 (1990).

Ziemiecki, A., M.G. Catelli, I. Joab and B. Moncharmont. "Association of the Heat Shock Protein HSP 90 with Steroid Hormone Receptors and Tyrosine Kinase Oncogene Products," *Biochem. Biophys. Res. Comm.* 138:1298–1307 (1986).

Index

Acetylcholinesterase
  enzyme inhibition, direct, 15–17
  and vertebral abnormalities, 173
N-Acetyl-β-D-glucosaminidase (NAG), 40
Acid phosphatase (ACP), 40
Adaptive responses, 13, 236
Adenylate energy charge, 21–22
Adrenal corticosteroids
  cytochrome P450 and, 100
  and immune system, 226
  as markers, 32–33
  metallothionein induction, 261
Age
  and antioxidant status, 109
  and histopathological markers, 192
Agglutination assays, 223, 226
Alanine aminotransferase, 40–42
Amino acid analysis, 101
Amino acids
  free, 58–60
  glucagon and, 37
  growth hormone and, 38
  plant enzymes, 20
δ-Aminolevulinic acid dehydratase (ALAS),
  17–18, 297
δ-Aminolevulinic acid excretion, 301
Aminotransferases, 40–42
Anamnestic immune responses, 216
Androgens, 34–35, 104, 110
Antibodies, *see also* Immunoglobulins;
Immunological markers
  agglutination assays, 226
  circulating levels, 218
  cytochrome P450, 239–240, 242
  humoral immunity, 216–217
Antibody-dependent cell cytotoxicity
  (ADCC), 216

Argyria, shellfish, 178
Aryl hydrocarbon hydroxylase (AHH), 239,
  241, 242, 246
Ascorbate, *see* Vitamin C
Ascorbate glutathione system, 19–20
Aspartate aminotransferase, 27, 28, 40–42
Asynchronous spawning, 34
ATP, direct enzyme inhibition, 18–19
ATPase
  organochlorines and, 55
  sodium pump, 18–19, 43, 157
Avian eggshell thinning and reproductive
  impairment, 54–55

Bile analysis, 97–98, 102
Bile ducts
  ductular hyperplasia, 164–165
  tumors, 170–171
Biliverdin, 300
Bioaccumulation in fish, 100
Biochemical markers, *see* Enzyme and
  protein systems
Bioconcentration in fish, 98
Biologically effective dose, 92
Biological specificity criterion, 2
Biotransformation products, *see* Metabolic
  products as markers; Phase II
  (conjugating) enzymes
Birds, *see also* specific markers
  cytochrome P450, 245–246
  immune function assays, 214–215
  phylogeny of immune system, 213
Blood chemistry, 39–47
  ion levels, 43–44
  serum enzymes, 39–43
  serum glucose and lipids, 44–47
B lymphocytes